Rebels within the Ranks

During the 1930s, psychologists Gordon Allport, Gardner Murphy, and Lois Barclay Murphy emerged as challengers to the neobehaviorist status quo in American social science from within the fields of social and personality psychology. Willing to experiment with the idea of "science" itself, these "rebels within the ranks" contested ascendent conventions that cast the study of human life in the image of classical physics. Drawing on the intellectual, social, and political legacies of William James's radically empiricist philosophy and radical Social Gospel theology, these three psychologists advanced critiques of scientific authority and democratic reality as they worked at the crossroads of the social and the personal in New Deal America. Appropriating models from natural history, they argued for the significance of individuality, contextuality, and diversity as scientific concepts as they explored what they envisioned as the nature of democracy and the democracy of nature.

Cambridge Studies in the History of Psychology

GENERAL EDITORS: MITCHELL G. ASH AND WILLIAM R. WOODWARD

This series provides a publishing forum for outstanding scholarly work in the history of psychology. The creation of the series reflects a growing concentration in this area by historians and philosophers of science, intellectual and cultural historians, and psychologists interested in historical and theoretical issues.

The series is open both to manuscripts dealing with the history of psychological theory and research and to work focusing on the varied social, cultural, and institutional contexts and impacts of psychology. Writing about psychological thinking and research of any period will be considered. In addition to innovative treatments of traditional topics in the field, the editors particularly welcome work that breaks new ground by offering historical considerations of issues such as the linkages of academic and applied psychology with other fields, for example, psychiatry, anthropology, sociology, and psychoanalysis; international, intercultural, or gender-specific differences in psychological theory and research; or the history of psychological research practices. The series will include both single-authored monographs and occasional coherently defined, rigorously edited essay collections.

Also in the series

Gestalt psychology in German culture, 1890–1967
MITCHELL G. ASH

Constructing the subject: Historical origins of psychological research
KURT DANZIGER

Changing the rules: Psychology in the Netherlands, 1900–1985
TRUDY DEHUE

The professionalism of psychology in Nazi Germany
ULFRIED GEUTER

Metaphors in the history of psychology
edited by DAVID E. LEARY

Inventing our selves: Psychology, power and personhood
NIKOLAS ROSE

Crowds, psychology, and politics, 1871–1899
JAAP VAN GINNEKEN

Measuring minds: Henry Herbert Goddard and the origins of American intelligence testing
LEILA ZENDERLAND

Rebels within the Ranks

Psychologists' Critique of Scientific Authority and Democratic Realities in New Deal America

Katherine Pandora
University of Oklahoma

PUBLISHED BY THE PRESS SYNDICATE OF THE UNIVERSITY OF CAMBRIDGE
The Pitt Building, Trumpington Street, Cambridge, United Kingdom

CAMBRIDGE UNIVERSITY PRESS
The Edinburgh Building, Cambridge CB2 2RU, UK
40 West 20th Street, New York NY 10011–4211, USA
477 Williamstown Road, Port Melbourne, VIC 3207, Australia
Ruiz de Alarcón 13, 28014 Madrid, Spain
Dock House, The Waterfront, Cape Town 8001, South Africa

http://www.cambridge.org

First published 1997
First paperback edition 2002

Typeface Times Roman

A catalogue record for this book is available from the British Library

Library of Congress Cataloguing in Publication data
Pandora, Katherine, 1958–
Rebels within the ranks: psychologists' critique of scientific
authority and democratic realities in New Deal America / Katherine
Pandora.
p. cm. – (Cambridge studies in the history of psychology)
Includes bibliographical references.
ISBN 0 521 58358 6 (hardback)
1. Psychology – United States – History – 20th century. I. Title.
II. Series.
BF105.P36 1997
150'.973'09043–dc21 97-3472

ISBN 0 521 58358 6 hardback
ISBN 0 521 52494 6 paperback

For my family –
the Barneses, the Howells, the Kablers, the Mihaljeviches,
the Pandoras, the Pirišins, and the Rosenbaums –
who first taught me that there are many pasts before us

Contents

to this fascinating phase of the insect life cycle. We have thus aimed the contents of the book at postgraduate entomologists in all stages of their careers. However, we hope that this will not deter the more advanced undergraduate from finding something of interest within these pages.

S. R. Leather
K. F. A. Walters
J. S. Bale
1992

Foreword

The study of insect overwintering has largely concentrated on a few well defined areas of study. Whether this has been the result of the understandable reluctance of entomologists to expose themselves to the rigours of the great outdoors during what is one of the more taxing times of the year or to the fact that the importance of the overwintering stage in relation to the population dynamics of a particular species has been underestimated, is a moot point.

Be that as it may, it is a fact that a large proportion of a temperate or polar insect's life cycle is spent in the overwintering stage and recent work within allied groups, e.g. red spider mite, and within insect groups such as the Aphidoidea, has highlighted the advantages, in terms of control and prediction, to be gained from a detailed knowledge of insect overwintering habits. The contents in each chapter of this book have been largely determined by the availability of published work and by what we have considered to be the more important aspects of this subject. However, we have included previously unpublished material in an attempt to make this a comprehensive and enlightening addition to the field. The overwintering habits of some insect groups have been largely ignored by the entomological world and we have tried, where possible, to point out areas where future research would be profitable. One of the great problems that has hampered the study of overwintering in insects has been as Danks (1978) points out, the tendency of researchers to consider ecology and physiology as two separate and unrelated disciplines. We have tried wherever possible to adopt an ecological and physiological approach to the problem and feel that by so doing, we have thrown new light on this subject.

We hope that this book, by emphasising the importance of insect overwintering, will induce other entomologists to pay more attention

Contents

Acknowledgments

I would never have reached this point without the support of many people. As a graduate student in psychology in the interdisciplinary department of Human Development and Family Studies at Cornell University, I was first exposed to primary source research in a course offered by Joan Jacobs Brumberg, who subsequently supervised my master's thesis. Her oversight of my forays into historical and cultural analysis were of profound importance. It was this experience, along with the compelling teaching of sociologist Glen Elder and historian Michael Kammen, that would lead me to pursue a doctorate several years later in the history of American science.

Upon entering the program in history of science at the University of California, Los Angeles, I received crucial initial advice and direction from Robert Westman. He not only helped me formulate my topic and begin research, but he also encouraged me to take seriously a number of key issues regarding cultural history and science and religion. Both he and M. Norton Wise helped me work through a wide array of historiography during this first phase of my training. At UCLA I was also fortunate in being able to further develop my ideas through extended discussions with Richard Weiss in history – who imparted to me his own deep fascination with the 1930s and encouraged me to look beyond conventional categories – and Jeffrey Prager in sociology, who shared with me his thoughts on the psychological and sociological issues inherent in literature on the individual and society. My discussions with visiting Professor Marian Lowe in Women's Studies were also invaluable.

When I transferred to the University of California, San Diego, Science Studies program upon its inception, I entered the second phase of my training, in which the teaching of Robert Marc Friedman, Philip Kitcher, Sandra Mitchell, Chandra Mukerji, Martin Rudwick, and Steven Shapin helped me expand my grasp of the interdisciplinary literature in science studies and add new layers of analysis to the issues that I would examine in this book.

As dissertation committee members, Friedman, Mitchell, Prager, Shapin, Weiss, and Westman read earlier versions of these chapters, and their critique of my work in its formative stages was most helpful. My studies were supported by a Regents' Fellowship from the University of California, University of California graduate fellowships, and fellowships from the Science Studies program and History department of UCSD.

During the third phase of my training, as an Andrew Mellon Postdoctoral Research Fellow at the University of Oklahoma's History of Science Department – and then as a member of the department – I acquired further debts in the development of this project. I am particularly grateful for the collegiality and support of Peter Barker, Steven Livesey, Gregg Mitman, Marilyn Ogilvie, Jamil Ragep, and Kenneth Taylor as I revised this work. Above and beyond this assistance was the intellectual comradeship extended by Gregg Mitman, who also read and commented on parts of the book and generously shared his own research expertise with me, as well as his work in progress. Thanks to the generous support provided by Cheiron, the History of Science Society, the UCSD Science Studies program, the Andrew Mellon Foundation, and the University of Oklahoma, I was able to attend annual meetings of the History of Science Society and of Cheiron. There I presented papers and participated in discussions that helped focus and refine my thinking. I also appreciate the opportunity to present my work at Northwestern University and the University of New Hampshire.

I owe a special debt to Lois Barclay Murphy, who generously shared her recollections and understandings of her life and that of Gardner Murphy, and of psychology during the 1930s. She responded kindly and patiently to my questions by letter, over the telephone, and in two personal interviews in November 1988. Eugene and Ruth Hartley were also gracious enough to spend an afternoon giving me a sense of psychology during the 1930s. Professors Marian McPherson and John Popplestone at the American History of Psychology Archives, University of Akron, offered guidance during my extended visit there to consult the Gardner and Lois Barclay Murphy Papers and the Archives of the Society for the Psychological Study of Social Issues. Archivist Sharon Ochsenhirt provided indispensable advice and expert assistance, and both she and archivist John Miller made the visit a pleasure. I owe thanks as well to the on-site staffs of Pusey Library at Harvard University, where I consulted the Gordon Allport Papers, the Edwin G. Boring Papers, the James Ford Papers, and the S. S. Stevens Papers; of the National Library of Medicine in Washington, D.C., where I consulted the Lois Barclay Murphy Papers and the Lawrence K. Frank Papers; and the Concord Free Library in New Hampshire, where I consulted the Murphy Family Papers. In addition, the staffs of Sarah Lawrence and Vassar universities greatly assisted me by mail, and the interlibrary loan departments at UCLA, UCSD, and the

University of Oklahoma made my investigation of the literature of early-twentieth-century America much easier.

This project would never have been completed without the enthusiasm and counsel of Mitchell Ash and William Woodward, coeditors of this series, who provided insightful critiques of my work, as did two anonymous readers. Editor Alex Holzman and the production staff of Cambridge University Press have displayed great patience in guiding a first-time author through the process of turning a manuscript into a finished book, and I thank them as well.

My greatest debt is to my husband, Ben Keppel, who experienced with me both the insanity and the benefits of writing dissertations and first books simultaneously. Ben provided cheer when my own spirits flagged, advice when I got stuck, and his own considerable insights into American history, which he shared with me at all hours of the day and night. My thinking on the cultural history of science has been enlarged by attending to his explorations of the politics of memory, and by sharing with him my own.

Introduction

> It is [the] combination of personal values without deep social underpinning which doubtless explains . . . the fact that we allow economic disorganization, which produces fourteen or fifteen million unemployed and their families to suffer; then, instead of letting them die off, we feel sorry for them to the extent of giving them "relief," maintaining them at a starvation level. It is a question whether such a combination of emphasis upon and violation of social values can be long sustained.
>
> Lois Barclay Murphy (1937)

> As . . . psychology goes beyond sheer common sense, and becomes dynamite to society, those dominant in society will try to protect themselves against the explosion. It seems to me that this is a sufficient answer to the ivory-tower remark that we should stick to science and let public practice alone. The answer is that public practice will not and cannot let psychology alone.
>
> Gardner Murphy (1939)

> To a large degree our division of labor is forced, not free; young people leaving our schools for a career of unemployment become victims of arrested emotional and intellectual development; our civil liberties fall short of our expressed ideal. Only the extension of democracy to those fields where democracy is not at present fully practised – to industry, educational administration, and to race relations for examples – can make possible the realization of infinitely varied purposes and the exercise of infinitely varied talents.
>
> Gordon Allport (1940)[1]

During the 1930s, the United States underwent one of its greatest periods of economic, social, and political crisis, as vocal groups of Americans engaged in intense debate over truths that no longer appeared to be self-evident. In an improvised chautauqua of national proportions, long-established patterns of authority were questioned, the nature of reality disputed, and the means and ends of the search for knowledge subjected to critique. Of the many questions reverberating through a diverse array of cultural arenas, none were so powerful as those placing the meaning of "democracy" under scrutiny. Given the scope of these cultural contests, it is not surprising that the moral, intellectual, and social relations of science became issues for public debate and professional soul-searching.[2]

Although the unsettling dynamics of life in 1930s America gave specific form and urgency to the questions put to the scientific community, such challenges existed prior to the great stock market crash of 1929, most notably in the radical theories of knowledge that preceding waves of thinkers had been developing since the closing decades of the nineteenth century.[3] Many of these theorists, impressed as they were by scientific achievements, were nevertheless disturbed by triumphalist pronouncements that seemed to be closing off debate concerning the nature of science. Willing to experiment with the idea of science itself, such critics rebelled against restrictive definitions of what properly constituted the boundaries of scientific life. This view was aptly expressed in 1925 by philosopher Alfred North Whitehead, when he asked in *Science and the Modern World:* "Is it not possible that the standardised concepts of science are only valid within narrow limitations, perhaps too narrow for science itself?"[4]

The arguments of these intellectual provocateurs and the depression era's debates over democratic ideals converged in the work of a cohort of social and personality psychologists who gained increasing visibility as the decade unfolded. Among those figuring prominently in this dissent from the status quo are the three psychologists – Gordon Allport (1897–1967), Gardner Murphy (1895–1979), and Lois Barclay Murphy (1902–) – whose work, circumstances, and communities of affinity I discuss in the following chapters.[5] The cultural critique articulated by Allport and the Murphys during the 1930s placed them at a tangent to the intellectual mainstream of scientific and political life, but not outside of it: they were rebels *within* the ranks of academic science.[6] They dissented from orthodox views that endorsed the status quo, but they gave voice to that dissent as fully credentialed members of the guild.

As the 1930s progressed, each of these psychologists acquired significant authority, as manifested in various ways. Allport, for example, served as editor of the *Journal of Abnormal and Social Psychology* beginning in 1937, was awarded a starred entry in the sixth edition of *American Men of Science* in 1938, and was elected president of the American Psychological Association (APA) the following year. Gardner Murphy was elected to both the APA council of directors and the presidency of the Society for the Psychological Study of Social Issues (SPSSI) in 1937; in 1939 he missed being elected APA president by eleven votes. He would receive this honor in 1944, the same year he would receive his "star" in the seventh edition of *American Men of Science.*[7] Murphy's influence was also furthered by his editorship of a monograph series for Harper's. In contrast, Lois Barclay Murphy chose, for the most part, to avoid organizational entanglements and committee obligations – she, never, for example, joined the APA. Murphy did, however, have the enthusiastic support of influential foundation officials such as Lawrence K. Frank during his tenure with both the Rockefeller and the Macy foundations, receiving substantial financial assistance to conduct her dissertation research and to establish a nursery school laboratory at Sarah Lawrence in 1937.[8] Such support enabled her and her collaborator, psychologist Eugene Lerner, to assemble a multifaceted consulting team to launch their research program.[9] If, by the decade's end, the work of these three psychologists was still considered to be at odds with ascendent conventions, it could hardly be characterized as marginal, and their dissent had, in fact, gained considerable momentum.

Social activists as well as social scientists, Allport and the Murphys drew on a densely interrelated set of resources in constructing their scientific practices and in articulating their cultural critique. In assaying the fields of social and personality psychology, these three individuals sought footholds from which they could simultaneously argue for reconstructing American society and American science: indeed, they believed that unless both efforts advanced together, neither would be successful.

Unlike the majority of their colleagues, Allport and the Murphys rejected the image of the laboratory as an ivory tower, contested the canons of objectivity that characterized current research practice, and argued against reducing the natural and the social worlds to the lowest possible terms. They realized that, in giving priority to questions about the "individual" and the "social" in their psychological research, they were manipulating cultural categories laden with political implications. Choosing to work at the intellectual intersection of the social and the personal in 1930s America was to plunge oneself immediately into the midst of a cacophonous crowd of communities struggling to come to grips with the meaning of "America," as customary configurations of values were questioned in the wake of the economic collapse.[10]

That historical, economic, and political questions played themselves out within personal and social worlds suggested to Allport and the Murphys that psychological research could be used to critique American culture and thus to help create a more democratic polity. That such a possibility existed did not change their judgment that psychology as then currently constituted displayed serious flaws as an analytical discipline. In a 1937 letter, for example, Allport characterized psychology as "a crude and arrogant discipline" and counted among "the worst of its historic blunders" such defects as its "excessive empiricism, grotesque nativism, traffic in boggled ethics, superficiality, and undue abstractness."[11] Gardner Murphy, for his part, was disturbed that

"17th-century naive mechanism" still held sway in so many of the sciences, and that psychology had "been content to model itself upon physics and biology and instead of challenging their tenets has felt that its own scientific status depended in large part upon acceptance of the standard world view."[12] Lois Barclay Murphy likewise complained that "research on children was largely dominated by criteria of objectivity whose influence it was not easy to transcend."[13] She suggested that "experimental human psychology" might do well to ponder the fact that "long ago a seer who was concerned with control of behavior remarked that the Sabbath was made for Man, not Man for the Sabbath."[14]

Despite finding themselves at odds with disciplinary dogma during this period, Allport and the Murphys judged that they could successfully challenge the status quo by furthering a scientific philosophy that was unafraid, in the words of William James, of lying "flat on its belly in the middle of experience."[15] The "narrow limitations" of the "standardised concepts of science" – to appropriate Whitehead's terms – became especially apparent in the attempt to study personal and social experience. To overcome these limitations, one could either advocate abridging the experiential world so as to fit scientific conventions or urge that scientific practice be recast in an attempt to capture larger and more complex dimensions of experience. Allport and the Murphys pressed the latter view, arguing that there was greater need to refigure scientific reality than to uphold scientific custom.

In discussing the work of these psychologists I pursue two ends. First, I try to elucidate the nature of the dissent advanced by Allport and the Murphys, by examining how they used various intellectual legacies, how they worked and argued, and what perspectives they developed. Here, I also explore the relationships that existed between their work within social and personality psychology and the cultural politics of this pivotal period. Second, I use this episode of psychological discontent to point to the existence within the scientific enterprise of robust critiques of the rules of the game, and to demonstrate the need for historians of modern science to examine the contours of dissent, as well as the structures and products of assent. In the conclusion, I address this issue further, suggesting why a critique that emerged from within *social* science possesses relevance for reflecting on the nature of the sciences in general.

The "Intellectual Commons" of 1930s America

Although it is true that psychology was Allport's and the Murphys' primary discipline, the nature of their thinking and the significance of their ventures cannot be understood without some idea of the extent to which their activities repeatedly traversed an "intellectual commons" made possible by the search for new structures of meaning in the 1930s. One of the reasons that the work of these three scientists is of historical interest is precisely because they chose to place themselves directly within this crowded intellectual crossroads, actively entering into the cultural issues at play during this period.

Within the frame of this trio's discourse, representations of science and representations of America were intimately interwoven, mediated by a core set of terms: the "individual" and the "social."[16] The sense in which they used these terms is similar to that given in a statement of John Dewey and John Childs, from a 1933 essay: "*Social* cannot be opposed in fact or in idea to *individual.* Society *is* individuals-in-their-relations. An individual apart from social relations is a myth – or a monstrosity."[17] Attempts to conceptualize the mutuality of the "individual" and the "social" were expressed in numerous ways during this period: for example, Lois Barclay Murphy spoke of "the total personality in its cultural setting" and of "the broader problem of personality development as a whole and the social context in which it appears"; Allport referred to "the social framework within which personality develops" and also used the expression "personality and culture," a shorthand tag that gained wide currency.[18] When referring in a general way to these perspectives, I will use the term "the-individual-in-social-context," a standardization of these various locutions.

As an overarching idea, there could scarcely be anything more prosaic than the concept denoted by the term "the-individual-in-social-context," and yet, as elaborated by activist psychologists such as these three, this framework embraced a variety of destabilizing assumptions. In the hands of Allport and the Murphys, the study of the "individual" became a consideration of "individuality," a concept at odds with the widespread belief that questions of uniqueness and singularity belonged to the arts, not to the sciences. Alternatively, attention to "contextuality" sat uneasily beside methodological prescriptions to abstract the objects of scientific scrutiny from the fields in which they were embedded in the quest for universal laws; in turn, such revisionist thinking brought into question laboratory practices assumed to be unproblematic. And, perhaps most disquieting of all, it required only a small leap of logic to move from the more general sense of "the-individual-in-social-context" to a more sharply defined category, such as that of "the-*scientific*-individual-in-social-context," and to challenge the presumption that scientific investigation existed apart from the larger polity. For the "pure" science contingent whom they were debating, the scientific dubiety of such constructs is indicated by the elaborate statistical and experimental lengths to which they went to "eliminate" them.[19]

In striving to work through the ramifications of the individual and the social from their positions within personality and social psychology, Allport and the Murphys participated in, and helped to further, arguments that cut across a number of disciplinary spheres of discourse, including anthropology, biology, history, philosophy, progressive education, religion, sociology, and the arts. One such argument, voiced by anthropologist Ruth Benedict in *Patterns of Culture* (1934), was that, "at different points in the interpretation of culture forms, both history and psychology are necessary; one cannot make the one do without the service of the other."[20] Similarly, the 1940 manifesto, *The Cultural Approach to History,* opened with a section on "techniques of cultural analysis" comprised of chapters by an anthropologist, a social psychologist, and a psychoanalyst. Historian Caroline Ware, editor of the volume, observed that the anthropological and the psychological "meet in the concept of 'personality,' the 'individual-in-society,'" and she maintained that historians had much to learn from the ongoing efforts of social scientists to deploy this framework.[21] In a like manner, Gardner Murphy called upon his psychological colleagues "to see the individual clearly in his full cultural context" by integrating anthropological and historical materials into their psychological analyses.[22]

The forging of such interdisciplinary sensibilities was a product of both communal discourse and personal encounters. As a young married couple, Columbia instructor Gardner Murphy and Sarah Lawrence faculty member Lois Barclay Murphy were friends with anthropologist Margaret Mead and discussed her fieldwork experiences with her; they also sat in on one of Benedict's anthropology courses, on "primitive religions."[23] Allport had close ties to colleagues in Harvard's sociology department, chief among them Pitirim Sorokin, who published a four-volume study, *Social Dynamics and Culture,* in the latter 1930s. Allport's course in social psychology was open to students in the sociology department and included such readings as Benedict's *Patterns of Culture,* Gregory Bateson's *Naven,* Karl Mannheim's *Ideology and Utopia,* and Thurman Arnold's *Symbols of Government.*[24] Sociologists were also part of the Murphys' New York intellectual circle, with Robert and Helen Lynd, authors of the sociological community studies *Middletown* (1929) and *Middletown in Transition* (1935), being especially close friends. It was Robert Lynd, in fact, who arranged for Gardner Murphy to attend an interdisciplinary conference organized by anthropologist Edward Sapir for the Social Science Research Council (SSRC) on the theme of "personality and culture" in 1930, which is where Murphy first met Allport.[25] As members of an SSRC subcommittee, Allport, along with Murphy and Mark May, worked with an interdisciplinary group of scholars to prepare an SSRC bulletin entitled *Memorandum on Competition and Cooperation.*[26] In the published report, Allport, Murphy, and May stated that the area being designated "for the want of a better name, 'Personality and Culture,' seems to represent one more attempt to find a field of inquiry that will enlist the active interest and joint support of psychologists, psychiatrists, anthropologists, and sociologists."[27]

Their literary output during the 1930s was crowned by the Murphys' *Experimental Social Psychology,* which appeared both in 1931 and in a revised edition in 1937, and by Allport's 1937 volume, *Personality: A Psychological Interpretation.*[28] Both projects were ambitious attempts at agenda-setting, as the authors made clear. Allport remarked, in regard to the psychology of personality, that he had answered "an insistent demand for a guide book that will *define* the new field of study – one that will articulate its objectives, formulate its standards, and test the progress made thus far." The Murphys stated, in regard to experimental social psychology, that they were responding to the "need for a volume suggesting what it is and what it may hope to become."[29] These massive tomes (*Personality* had 588 pages, *Experimental Social Psychology* 709, the revised edition 1,121) broadcast their authors' intention to place themselves at the forefront of debates in the still-developing fields of personality and social psychology.

That Allport made his presence felt most explicitly in the realm of personality psychology and the Murphys in social psychology should not be construed, however, as indicating that any of these three restricted themselves to one area of emphasis; quite the contrary. Allport contributed a review chapter entitled "Attitudes" to the 1935 *Handbook of Social Psychology,* while Gardner Murphy produced the survey *Approaches to Personality: Some Contemporary Conceptions Used in Psychology and Psychiatry* five years before Allport's interpretive treatise on the subject.[30] Their overarching goal was to find ways to juxtapose the individual and the social, and in some of their works this commitment is evident on the title page: Lois Barclay Murphy's 1937 study, *Social Behavior and Child Personality: An Exploratory Study of Some Roots of Sympathy,* is one expression of this concern, as are such analytic turns as Allport's essay "Dewey's Individual and Social Psychology" and Gardner Murphy's article "Personality and Social Adjustments" (or, indeed, the remarks under the section "The Individual in Relation to His Culture" in the revised *Experimental Social Psychology*).[31]

Nor were these psychologists' efforts restricted to professional audiences during this period. Allport sought a wider public with *The Psychology of Radio,* a work indicting corporate control of the "ether."[32] Gardner Murphy produced *Public Opinion and the Individual,* an inquiry into attitudes on social issues. He also collaborated with author Paul Grabbe on a book for a broad audience, entitled *We Call It Human Nature.*[33] Allport and the Murphys also brought their thinking on various psychological topics before the public in such popular magazines as *Understanding the Child* and *The Family* (Allport), *Harper's Magazine* and *American Magazine* (Gardner Murphy), and *Good Housekeeping* and *Parents' Magazine* (Lois Barclay Murphy).[34]

The "Disciplinary Commons" of 1930s American Psychology

The scope, form, and content of the work of Allport and the Murphys have received little notice in histories of the social sciences. To some extent, research such as theirs has been overlooked because historians interested in the 1920s and 1930s have been focusing on the rise and elaboration of behaviorist ideologies of science. The "behaviorist standpoint" in psychology was most forcefully articulated by John Watson in his manifesto of 1913; at heart, as Stephen Toulmin and David Leary observe, Watson sought "to make psychology as close to experimental physics as he knew how."[35] Watson's "classical" behaviorism soon yielded to the efforts of revisionists, with the result that a number of neobehaviorist and neopositivist stances contributed to the establishment ethos of the 1930s.[36] The reigning orthodoxy on the question of studying the individual-in-social-context, as expressed by John Dashiell in his 1928 text, *Fundamentals of Objective Psychology,* struck a decidedly chastening tone. Proper scientific procedure – that of "natural science" – Dashiell affirmed, did not allow the observation of "concrete persons as wholes in complex social situations."[37]

Dashiell's insistence that the legitimacy of psychological science lay in its adherence to principles derived from natural science represents one of the key tenets of what can be loosely characterized as the neobehaviorist community. This trope was stirringly rendered in the 1936 APA presidential address of Yale psychologist Clark Hull, a neobehaviorist standard-bearer. Hull exhorted his audience to adopt "strictly orthodox scientific methodology," which he illustrated with allusions to the work of Galileo. He then bracingly reminded his audience that Galileo had practiced "this methodology at the imminent risk of imprisonment, torture, and death," while his twentieth-century colleagues had only to wrench themselves free from "the thrall of the Middle Ages" by throwing off "the shackles of a lifeless tradition."[38] Jill G. Morawski, in her perceptive analysis of Hull and his associates at Yale's Institute of Human Relations, argues that Hull envisioned scientific research as proceeding "most economically if structured like a psychic machine, an automatic mechanism free from subjectivity." Not surprisingly, Hull inculcated scientific orthodoxy in seminar participants by requiring them to study Isaac Newton's *Principia Mathematica.*[39]

Such neobehaviorist exhortations existed alongside psychology's increasing emphasis on laboratory experimentation, mental testing, and statistical procedures.[40] These methodological imperatives – aptly described by Toulmin and Leary as the foundation of a "cult of empiricism" – were proposed as prerequisites for the establishment of rigor and objectivity within psychological science.[41] Since the mid-1970s, these interwar trends have been increasingly scrutinized by historians of psychology from a number of vantage points. Such work has detailed the pervasive reach of neobehaviorist initiatives and has illuminated the diverse array of commitments these initiatives advanced.[42]

My use of neobehaviorism as an umbrella term for the kinds of disciplinary moves advocated by psychologists such as Dashiell and Hull inevitably homogenizes a diverse sphere of discourse that displayed idiosyncrasies at every level, from foundational assumptions and methodological prescriptions to guiding metaphors and final conclusions. Psychologists who found themselves at odds with the general trend of pledging allegiance to the physical sciences, such as Allport and the Murphys, understood that scientists such as Karl Lashley, Clark Hull, B. F. Skinner, and Edward Tolman did not hew to a neatly categorized set of principles that rendered their work interchangeable.[43] Nevertheless, such neobehaviorist efforts often shared a constellation of values intended to channel academic research into increasingly restrictive paths, by assigning priority to what were declared to be rules derived from physical science in framing research questions. Such strictures (whether characterized by participants as objectivist, positivist, logically empiricist, physicalist or behaviorist) frequently sanctioned reducing human phenomena to animal analogues, physiological substratum, statistical distillates, or mechanistic systematics.[44]

In 1931 Robert S. Woodworth, one of psychology's elder statesmen, remarked that "behaviorism is a spirit or attitude rather than any fixed theory."[45] My use of neobehaviorism as a collective term is meant to evoke this sense of common spirit as well as to refer to a set of academic ventures, for figures like Allport and the Murphys objected not only to aspects of the methods and theories offered by "austere" empiricists but also to the restrictionist tone in which the debate was being waged. If a certain degree of latitude was allowed within the neobehaviorist circle – as in, say, Tolman's nonconventional insistence on the purposive and cognitive nature of behavior – those who worked outside of it were not strangers to feeling, in Allport's words, "the scorching displeasure of behaviorists and objectivists" turned upon them.[46] When Allport publicly attacked what he saw as a mood of coercion and intolerance being promoted within psychology, he was scored by colleague E. G. Boring for presumably displaying a parallel dogmatism in objecting to the move to establish neobehaviorist principles as normative. Allport replied, "In so far as current trends displease me, I reserve my right to fight them, even though I am in the minority. But I think I can fundamentally tolerate the views of those I am fighting." Stating that he realized "the danger of saying that the intolerance lies on the other side," he nevertheless asked "is it not in fact likely to lie on the side of the majority (meaning by majority the distinguished and articulate members of the profession)?"[47]

Although Gardner Murphy avoided the head-on confrontation Allport mounted, he also pointed to aspects of behaviorism that gave him pause. In offering a definition of behaviorism in his 1929 text, *An Historical Introduction to Modern Psychology,* Gardner Murphy observed, "Behaviourism has become in many quarters simply a name for mechanistic psychology, or has been reduced to a mere *emphasis* upon objective, as opposed to subjective, data." In Murphy's view, the essence of behaviorism was its promise that psychology would one day achieve the objectivity held to be characteristic of physical science, with all psychological data being "verified in the manner of the physical experiment."[48] Yet in his concluding remarks Murphy found reason to counsel caution regarding the fruitfulness of experimentation. Although he allowed that experimental methods had gained "in variety and in reliability," he nonetheless claimed that they had proven "hopelessly unable to keep pace with the imperative demand for more factual material upon the emotional and volitional life, the nature of suggestibility and imitation, the relative importance of heredity and environment in the causation of individual differences, the manner in which social likes and dislikes, ambitions and ideals, are acquired, and a host of equally pressing questions." Murphy recommended an emphasis upon developmental perspectives as a remedy for this deficiency in experimental efficacy.[49] Indeed, he asserted that one hundred years into the future, "when the laws of physiology and of quantitative psychology have merged," that psychology's major methodological challenge would have yet to be conquered: the struggle "to devise reliable methods for the direct study of experience, methods which we cannot at present even dimly outline."[50]

In his surveys of the psychological discipline at the end of the 1930s, Allport tracked the growing influence of experimentation based on "the self-discipline of mathematics and of the natural sciences." The current popularity of animal research, Allport argued, was due to "its delightful suitability for the exercise of objective and approved methods. By studying rats, not men, we gain status as scientists, for like the natural sciences we can, in this line of investigation, employ precision techniques and operational modes of communication."[51] Allport remained unconvinced of the relevance of such ideals, maintaining that "most of the vital and practical questions of psychology are difficult to approach with the sterilized forceps of strict physical science."[52] In his APA presidential address, Allport underscored this point to the assembled audience by detailing his bemusement when a close colleague challenged him to identify a single psychological question that could not be solved by using rats as subjects.[53] Somewhat taken aback, Allport managed to murmur something about the psychology of reading disabilities. Upon reflection, there came flooding to mind "the aesthetic, humourous, religious and cultural behavior of men. I thought of how men build clavichords and cathedrals, how they write books, and how they laugh uproariously at

Mickey Mouse; how they plan their lives five, ten or twenty years ahead." Could the study of rats, he asked, clarify how some individuals "by an elaborate metaphysic of their own contrivance . . . deny the utility of their own experience, including the utility of the metaphysic that led them to this denial"?[54] He ended his address by accusing those who sought to confine psychological inquiry to the empirical, the mechanistic, the quantitative, the nomothetic, the analytic, and the operationist of "authoritarianism," an especially loaded term in light of global political developments in the 1930s.[55]

During the interwar years, prominent members of the neobehaviorist community asserted that the local authority that held among those working within that community's parameters should be extended into a collective authority that would be mandatory and irrevocable in regard to the discipline as a whole. Psychologists' claims to be legitimate heirs to the scientific legacy of Galileo and Newton were in fact presented as depending upon the acceptance of neobehaviorist research practices. Where scientists such as Allport and the Murphys looked to extend the disciplinary commons of psychology into the intellectual commons of 1930s America, influential neobehaviorists sought to institutionalize their research community as the disciplinary commons itself. The consequences of opting for restrictionist modes of scientific investigation did more than pare back the contents of psychological toolkits. Choices in methodology, as Danziger argues, place limits upon "the kind of reality that can be represented in the products of scientific investigation."[56] This is precisely what was at stake in the debates between those making up what I have termed the neobehaviorist status quo and psychologists such as Allport and the Murphys, who considered themselves to be "rebels within the ranks."

Explorations of such periods of contestation offer insight into issues specific to psychology, but they can offer something more as well, for these telling descriptions of discipline regulation also speak to historical matters that reach beyond the field of psychology itself. As historian Mitchell Ash has astutely pointed out, the struggles of academic psychologists to gain legitimacy as scientists provide "a counterimage marking the limited reach of such scientific ideals as objectivity, measurability, repeatability, and cumulative knowledge acquisition."[57] The problematic nature of these ideals is an issue that dissenting scientists such as Allport and the Murphys began to explore during the 1930s.

Even though neobehaviorist ideologies had become orthodoxy by the 1930s, the scientistic moves made by devotees of the "cult of empiricism" continued to encounter opposition, disruptions, and impediments. In this book, I examine how and why some contemporaries challenged the commitments that underlay the quest for scientific purity that preoccupied so many researchers. In doing so, my aim is not only to describe what these dissenting scientists believed that they were fighting against but also to examine the vision of scientific knowledge that they sought to advance, and the ways in which their efforts converged with other cultural disputes over representing reality. In approaching this topic, I have kept in mind historian Fred Matthews's admonition that "one of the dangers of the approach to intellectual history through ruling paradigms is that it leads us to assume a deeper and more sweeping consensus than actually existed."[58] And, as Allport himself warned, "To trace the history of any department of science rapidly is a dangerous thing, for epochs intertwine and each age has its dissenting voices."[59]

To what extent were the dissenting voices of Gordon Allport, Gardner Murphy, and Lois Barclay Murphy raised as a group? It is certainly the case that they admired each others' work and saw their values as compatible. In an autobiographical essay, Gardner Murphy stated: "It is self-evident to anyone who knows me that I have learned a great deal more psychology from Lois than from any other living person," while Lois Murphy described their marriage as "a profound, close, enduring, and creative relationship growing out of broad and deeply shared loves and values."[60] In turn, Lois Barclay Murphy described Allport as "a kindred spirit," and Allport characterized Gardner Murphy as one of the "first cousins of my thinking."[61] Indeed, Allport argued in 1966 that the APA's highest award should be jointly granted to both the Murphys.[62]

More particularly, in regard to political engagements, the three display parallel commitments. Gardner Murphy was an energetic member of the Columbia University Faculty Committee for Aid to the Spanish People, which raised funds and pressured the Roosevelt administration to support the Spanish loyalists against Franco's fascist forces. The Federal Bureau of Investigation (FBI) records that Murphy was active on behalf of civil liberties, supporting, for example, the American Committee for the Protection of the Foreign Born, the Citizen's Committee for labor leader Harry Bridges, and signing a 1940 statement warning against following the example "of many European countries where suppression of the Communist Party was but a beginning, followed by a campaign against trade unions, cultural groups, Jews, Catholics, and Masons, ending with the destruction of all freedom."[63] Both Murphy and Allport were instrumental in guiding the early years of SPSSI, an organization that aimed at using psychological research to counter fascism and racism and to support such groups as labor unions.[64] Allport was a prominent member of the American Federation of Teachers at Harvard during the 1930s and actively recruited faculty as members.[65] Lois Barclay Murphy was fired after her first year at Sarah Lawrence for "fomenting" faculty demands for democratic participation in the governing of the college (she was rehired a year later when the president was replaced).[66] A report from an FBI agent remarked that Sarah Lawrence had the reputation of being extremely progressive, with no color line, and that many of the faculty members appeared to be of foreign birth; Murphy was classed with those who were either "starry-eyed idealists" or members of the Communist party.[67] Murphy insisted that her students learn about social issues firsthand in fieldwork courses, taking her Sarah Lawrence students into New York City to observe homelessness, for example, or to sit in on a club meeting of unemployed men.[68]

Although they agreed on a great deal, these three psychologists did not always speak with one voice. Allport, who chose to present himself as a latter-day Martin Luther, was clearly the most polemical and outspoken of the three, especially on the scientific significance of "individuality." Gardner Murphy opted most often for a tone of sophisticated tolerance, and a less conspicuous strategy of "boring from within," seeking to turn conventional practices such as experimentation and quantification to unconventional ends. Embracing the stance of the "explorer," Lois Barclay Murphy shared Allport's more forthright style in addressing methodological and theoretical questions, although not his most provocative rhetoric.[69] Such divergences in form relate to dissimilarities that existed in the disciplinary challenges that confronted each of these three psychologists, to differences in the nature of their local social and professional locations, and to the varying weight that each assigned to common priorities (distinctions that I discuss at greater length in chapter 2).

In the particulars of their lives, there is as much to mark them as different, perhaps, as there is to designate them as confederates. Within the framework of this book, however, I have emphasized their shared commitments, which infused their separate undertakings with related patterns of meaning, and which rendered them dependable allies in the pursuit of revisionist ends. In treating these three as joined together in identifiable ways, I do not mean to imply that they alone are of interest in studying the currents of dissent that gained strength in psychology during this period. Others affiliated with the goals and values of Allport and the Murphys at this time were such figures as Barbara Biber, Kenneth and Mamie Clark, Eugene and Ruth Hartley, Otto Klineberg, Kurt Lewin, Lucy Sprague Mitchell, Henry A. Murray, Theodore Newcomb, Muzafer Sherif, and Goodwin Watson. I have chosen to concentrate on Allport and the Murphys because they actively sought to define the emerging fields of social and personality psychology, and because the scope of their collective research interests opens doors to numerous questions of historical significance.[70]

In focusing on the work of Allport and the Murphys during the 1930s, neither am I suggesting that their careers are somehow reducible to this time span, for this is clearly not the case. First of all, much of the work for which each of these three individuals would win their widest renown lay before them: Allport's *The Nature of Prejudice,* which appeared in 1954; Gardner Murphy's 1947 theoretical study, *Personality: A Biosocial Approach to Origins and Structure;* and Lois Barclay Murphy's longitudinal study of child life, published as *Vulnerability, Coping and Growth* in 1976.[71] Furthermore, during the 1950s and 1960s each would also be involved in psychological issues of national import. Each assisted, for example, in the efforts of those presenting the social science evidence cited in the Supreme Court's *Brown v. Board of Education* school desegregation decision in 1954; Lois Barclay Murphy participated in the planning and implementation of Head Start and Parent-Child Centers in the 1960s.[72]

All three continued to seek out institutional locations that would allow them to expand on their professional commitments: Gordon Allport at Harvard's interdisciplinary experiment, the Department of Social Relations, which was formed in 1946; Gardner Murphy as director of graduate studies at the City University of New York in the 1940s, and then as director of research at the Menninger Foundation during the next two decades; and Lois Barclay Murphy as director of a multidisciplinary research group that carried out a longitudinal study on coping behavior in children from 1953 to 1969, funded primarily by the National Institute of Mental Health. Allport and the Murphys would also receive various honors and forms of recognition for their professional achievements. In a 1951 survey requesting clinical psychologists to name the theorist who had influenced them most in their daily work, Allport was cited second to Freud; in a 1957 survey asking psychologists to state who had inspired them to go into psychology, Gardner Murphy similarly ranked second after Freud.[73] Both men would receive gold medals from the APA for their lifetime achievements, and Lois Barclay Murphy would receive that organization's G. Stanley Hall Award for Distinguished Contributions in Developmental Psychology.[74]

Although I hope to shed light on the foundational years of these three psychologists' careers, that is not my only purpose in writing this book. In truth, I have chosen to "borrow" their work and lives in order to explore critical dimensions of the cultural history of American science that have fallen outside the main lines of development presented in disciplinary histories.[75] The depression years were pivotal ones for social theory, as intellectuals such as Allport and the Murphys reworked the modernist legacies of European and Anglo-American thinkers in a world that was besieged by the failures of capitalism, unnerved by the success of new political philosophies such as fascism, and seeking to come to grips with the increasingly interconnected nature of global life in the wake of technological developments. In their responses to these transformations in the social order, Allport and the Murphys challenged prevailing understandings of the nature of scientific knowledge and of the place of science in American society. This book is a study of the resources that these scientists used to contest the status quo, the alternatives they sought to legitimate, and the ways in which their critique of scientific authority and their pursuit of democratic realities contributed to the improvisatory chautauqua under way in 1930s America.

In chapter 1, I discuss the intellectual frameworks that served as philosophical and political resources for Allport and the Murphys as they constructed their scientific lives: William James's radical empiricism and insurgent versions of the Social Gospel. As second-generation Jamesians, these three psychologists brought the philosopher's postpsychological thinking to bear on questions of experiential reality that many of their colleagues either sidestepped or ruled out-of-bounds; in doing so, they challenged orthodox understandings of the nature of scientific authority. As adherents of theological perspectives that placed social salvation above individual salvation, they brought values derived from the sphere where radical Progressivism and Christian socialism converged to bear on both the form and content of their research in personality and social psychology. In other words, they contested scientific conventions that placed religion and politics outside of science. In chapter 2, I consider the institutional contexts in which these three psychologists worked and the larger networks of which they were members, pointing to their ties to such areas as progressive education, social ethics, and parapsychology.

In chapters 3 and 4, I consider their attempts to study the "individual" and "social" aspects that make up the "individual-in-social-context" construct. I pay particular attention to the political and intellectual ramifications embedded in their use of the concepts of individuality and contextuality, and to the ways in which these concerns informed the scientific arguments they made and the research questions they framed. I note that a critique of capitalist economics underlay the concept of individuality in 1930s discourse, as well as a concern about how the idea of individuals-in-social-context could be extended to that of scientific-individuals-in-social context. I also consider Allport's and the Murphys' deep interest in conceptualizing the dynamics of resistance to social change and their concern that domestic fascism was on the rise.

In chapter 5, I turn to the ways in which natural history served as a countermodel to classical physics in the research practices of Allport and the Murphys, allowing them to bring questions of subjectivity and emotion into academic psychology. I describe the radically empiricist cast of their natural history-based research ventures as a form of "experiential modernism," that is, as a search for scientific forms of knowing that would unsettle conventional ways of thinking without simultaneously divorcing reason from feeling, and thus from the realm of moral sentiments.

In chapter 6, I elaborate on the wider patterns of cultural critique with which their work intersected during the depression era. I examine the relativistic nature of their scientific discourse, in which questions of diversity, unpredictability, and change were highlighted. I refer to these stances as "exploratory relativism," to indicate that this framework was not intended to sanction an ethic of "anything goes" but instead was to serve as a means of bringing unexamined assumptions into the open as the patterns of American culture became a topic of public debate in New Deal America.

1

The Deep Context of Dissent: Jamesian Philosophy and Social Gospel Theology

> In sum, James is a metaphysical democrat.
>
> Horace Kallen (1925)

> Political democracy without economic democracy is an uncashed promissory note, a pot without a roast, a form without substance.
>
> Walter Rauschenbusch (1912)[1]

As children of the Progressive Era who would later choose to work in the social sciences, Allport and the Murphys came of age in social and intellectual worlds that have been well scrutinized by scholars.[2] Indeed, increasing historical attention has been directed at the fact that the attitudes characteristic of their parents' generation – reformist zeal, faith in scientific knowledge, and passion for organization building – powered the professionalization and institutionalization of the social sciences at the turn of the century.[3] In many ways, the historiography of the social sciences in this period of disciplinary formation is a province of the larger realm of the historiography of Progressivism.

The most insightful set of interpretations that exist regarding the growth of the social sciences in the early twentieth century relate to the increasing appeal that scientism held for many practitioners. Dorothy Ross, for example, in *The Origins of American Social Science,* finds that "during the 1920s the cultural authority and social power of scientism pushed the center of gravity in the social science professions toward a harder and more technocratic conception of social science."[4] But as Ross herself has observed elsewhere, "The wave of scientism launched before and after World War I did not wholly prevail," and, in fact, "scientism set in motion substantial countercurrents in defense of social science methods that took into account the uniquely human attributes of subjectivity and choice," especially during the 1930s and 1960s, "when reformist attitudes arose in the society at large."[5] In comparison with the considerable historiographic literature that has developed regarding the Progressivist agendas that launched the scientistic wave, historical study of the Progressivist traditions that informed the "substantial countercurrents" noted by Ross has been less extensive. To assay what Robert Westbrook has called the "radical wing of progressivism" is to work against the grain of interpretive conventions; however, recent studies have begun to remedy this gap in our historiographic understanding, providing interpretive perspectives on the particular intellectual legacies upon which individuals such as Allport and the Murphys drew.[6]

Foremost among these efforts is James Kloppenberg's revisionist analysis, in *Uncertain Victory: Social Democracy and Progressivism in European and American Thought, 1870–1920,* of the transatlantic intellectual transformation that gave birth to a diverse community of thinkers elaborating a *via media* between revolutionary socialism and laissez-faire liberalism.[7] Kloppenberg disputes characterizations of *via media* philosophies as anemic compromises devised by individuals who, allegedly lacking the courage of their doctrinaire contemporaries, could offer nothing more than "unstable accommodations to the new realities of organized capitalism in an urban-industrial environment."[8] On the contrary, Kloppenberg presents a compelling rendering of a company of "restless challengers [who] jolted philosophical and political discourse from standard categories and made possible new ways of thinking."[9]

Kloppenberg argues that what distinguished these new ways of thinking from inherited patterns of thought was an insistence "that ideas emerge from, and must be validated in, neither language nor logic but life." Proponents of these new intellectual frameworks emphasized contingency as opposed to certainty, highlighted the interpretive nature of knowledge making, and supported perspectives that started from human experience, thus collapsing dualistic categories such as those of mind/body and subject/object.[10] Within these radical theories of knowledge, Kloppenberg relates, were the "seeds of a new political sensibility," one that received its stimulus from the belief "that knowledge begins in the uncertainty of immediate experience and that all ideas must remain subject to continuous testing in social practice."[11] Kloppenberg's "renegade philosophers" saw "the unpredictability of the democratic project" as an intellectual imperative to be taken seriously, not as an irksome liability to be pushed aside in favor of the certitudes of historical materialism, or to be kept at bay by the rationalized bureaucracies of paternalistic elites.[12]

In bringing their democratic critique to psychology during the 1930s, scientists such as Allport and the Murphys were not simply responding to transitory rumblings of discontent thrown up by the current economic crisis, although the shifting landscape of the depression era polity certainly presented new prospects for carrying their arguments forward. Working within the idioms of the *via media,* they belonged to a new cohort of radical Progressives and social democrats who were elaborating a perspective that judged the democratic credentials of American society to be in question owing to its economic and political arrangements, and that considered the authoritative claims of science to be compromised by metaphysical, social, and ethical commitments that turned away from the experiential world in favor of abstract purity. In their depictions of the "scientific," Allport and the Murphys indicated their commitments to what the word "American" should signify; likewise, their vision of America informed their interpretations of scientific reality. Through the representations that they developed, they explored what they took to be the nature of democracy, and the democracy of nature.

The radical bent of Allport and the Murphys does not conform to the categories of political dissent most familiar to American historians, who typically associate such dissent with membership in the Socialist or Communist parties. Many generative tensions present during the 1930s remain unanalyzed because American patterns of dissent have been considered inferior to European Marxism and therefore discounted, no matter how energizing the "natives" may have found them.[13] To understand the homegrown version of socialist thought that Allport and the Murphys adopted as members of the radical wing of Progressivism, it is necessary to see it in the light of the indigenous intellectual legacies of the *via media* on which they drew in taking their bearings, and which served as constitutive resources for their science: the radically empiricist philosophy of William James and the tenets of a radicalized Social Gospel.

William James's Radical Empiricism

In his 1935 biography of James, Harvard philosopher Ralph Barton Perry suggested that, although there existed "very few pure Jamesians, in the sense of direct descent," the world was nonetheless full of "mixed Jamesians."[14] Allport and the Murphys certainly belong to this tribe of "mixed Jamesians," for they saw the "hospitable framework" of William James as an especially rich intellectual bequest.[15] David E. Leary judges Allport as "perhaps the most 'Jamesian' psychologist of his generation," and Eugene Taylor argues that, among James's "intellectual descendants," Allport, Gardner Murphy, and Henry Murray form a "great triumvirate."[16] Lois Barclay Murphy speaks of her husband as being "in the William James tradition," and of Allport as "a kindred spirit" for the same reason.[17] Murphy herself states that, as a college student, she "tended to keep my distance from psychology and psychologists," because "the current psychology generally seemed to be so far from William James and what I had absorbed from both my mother and father with their James and John Dewey orientation."[18]

If it is a relatively straightforward matter to identify the intellectual frameworks of Allport and the Murphys as being "Jamesian" in outlook, historiographic traditions make it somewhat more difficult to specify what this signifies. James, along with Charles Sanders Peirce and John Dewey, is most well known as one of the founders of "pragmatism," which is a philosophical theory positing a new conception of truth, opposing the view that "the truth of an idea is . . . a stagnant property inherent in it" with the argument that truth is "in fact, an event, a process."[19] Identifying Allport and the Murphys as "pragmatists" would follow historiographic convention and offer some guidance in placing their work within the context of their times. But if "pragmatism" is a useful label for identifying certain features of James's thinking, it does not fully apply to the views of Allport and the Murphys. In current historiography, "pragmatism" denotes an intellectual movement as well as a set of particular theoretical statements, and although Allport and the Murphys would certainly have found much with which to agree in the work of those attempting to analyze, clarify, and correct the philosophical claims of the pragmatist tradition, such technical concerns were not theirs.

More problematic, however, is the fact that the pragmatist label obscures those aspects of James's intellectual inquiries that focused less on describing a method for determining truth, and more on exploring the nature of experience – inquiries such as *Varieties of Religious Experience: An Inquiry into Human Nature, A Pluralistic Universe,* and the articles collected posthumously as *Essays in Radical Empiricism.* At heart, the work of such "rebels within the ranks" as Allport and the Murphys drew less from pragmatism than from this second constellation of Jamesian concepts, that is, the "radically empiricist" aspects of his philosophy.[20]

Recent philosophical discussions regarding the slighting of James's radically empiricist explorations by subsequent commentators have begun to outline why it is necessary for historians to begin developing a historiography of radical empiricism. Philosopher John J. McDermott, for example, argues that James's identification with pragmatist philosophy "has turned out to be an unfortunate development, for although a pragmatic epistemology is an important strand in James's philosophy, it does not occupy the center of his vision." To the contrary, McDermott contends that the center of James's vision is to be found in those aspects of his work that he "referred to as radical empiricism."[21] Indeed, when Alfred North Whitehead declared that James had inaugurated "a new stage in philosophy," one that marked the end of 250 years of reliance on categories that Descartes had introduced in his *Discourse on Method* in 1637, Whitehead was not referring to James's pragmatist lectures, but rather to his pronouncements on radical empiricism.[22] As philosopher Richard Bernstein has pointed out, James's radically empiricist vision of experience "is nothing less than a critique of Western philosophic thought."[23]

To some extent, the pragmatist aspects of James's work have received undue attention because, as McDermott notes, "after James's death, pragmatism dominated the philosophical climate and James's thought was dragged, posthumously, into a fray which did not represent" the larger sweep of his intellectual efforts.[24] Similarly, historian David Hollinger has suggested that the "distinctive intellectual tradition" known as pragmatism that James helped to create was in fact "largely put together by others making use of James's work in the context of the writings of John Dewey."[25] In this earlier period, as in our own time, with the so-called pragmatist revival initiated by philosopher Richard Rorty, interest in James is linked with pragmatism not because of a critical engagement with the entire scope of his thought, but in large part because of an interest in how Dewey later interpreted pragmatist conceptions of truth; as James Livingston has recently remarked, many writers deal only in passing with James "on their way toward the more tough-minded Dewey," delivering condescending interpretations of his work that "patronize a pragmatism he would not recognize."[26]

In 1897 James wrote that it was his intention to make visible the presence of a radically empiricist attitude that had been "eclipsed from sight" even though it had always existed alongside "the higher and lower dogmatisms" of dominant philosophic traditions.[27] Nearly one hundred years later this philosophical challenge has become a significant historiographical opportunity as well. This opportunity, while intersecting with philosophers' attempts to reconstruct and analyze James's radical empiricist statements, has instead a different aim: that of considering how radically empiricist theories and research practices have constituted a discourse that has contributed to twentieth-century science even as it has critiqued mainstream frameworks, and of identifying the relationships that exist between this discourse and other cultural forms of thought. To be sure, a radically empiricist discourse is one that draws from and feeds into various pragmatist traditions in diverse ways. My point here is that radically empiricist thinking is neither reducible to pragmatism nor merely subsumable by it.

In positing that Allport and the Murphys are second-generation Jamesians whose psychological research during the 1930s can best be understood as emerging from a radically empiricist framework, my intention is to highlight their interest in the phenomenon of immediate experience. Allport, for example, stated that in his later years James moved away "from the positivistic empiricism to which he himself had given such impulse," and that he came to defend "that special form of subjectivism which he chose to call 'radical empiricism.'"[28] Allport defined radical empiricism as "a tentative theory of knowledge, admitting all experiences of fact as hypotheses to be verified in the course of future experience." Unlike the psychologies elaborated by contemporary neobehaviorists, Allport found that James's radical empiricism pointed to the significance of immediate experience as "the essence of psychological truth."[29]

James's move to radical empiricism is one that Gardner Murphy would comment on as well, especially in regard to James's interest in psychic phenomena. Murphy pointed out that when James was asked in his later years "to tell what 'pragmatism' was really about," he would stress "the fact that it dealt with the practical *and the concrete,* and that if one must choose between the two, to be concrete was even more important than to be practical." Murphy observed that, during "those last extraordinary ten years of his life from 1900 to 1910," James was "moving toward 'radical empiricism,' the habit of thrusting oneself forward into the world of experience, to make the richest possible contact with the concrete, the immediate, the real." Murphy pointed to James's *Varieties of Religious Experience* as the work offering the clearest and most expansive rendering of James's radically empiricist commitments, that is, his concern with making "real contact with the tough, vital, throbbing, everyday realities with which our immediate life is concerned."[30] In Murphy's assessment, James's interest in psychical research was a natural extension of the radically empiricist frame of mind.

In his analyses of the direct apprehension of concrete experience, James challenged what he saw as the thin and superficial depiction of reality offered by a "half-way empiricism." James proposed instead his own vision of a "thicker and more radical empiricism" that would confront the fact that "reality, life experience, concreteness, immediacy, use what word you will, exceeds our logic, overflows and surrounds it."[31] For James, the world of "half-way empiricism" is one in which "we carve out order by leaving the disorderly parts out."[32] This is the classical mode of thinking: James noted that "philosophers have always aimed at cleaning up the litter with which the world apparently is filled. They have substituted economical and orderly conceptions for the first sensible tangle; and whether these were morally elevated or only intellectually neat, they were at any rate always aesthetically pure and definite, and aimed at ascribing to the world something clean and intellectual in the way of inner structure."[33] The classic-academic view of the world "explains things by as few principles as possible and is intolerant of either nondescript facts or clumsy formulas."[34]

James styled his empiricism as radical because his exploration of experiential reality persuaded him to treat the belief that the world is a single system as a theoretical conjecture rather than as a matter of fact. A radical empiricist, James contended, views "the doctrine of monism itself as an hypothesis, and unlike so much of the half-way empiricism that is current under the name of positivism or agnosticism or scientific naturalism, it does not dogmatically affirm monism as something with which all experience has got to square." To treat monism as an hypothesis was to open the door to the possibility of a pluralistic universe. James asserted that "the difference between monism and pluralism is perhaps the most pregnant of all the differences in philosophy. Prima facie the world is a pluralism; as we find it, its unity seems to be that of any collection; and our higher thinking consists chiefly of an effort to redeem it from that first crude form."[35] As James saw it, then, the difference between monism and pluralism exemplified the difference between seeking to account for the world from the vantage point of concepts achieved by intellectualization, or by the percepts that arise from our immediate experience. Contending with this distinction was more than a speculative exercise for James, for in identifying himself as a radical empiricist he was throwing his weight behind those who hypothesized that pluralism "is the permanent form of the world." For those who adopted this perspective, James explained, "the crudity of experience remains an eternal element thereof."[36]

Basing one's understanding of the world by starting from experiential reality as opposed to intellectualized reality was to apprehend a world conceived not "statically as a geometric structure" but one that was, James commented, "a turbid, muddled, gothic sort of affair, without a sweeping outline and with little pictorial nobility."[37] Elsewhere, James elaborated on this point: "The ultimates of nature – her simple elements, if there be such – may indeed combine in definite proportions and follow classic laws of architecture; but in her proximates, in her phenomena as we immediately experience them, nature is everywhere Gothic, not classic. She forms a real jungle, where all things are provisional, half-fitted to each other, and untidy."[38] It is this consideration of nature as immediately perceived rather than as intellectually conceptualized – or, in James's terms, of a "Gothic" rather than a "classic" sense of reality – that represents one of the key aspects of a radical, as opposed to a halfway, empiricism. In arguing for the legitimacy of the radical empiricist perspective, James challenged his audiences to acknowledge a world more saturated, more complex, more various, and more mutable than they were accustomed to conceiving. The world that exists as a result of our efforts to categorize its contents is always a selection made from within a plenum of the unselected. As Arnold Metzger remarked in a commentary from 1942, this selected world "cannot pretend to be the true world; is not a copy of the auto-existent, but is rather the result of 'attempts to get the chaos of the crude individual experience into a more manageable shape' – 'a selection detached by some interest, or defined by some point of view.'"[39]

Many of those who found James's vision of a pluralistic universe plausible perceived that the radical empiricist critique of experience challenged current scientific conceptions regarding the nature of reality. Horace Kallen, one of James's most influential acolytes during the interwar years, charged that scientific investigation was currently structured so that "the inevitable is the only trail it hunts; the formal equivalences and identities of mathematics are its goal. The biological sciences, consequently, look structurally backward toward the formal simplifications of chemistry and physics; psychology retreats into physiology; the social sciences into statistical averages."[40] As a result, Kallen argued, modern science, "at its core, where it carries most weight and influence, where it compenetrates with the tradition of thought in the western world," has become as "increasingly centralized, monistic, determinist, necessitarian, as industry." In Kallen's opinion, the consequences for society of these practices were both real and immediate: "The folkways of a society regimented by scientific generalization and machine-made uniformities in its work, its play, its arts, its religion, are upon us."[41] Kallen held out the hope that James's views could serve as "an anodyne and a correction" to current scientific and political practices, for his alternative vision took its cues not from "industry" but from the principles of American egalitarianism: "In sum, James is a metaphysical democrat."[42] Edna Heidbreder, in her 1933 survey, *Seven Psychologies,* characterized James in much the same way as did Kallen. Heidbreder saw "a radical democracy in James's nature," which she traced to James's insistence that "every man, every idea, every moment of experience, must have its chance to speak out and be recognized. This democracy, this unfeigned desire that all be counted somehow – all experiences, all men, no matter how disreputable in the eyes of the savants – was probably his fundamental prejudice." Or, as a young Walter Lippman wrote in his éloge of James, "I think he would have listened with an open mind to the devil's account of heaven, and I'm sure he would have heard him out on hell."[43]

This emphasis on democratic precepts in James's thinking has been recently highlighted by historian George Cotkin, who argues that James, as a *public* philosopher, believed that intellectuals bore a fundamental responsibility for bringing their academic work to bear on public problems.[44] Compelling evidence that James was indeed successful in this regard is demonstrated by the fact that his contemporaries never evaluated his later philosophical writings simply in terms of their internal logic but concentrated upon their political and social import.[45] At the heart of James's venture into the public arena was a desire to combat imperialist ambitions in intellectual and political life – modes of thought and action that he was certain did "violence to the bedrock of American traditions."[46] As James expressed himself in a letter commenting on nineteenth-century American politics, "I think that the manner in which the McKinley administration railroaded the country into its policy of conquest was abominable, and the way the country puked up its ancient soul at the first touch of temptation, and followed, was sickening."[47] James's characteristic melding of metaphysics and political ethics in support of his democratic ethos is one that second-generation "Jamesians" such as Allport and the Murphys appropriated and imported back into psychology.

Recent commentators such as Seigfried have argued that "the time is right for undertaking the task of reconstructing [James's] psychology . . . in the light of his radically empiricist insights."[48] Eugene Taylor, in turn, has pointed out that the possible scientific implications of James's radical empiricism were ignored in the early twentieth century in the wake of an anti-Jamesian backlash led by psychologists who invoked German experimentalism as the scientific ideal, working "under the banner of quantification, laboratory apparatus and positivist rhetoric," and he has also called for a resurrection of radically empiricist investigation in psychology.[49] In important respects the dissenting discourse of such second-generation Jamesian psychologists as Allport and the Murphys represents just such an impulse: a reconsideration of James's *Principles of Psychology* in the light of his later elaboration of radically empiricist philosophy. In appropriating these Jamesian concerns, Allport and the Murphys also absorbed the lessons of doing public philosophy, and their pursuit of a reconstructed science is one that simultaneously had in view the reconstruction of American society.

At the conclusion of James's last complete major work, *A Pluralistic Universe,* he offered an italicized injunction: "*It is high time for the basis of discussion in these questions to be broadened and thickened up.*" James specifically urged this task on "the younger members" of his audience, charging them, he said "with the cheerfullest of hearts, [to] 'Ring out, ring out my mournful rhymes, but ring the fuller minstrel in.'"[50] Allport and the Murphys accepted James's challenge and set out to "broaden and thicken up" scientific discourse in the search for new realities.

The Radical Social Gospel

Although Jamesian radical empiricism represents one source on which Allport and the Murphys drew when thinking in terms of public philosophy, the form and content of their political challenge to the scientific status quo will be misunderstood unless attention is also given to the ways in which religious convictions informed their scientific practices. The growing elaboration of the subfields of social and personality psychology during the 1930s, though fed by many sources, owes a special debt to activist versions of the Social Gospel.[51] The teachings of this religious movement provided both a moral imperative and a theoretical framework for constituting the subject matter and research practices of social and personality psychology, especially because these subfields were intertwined with attempts to use studies of "the-individual-in-social-context" to critique the current social order. Resacrilizing the world under the investigation of science according to the Social Gospel was thus a repoliticizing of it as well.

That religion may be a *constitutive* resource in modern-day science is a question that scholars have investigated only irregularly. The conventional wisdom that ours is a "secular" age, accompanied by a presumption that science and religion represent mutually exclusive ways of thinking, is still widely accepted in academic circles.[52] Historian of science Martin Rudwick has observed that "the strength of the historian's empathy for religious beliefs often seems to be directly proportional to the space of time that separates him from them, fading away as one approaches the present day."[53] But if a presumption regarding the lack of religiosity on the part of scientists has hampered recognition of the expression of theological concerns in twentieth-century science, it is also true that many intellectual historians have similarly framed questions regarding twentieth-century political activism in terms that underplay the possibility of religious considerations. As Henry F. May points out, however, closer attention to religious history can help identify wellsprings of social reform that have previously escaped notice.[54] The politically activist vision at the core of Allport's and the Murphys' psychological research is one that drew on a radicalized Social Gospel.

What marked the Social Gospel as a distinctive movement, as William R. Hutchison comments, was its proponents' belief that "social salvation precedes individual salvation both temporally and in importance."[55] Falling somewhere in the realm lying between "liberalism" and "revolutionary socialism," the insurgent status of the Social Gospel has often been viewed with suspicion by scholars. William McGuire King observes that the term "Social Gospel" became "a vague catchword for the supposed sins of liberal Protestantism," with critics frequently charging that "the social gospel represented a superficial moralism and theological naïveté."[56] As early as 1949, however, May had pointed out that proponents of a radical social gospel could be easily distinguished from their moderate counterparts, in that they "did not confine themselves to demanding a 'new social spirit' or a few limited reforms. They did not believe that everything was basically all right. The remedies they proposed, though they were Christian, nonviolent, and often unrealistic, were sweeping."[57] Indeed, many of the political and economic critiques embedded within these radical doctrines were socialist in nature. Walter Rauschenbusch, for example, who set the Social Gospel's tone in the early part of the century, maintained that "the most important advance in the knowledge of God that a modern man can make is to understand that the Father of Jesus Christ does not stand for the permanence of the capitalist system."[58]

The religious odysseys of Gordon Allport, Gardner Murphy, and Lois Barclay Murphy began in their childhoods, within earnestly devout families. Allport's early religious training was intense; his maternal grandmother had been a preacher at the Free Methodist church she founded in Fulton, New York, and his mother was strenuously pious as well, insisting that her children heed the "importance of searching for ultimate religious answers," and overseeing their attendance at numerous camp meetings and revivals.[59] The turmoil of the young Allport's religious home life was echoed in the political activity of Cleveland, where he spent much of his youth. Ohio's progressive leaders in the first two decades of the twentieth century – especially Tom L. Johnson of Cleveland, who presided over "the best governed city in the United States," according to Lincoln Steffens – set the agenda for the United States' "gas and water" municipal socialists.[60] The ideology of the Ohio Progressive movement had a distinctly religious tone, drawing both from the Christian socialist ideas of Henry George and from the Sermon on the Mount.[61] Allport's father, a physician who founded a cooperative drug company, often expressed a personal philosophy to his sons that was in keeping with this ethos: "If every person worked as hard as he could and took only the minimum financial return required by his family's needs, then there would be just enough wealth to go around."[62]

As a Harvard undergraduate, Allport majored in both psychology and social ethics. The aim of the social ethics department during these years was to train students in the scientific investigation of social problems. In social ethics, Allport worked most closely with the Progressive reformer James Ford, who supervised Allport's fieldwork in social service.[63] During his undergraduate years, Allport participated in such Progressive Era activities as working with foreign students, locating housing for industrial workers, serving as a probation officer, leading a boys' club, and holding down a job with the Cleveland Humane Society while on summer vacation.[64] Allport was also an ardent internationalist: his Harvard scrapbook is replete with literature from various conventions and meetings of the Cosmopolitan Club – an organization whose aim was "to stimulate a sympathetic appreciation of the characters, problems and intellectual currents of other nations" – and tributes to Woodrow Wilson and the League of Nations. Allport took his studies seriously as well. An item in Allport's scrapbook shows that the hard-working youth found favor with the Harvard aristocracy: as an honor student in philosophy, Allport was extended an invitation to Ralph Barton Perry's home "to meet Mrs. Perry and to hear Dr. [George] Sarton talk on 'The New Humanism.'"[65]

As a senior in 1917, Allport traveled to New York to attend the 13th Annual Conference of Eastern College Men on the Christian Ministry, held at Union Theological Seminary (UTS). The purpose of the conference was to "assist college students, who may be considering their choice of career, to weigh the claims and opportunities of the Christian ministry as a life-work"; the program included addresses by such Social Gospel stalwarts as Harry Emerson Fosdick on the personal religious needs of men, and Harry Ward on the minister's opportunity.[66] Torn between his interests in psychology and in social ethics, Allport chose, upon graduation, to accept a position to teach social ethics at Robert College in Constantinople (an American-run, nondenominational college founded in 1863), under the sponsorship of the "Harvard Mission."[67] After teaching for a year – during what turned out to be the final days of the sultanate – Allport returned to Harvard to earn a doctoral degree in psychology, working under Herbert Langfeld.

Allport continued his association with social ethics even after earning his Ph.D. in psychology. After spending two years in Europe on a postdoctoral fellowship, Allport returned to Harvard as an instructor in social ethics for two years, under the chairmanship of Richard Clarke Cabot, an internationally renowned physician and social philosopher. Allport would come to hold Cabot in heroic esteem, describing him as a "rebel," an "idealist," and a "visionary."[68] Cabot was a radical critic of the medical profession, urging the adoption of socialized medicine and nearly being thrown out of the Massachusetts Medical Society for his articles alerting the public to medical malpractice.[69] Cabot's ethics and politics merged in his religious vision of socialism; he proclaimed Christian socialism to be a "more effective American" manifestation of social critique than "Marxian socialism."[70] After Cabot died, the *Christian Century* reflected that "one hardly knows which was more important in Dr. Cabot's career – his crusading zeal for socialized medicine or his physician's recognition of the spiritual basis of the good life. . . . When socialized medicine becomes generally established in the not distant future, Richard C. Cabot will be honored as one of its greatest prophets."[71]

Gardner Murphy's father, Edgar Gardner Murphy, was an Episcopal minister based in Montgomery, Alabama, and a Progressive reformer of national prominence. When his social activism began to attract public attention, Murphy resigned his clerical orders to more freely pursue racial and child labor reforms in the South. Murphy embarked on a massive public education effort regarding child labor in Alabama, indefatigably producing pamphlet after pamphlet until he hit on his most effective publication in 1903: a twelve-page leaflet entitled "Pictures from Life. Mill Children in Alabama." The documentary realism of the text proved a valuable propaganda tool, in the South and throughout the nation.[72] Edgar Gardner Murphy's work in establishing the National Child Labor Committee brought him into personal contact with other prominent Progressive activists, such as Jane Addams, Florence Kelley, and Felix Adler.[73]

Murphy was a theological modernist who regarded Scripture as "fallible, human but nevertheless inspired." On racial reform, Murphy's stance was mixed: he vigorously decried lynching and supported the educational and social advancement of African-Americans, although he accepted segregation as a social good, issuing apologias for its existence; in *An American Dilemma,* Gunnar Myrdal regularly drew upon Murphy's perspective to represent the view of "the enlightened and responsible Southerner."[74] In regard to his assault upon child labor, one of Murphy's biographers suggests that he "ranked with the ablest of the abolitionist and feminist reformers. As a social critic, he had a fine sense of the disparity between society's articulated values and its operative assumptions."[75]

As a youth, Gardner Murphy was staunchly evangelical, and he planned to become a missionary.[76] Murphy's undergraduate years at Yale were ones of emotional turbulence as he attempted to determine his true vocation. Such searching can be seen in a letter of academic advice that the elder Murphy sent to his son, in which he replied to a previous query by indicating that his "own preference would be to put psychology and 'such like' ahead of biology." Murphy stated that "we really know nothing of matter and motion except under the forms of mind; and the student who starts in on the merely physical or biological side of things is likely to carry false assumptions into all his later work." At the same time, Murphy warned his son that "psychology itself *can* be taught merely as a sort of brain-mechanics."[77] The question of vocation was heavy with religious implications.

Edgar Gardner Murphy would die from heart disease at the age of forty-three during Gardner Murphy's freshman year in 1913, and Murphy soon found his religious faith being shaken by scientific doctrines.[78] In a letter from 1916, his best friend's mother responded to a troubled letter from Murphy by stating that she knew that he was "eager to serve God and man," and that, in order to do so, he "must face these spiritual problems, which are unsettling the Christian faith of so many of our college young men, squarely and honestly." She applauded the "life purpose" he had expressed to "reconcile science and religion and be able to help struggling humanity," a desire that he reiterated in a letter of 1917 to his brother.[79] Murphy hoped to find a basis for demonstrating the independence of consciousness from matter, and to therefore justify belief in a mind–body dualism.[80] As a graduate student at Harvard, Murphy felt forced, as he became further committed to a scientific career, to declare himself once and for all on the matter of his religious faith. Admitting that the evidence was "not strong enough," he felt compelled to "give up my dualism, which meant my religious faith, and I did this with one clean, if bloody, stroke."[81] At the same time that he put aside his search for a way to ground God and religion in mind–body dualism, Murphy developed "an *intense* focus on psychical research."[82]

After earning his master's degree from Harvard in psychology, Murphy served in the Yale Mobile Hospital Unit in France during World War I; he returned to Columbia University after the war to obtain his Ph.D. in psychology. While in New York, Murphy also attended classes at UTS, recalling particularly "the extraordinary course by Harry Emerson Fosdick . . . entitled The Use of The Bible."[83] In 1924 he met a first-year seminary student named Lois Barclay, who was herself a clergyman's offspring. The two were married in 1926.

The young Lois Barclay grew up with table talk from her Methodist Episcopal minister father, Wade Crawford Barclay, about the work and ideas of such figures as Walter Rauschenbusch and Jane Addams. As a child she accompanied Barclay as he preached social justice at churches throughout the city of Chicago and recalls delivering her own "sermons" at home from the landing on the stairs.[84]

Barclay, along with fellow ministers Harry Ward, Harris Franklin Rall, and Bishop Francis McConnell, was a leader in the radical Methodist Federation for Social Service (MFSS), an activist organization "which was at the forefront of progressivism within the Methodist Episcopal Church in its confrontation with racism, the profit motive of capitalism, class exploitation, and American imperialism."[85] Styled as the Marxist Federation for Social Strife by its detractors, the MFSS was a frequent target in the pages of William Randolph Hearst's newspapers.[86] Barclay earned an entry in the section entitled "Who Is Who in Radicalism?" in Elizabeth Dilling's *The Red Network* for his religious activism. Lois Barclay Murphy herself, after graduating from Vassar in 1923, received a degree from the institution most distrusted by reactionary patriots such as Dilling: UTS, which they called The Red Seminary.[87]

As a Vassar undergraduate, Murphy majored in economics, a department headed by Herbert Mills. Vassar president Henry Noble MacCracken recalled that Mills "not only participated in social movements of his time, but his passionate partisanship of the right inspired his pupils to share in the social movements of their day. . . . When he occasionally led Sunday service, he preached social justice."[88] In many ways the classes offered under the rubric of economics were sociological in nature, such as the year-long course on socialism and labor problems begun in 1913.[89] Murphy's interest in these matters can be seen by her membership in the Socialist Club.[90] Another influential teacher was Mary Redington Ely, whose area of scholarship was the early Christian Gospels.[91]

Although Murphy considered going on to graduate school in economics, she instead decided that she would prefer to teach comparative religion and entered UTS. Among Murphy's teachers at UTS were noted Social Gospel figures such as Harry Ward, Eugene Lyman, Ernest Scott, and Julius Bewer; she also studied under Harris Franklin Rall during summer courses at Garrett Theological Seminary in Evanston.[92] Murphy chose comparative religion as her field of emphasis and wrote a thesis entitled "The Problem of Evil and Suffering in the Early Vedas." Murphy planned to use a "combination of historical, psychological and philosophical treatments" in order to explore "the relation of experience and ideas in the growth of fundamental attitudes toward the universe" in India.[93] Murphy began teaching comparative religion at Sarah Lawrence in 1928, the same year that she received her UTS degree.

Murphy's first exposure to psychological precepts, like her introduction to religious issues, had occurred at home, where her parents discussed the ideas of such thinkers as William James and John Dewey. During her college years, Murphy found little to her liking in current psychological theory, dominated as it was by varieties of behaviorism. But the young Sarah Lawrence instructor would find herself drawn into the world of psychology through her relationship with her husband Gardner. In Gardner Murphy, Lois Murphy believed she discerned "a completely different kind of psychologist from my image of the current hardnosed scientists who were not interested in human beings." She credited "the combination of Gardner's kind of psychology and the stimulation of two irresistible and delightful children" with opening "another door" – that other door being her pursuit of a doctorate in psychology at Columbia University's Teachers College.[94] Murphy was awarded her Ph.D. in 1937.

Murphy was hardly alone in joining the sacred to the secular through a dual apprenticeship at UTS and Columbia. UTS was, in fact, the breeding ground for a number of students who would find their "way to psychology by way of religion," making the move across the street to Columbia University to pursue their Ph.D.'s.[95] In addition to Lois Barclay Murphy, that number included at least Theodore Newcomb (who would coauthor the revised edition of *Experimental Social Psychology* with the Murphys in 1937) and Rensis Likert, who conducted research with Gardner Murphy that Murphy wrote up and published in 1938 under their joint authorship as *Public Opinion and the Individual.*[96] Goodwin Watson, one of the moving forces behind SPSSI during the 1930s – Watson served as president of SPSSI in its founding year of 1936, Gardner Murphy in 1937 – was another of those who first studied theology at UTS and then took a doctorate in psychology at Columbia. Psychologist Carl Rogers and sociologist Robert Lynd were two other figures who studied at UTS during this period.

By the 1930s, as Robert Moats Miller has remarked, "the cutting edge of Protestant idealism approached socialism." This observation is testified to by the frequent approval, at official denominational meetings at both the local and the national levels, of "resolutions which avowedly or tacitly advocated socialism." Miller wryly notes that such pronouncements were so prevalent "that the investigator receives the superficial impression that two clergymen could not meet each other on the street without one of them banging a gavel, calling the other to order, and then introducing a resolution damning capitalism."[97] Such descriptions aptly characterize the spirit of the MFSS during the 1930s, and of radical Social Gospel theology in general.

As an MFSS leader, Barclay compiled a devotional book in 1936 entitled *Challenge and Power,* which encompassed such topics as "prayer for the speedy coming of the new social order" and "prayer for equitable distribution." In words written in "Fellowship with Those Who Hunger," Barclay prayed that God "wilt so arouse the conscience of mankind that men and women no longer shall consent to eat the bread of oppression, but applying their knowledge and skill to social arrangements and processes of distribution they shall forevermore assure that those who wear out their lives in production may receive their just share of thy gifts."[98] Barclay was clear that science had served the interests of those who had contributed to this oppression. In praying for "the coming of a more just and equitable order," Barclay looked for a reconstructed world "in which scientific knowledge and technical skill shall not serve the private gain of the few but higher interests and welfare of all."[99] Clearly, such religious thinkers believed that social salvation could not be achieved without a fundamental reordering of social relations, for an economic order that sanctioned the treatment of individuals as objects to be exploited was an economic order based on an immoral foundation.

It was this egalitarian spirit of religious activism that Allport and the Murphys carried into their research in social and personality psychology, counting them among those whom Lois Barclay Murphy characterized as "apostles of democracy."[100] Such religious and political views were consonant, for example, with Allport's proclamation of the "basic ideological affiliations of democracy, socialism, and Christianity." Allport pointed out that, "when democracy persecutes socialism, as it frequently does, it denies its own essential creed, refusing to extend democratic rights into certain tabooed regions such as industry, finance, or perhaps social security or colonial policy." Likewise, Christianity was accountable for persecuting socialism, while continuing "to shelter Father Coughlin, the Klan, the Christian Front, and other race-hating, rabble-rousing, person-destroying travesties of Christianity." Dismayed at democracy's "half-heartedness" and at the church's "corruption," socialism, Allport stated, "sometimes turns upon both and repudiates kinship." Thus it was that "these three expressions of the doctrine of human liberty" were kept apart, Allport observed, "while in half the world the person is being suffocated by the poisonous vapors of totalitarianism."[101] Allport and the Murphys were deeply committed to their apostolic convictions regarding the democratic mission of the United States, and they sought to incorporate the teachings of a radicalized Social Gospel into their search for scientific knowledge.[102]

But Social Gospel activists were interested in more than the possibilities of putting science to work toward more egalitarian economic ends: they also spoke to the subject matter of the social sciences by arguing for the need to highlight the ways in which selfhood and society mutually constitute each other. During this period an avowedly Social Gospel was equally the gospel of "personality," with writers expressing concern regarding the deindividuating aspects of modern life.[103] In 1932 Boston University theologian and professor of philosophy Edgar Brightman observed that "we are living in an impersonal age," which was becoming manifest in numerous ways. Science, Brightman noted, sought to eliminate personal considerations in the search "for pure objective truth," while capitalism counted persons "only as means to the end of profit," and communism required every individual to "conform to the plan and the ideas of the state." Contemporary life, according to Brightman, was "an age of statistics and statistics are impersonal, obscuring the peculiarities of the individual instance in the interests of the curve or the percentage. This has rightly been called a machine age."[104] Within the terms of Social Gospel theology, the "individuality" of each person represented an aspect of the divine, and any diminishment of the individuality of persons was thus a diminishment of the spirituality of human life.[105]

Radical religion and democratic faith thus converged in a belief in the central ethical significance of individuality. According to Rall, what lay at the heart of the democratic faith was "the conviction as to the sacredness of human personality. Man is always to be treated as an end, never as property or tool, as Kant pointed out, and as the prophets and Jesus made plain centuries before."[106] Similarly, Allport wrote in 1939 that his "personal philosophy of life" lay close to what "theologians sometimes call the Dogma of the Person, the doctrine that ultimately nothing is sacred except the individual, and nothing so morally binding as our duty to permit and assist growth, freedom, and self-perfecting for each mortal man." This belief in the sacredness of personality, Allport stated, lay at the core of the democratic, socialist, and Christian doctrines, which all shared a belief in "the right of each man for unlimited growth however complex the social situation in which he must adaptively live, and however discordant the rival demands of other mortals who have privileges equivalent to his."[107]

The interest of Social Gospel adherents in individuality extended as well into a concern with the subjective nature of human personality. Historian John C. Burnham observes that, "largely under the influence of liberal or modernist writings, theologians were shifting the focus of their interests from the Bible as such to the life of Jesus . . . suggesting that humans might to some extent share that internal experience."[108] Popular theologians such as Fosdick reinterpreted the Bible as revealing a historical progression in humanity's understanding of God. In the beginning, religion was "the relationship of the social group with its heavenly chieftain," for Jehovah, Fosdick suggested, was not at first "thought of as caring for individuals one by one. He was the God of the nation and he regarded individuals only as they were incidentally affected by the national fortunes." It was under the influence of Jesus that religion became "the relationship of the soul with God," for by then "the personality of the individual" had finally "been shaken loose from its submergence in the mass."[109] In like fashion, the authors of one of the meditations in Barclay's book reflected: "The world into which Jesus came nineteen hundred years ago was a vastly darker world than that of today. He dared to teach the worth of human personality in a world that denied its worth much more persistently and successfully even than our own."[110] Within the idioms of progressive Protestantism, the gospel of social salvation drew inspiration less from the image of an other-worldly "Christ" than from the life of a charismatic personality named "Jesus," who displayed radically egalitarian views.

This theological interest in personality can be seen as well in the response of such figures as Brightman and Fosdick to Allport's writing on this issue. Brightman, for one, considered the publication of Allport's *Personality* to be an event of moment. Offering a testimonial to the publisher to be used on Allport's behalf, Brightman wrote that "it is difficult for me to restrain my enthusiasm regarding this book and to speak in properly measured terms. It is the first work in psychology that I have ever seen which does adequate justice to all aspects of personality." Brightman judged the book to mark "a turning point in the development of American psychology" and contended that Allport's text would "go down in history as a masterpiece," second only to James's *Principles of Psychology*.[111] For his part, in a letter to Brightman, Allport reported his deep gratification for "the friendly attitude you have taken toward my book. I have no hesitation in saying that among its readers known to me, you have shown the keenest understanding of the foundations I am trying to establish. Your sympathy has been a source of encouragement more than offsetting the chastising that I have received by my colleagues for my 'return to the genteel tradition.'"[112] In 1941 Wallace F. Abadie, a student of Fosdick's, wrote to Allport: "It may interest you to know that a number of us first became acquainted with your book through lectures by Dr. Harry Emerson Fosdick of New York City." Abadie stated that Fosdick had "drawn freely on you for many points in his preaching – and has stated publicly in his lectures that your work *Personality* is the best thing of its kind written in the past 25 years."[113]

In pursuing democratic realities within nature and within the American polity, Allport and the Murphys drew freely on concepts and values derived from Jamesian radical empiricism and radicalized versions of the Social Gospel as they constructed their psychological theories and practices. In doing so, they served as missionaries to the scientific realm, bringing the tenets of Jamesian philosophy to bear on personality and social psychology and opening up social science to the doctrines of radical religion. In turn, by incorporating these perspectives into their scientific work, Allport and the Murphys embedded elements of the democratic contestation alive in the larger culture within the structure of their psychological research.

2

Challenging the Rules of the Game

> The disturbing fact is that any persistent and searching inquiry into the criteria of *psychology as science* leads inevitably to a discussion of the essential principles of science itself . . . We cannot afford to forget the ideal of empirical science embodied in the words of Isaac Newton, "In this philosophy, propositions are based on phenomena, and laws are derived by induction."[1]
>
> Arthur Bills (1938)

By the close of the 1930s, arbiters of the status quo had begun to mount worried defenses of scientific method that sought to rebut what they saw as the pernicious effect of oppositional sensibilities such as those expressed by Allport and the Murphys. Defenders of right thinking such as psychologist Arthur Bills believed that the ground that had been "won by the efforts of Galileo, Bacon, Newton and others" was imperiled by "newer psychological systems which are clamoring for scientific status with all the vigor and scorn for tradition that characterize youthful movements."[2] C. C. Pratt, in his 1939 text, *The Logic of Modern Psychology,* defended the reigning orthodoxy, stating that science was "too serious and well established a game to be entered into lightly or altered, unless it can be shown either that a proposed alteration does not upset the essential features of play, or that the new game is better than the old one." This is why those who "played" science had "a right to insist that it be played according to rule."[3] This was precisely the challenge that Allport and the Murphys took up in their dissent from the status quo during the 1930s: that of upsetting the "essential features of play" in the hope of offering a "better game," one in which the grounds for scientific authority would be refigured.

In the late 1920s, when Allport and the Murphys were making their initial forays into academic psychology, they were joining a "small, fairly tightknit community" in which most of the members "knew many of their colleagues well and could speak of them as friends and coworkers, even if they sometimes disagreed with one another."[4] The American Psychological Association (APA) was increasing in numbers, from 307 in 1917 to 1,000 by 1929 and to 2,739 in 1940; there were twice as many Ph.D.'s in 1930 as there had been in 1920.[5] Between 1916 and 1938 there was a fivefold increase in the number of APA members in teaching positions (from 233 to 1,229) and a twenty-nine-fold increase in the number of members in applied jobs (24 to 694).[6] As is true for many other fields, Laurel Furumoto finds that the pattern that emerged was "for women psychologists to predominate in the lower status practitioner roles as mental testers and clinicians, while men psychologists retained their hegemony over the higher status academic areas – applied as well as pure – of the discipline."[7] The only substantial funding available for psychologists during this period came from private organizations such as the Rockefeller Foundation and the Carnegie Corporation.[8]

In short, Allport and the Murphys were entering the field of psychology at a time when it was just beginning to shift from a small nucleus of gentlemen-scholars to a more amorphous body of researchers with potentially divisive interests, during an era when institutional investments in psychology departments were still small. Psychology was a discipline in which leadership was being passed to a new generation of researchers whose influence would come less from convincing each other of their authority than from recruiting allies from the expanding membership rolls. Until the next generational shift, those setting the discipline's agenda would be individuals who were comfortable acting as intellectual entrepreneurs, risking their professional reputations on organizing and managing new subfields within the profession and seizing the initiative in bidding to restructure psychology's relationship to neighboring domains. Because psychologists were increasingly coming to accept varieties of behaviorism as modeling good scientific practice, the ventures launched by Allport and the Murphys during the 1930s contested ascendent conventions.

The dissent launched in the United States by psychologists such as Allport and the Murphys would not be the only existing challenge to the behaviorist ideologies gaining prominence, however, for the Gestalt movement initiated in Germany also began to make significant inroads in the American context during the 1930s.[9] Coming equipped as they did with ambitious theoretical goals and provocative experimental findings, the Gestaltists served as a potent disruptive force, placing many assumptions operative in American psychology under scrutiny.[10] For psychologists such as Allport and the Murphys, the growing presence of Gestalt psychology served as a powerful illustration that neobehaviorist frameworks were vulnerable to challenge by new perspectives.

Allport counted the Gestaltists among "the many rebels within the ranks of general psychology attacking its long-entrenched assumption that elements abstracted from individual experience are the proper data of the science."[11] Gardner Murphy described the Gestaltists' doctrines as a "revolution," one advocating that "the description of parts and their connections never tells the nature of the whole; that, in fact, the whole is no mere sum of its parts, but a totality, an integer, a form, the nature of which must be directly grasped. There must be an end to elements." As Murphy explained, "just as chemical analysis of living tissues destroys life, rendering the mere chemical description inadequate as a description of the total reality, so the reduction of mental states to elements destroys that unity, that organization, which is mental life itself."[12]

The radical implications of the Gestalt movement were clear to others as well. Robert Woodworth styled Gestalt psychology as "a revolt against the established order," as did Edna Heidbreder in her 1933 survey of schools of psychology, describing Gestalt as a "psychology of protest," and as a movement intent on reaching beyond the bounds of psychology proper. A school "that started as an attempt to do justice to a particular problem in visual perception," Heidbreder observed, had, by the 1930s, "ended by demanding a thoroughgoing revision of the fundamental principles of the science. It did not, indeed, stop with psychology itself but extended its principles into the physical sciences."[13] Harvard philosopher Ralph Barton Perry noted that Gestalt theory existed "on what might be called a war-footing. It enjoys the *esprit* of an armed revolution – against sensationalism, against associationism, against mechanism, against behaviorism, against authority and tradition generally."[14]

Gordon Allport and Gardner Murphy were among the first to take note of the Gestalt movement, and they had been following the developments in Germany with interest. When Murphy published his lectures on the history of psychology in 1929, he commissioned Heinrich Klüver to write a supplement addressing contemporary German psychology, including the Gestalt viewpoint; in his 1932 *Approaches to Personality,* Murphy presented his own analysis in the first chapter, "Gestalt and Type."[15] Allport had encountered Gestalt psychology for the first time in the early twenties, while doing postgraduate work in Germany, and wrote two early assessments of the movement, "The Standpoint of *Gestalt* Psychology" and "The Study of the Undivided Personality."[16] The Gestaltist framework was, Allport postulated, "only a partial expression of the more comprehensive *Strukturbegriff* which has caused an upheaval in psychology on the continent."[17] The distinction between the two, Allport pointed out, is that "the original conception of *Gestalt* relates rather narrowly to the sphere of perception," while "the doctrine of the *Struktur* has been extended in other directions."[18]

The central tenet of Gestalt theory, as historian Mitchell G. Ash indicates, is that "dynamic structures in experience *determine* what will be wholes and parts, figure and background, in particular situations." As scientists such as

Wertheimer and Köhler advanced Gestalt theory in the 1910s and 1920s, they argued that the dynamic self-organizing processes they observed in perception and brain physiology were similar to equilibrating physical systems, and that these processes were thus "properties of both mind and nature," and not simply conceptualizations brought to bear on experience.[19] The major Gestalt theorists, Ash argues, while critical of mechanistic and elementalist perspectives, saw themselves as working within natural science, bringing holistic lines of thought to bear on the search for objective order in experience.[20] The presumption of these German psychologists that their work was on an even footing with natural science was a viewpoint that startled many American researchers.

The Gestalt movement in fact emerged within a cultural context in which many of those who were developing holistic theories saw their work as containing implications for the sciences as a whole. In her study of the general contours of interwar German psychobiology, Anne Harrington, for example, points out that such theorists did not see their research as being confined to the attempt to develop "a superior approach to understanding or modeling the phenomena of life and mind."[21] The array of efforts aimed against mechanist-based science within psychobiology, Harrington suggests, challenged "the methodological and epistemological principles on which the entire international natural scientific enterprise was based."[22] That the opening decades of the twentieth century had seen great ferment across disciplinary lines in Europe suggested to individuals such as Allport and the Murphys that scientific life was in the process of emerging in new configurations.

Indeed, in 1931 Gardner Murphy argued that the Gestalt psychologists had not made their case strongly *enough.* In explicating what he felt was the crux of the Gestalt system, Murphy contended that "the Gestalters have made a very much bigger contribution than they realize; that their claims have been much too modest and in large part irrelevant; that an actual consideration of what they have done shows the movement to be of astounding significance, not only to psychology, but to *all* science and philosophy."[23] Murphy commented that "no, not even in physics or chemistry, not even in the geometry of the block in which you live or the time relations of the melody to which you listen, can the form, the organization, the pattern by which wholes are constituted ever be treated as mere parts of a total."[24] Murphy emphasized that the problem of whether wholes are made up of parts is "a problem as to the *nature of reality,* not merely a problem within the sphere of psychology." Indeed, Murphy goes on to proclaim that "this is no mere revolution in psychology; we can be content with nothing less than a new metaphysics, in which the part–whole relationship shall be irrevocably banished and the concept of unique characters, entities, or integers elevated to a position of genuine explanatory value."[25] For Murphy, the experienced reality of jumps, gaps, discontinuity, and discreteness pointed to aspects of fundamental significance in the natural world. Perhaps, Murphy suggested with irony, "the discontinuity between the very principles of continuity and discontinuity is itself the most perfect of discontinuities." At any rate, Murphy advised, "we must shut our eyes and jump as far as the structure of the human mind permits."[26] As evidenced by the efforts of the Gestaltists, work from within contemporary psychological science was fruitfully engaged with philosophical questions of the deepest consequence regarding the nature of reality and of the nature of science itself.

Because the actions of Hitler's regime made life in Germany untenable for many of the Gestaltists, the leaders of this psychological rebellion had, in the main, replanted themselves in America by the mid-1930s. Within a few years, such distinguished refugees as Rudolf Arnheim, Kurt Goldstein, Kurt Koffka, Wolfgang Köhler, Kurt Lewin, William Stern, Heinz Werner, and Max Wertheimer literally became part of the American scene, bringing Gestaltist perspectives in direct contact with American views of mind and science. Although no one knew at the time how this sudden influx of daring psychological theorists might affect the status quo of American psychology, dissenting figures such as Allport and the Murphys were excited at the possibilities. Prior to his presidential address to the APA, Allport confided to psychobiologist Kurt Goldstein, "Since my remarks at the meeting will be definitely contrary to the convictions of most American psychologists it is most encouraging and helpful to have the assurance that discriminating investigators such as yourself would probably not find my remarks as fantastic as will many of my colleagues."[27] As a self-professed leader of a hoped-for "revolutionary upset in the conception of psychology as *science*," Allport conveyed a double meaning in his address, when he saluted those "gifted emigrés who have come so recently to join their strength to ours."[28]

That strength was welcomed by Allport and the Murphys in their struggle against neobehaviorist trends in their day-to-day professional lives. By virtue of their associations with the psychology departments at Harvard University, Columbia University, and Columbia University Teachers College, Gordon Allport, Gardner Murphy, and Lois Barclay Murphy, respectively, were obliged to contend with three particularly vigorous varieties of strict methodism. Harvard, for example, was the site of S. S. Stevens's energetic promotion of "operationism," which sought to remake psychology in the image of the Vienna Circle. At Columbia, Robert S. Woodworth was restricting the causal basis of "scientific psychology," as he reinterpreted the act of experimentation to mean the manipulation of an independent variable, a conviction he disseminated in his influential 1938 textbook, *Experimental Psychology.* That Woodworth claimed to be an adherent of no school – indeed he emphatically insisted his position was that of a nonpartisan "middle-of-theroader" – does not negate the neobehaviorist implications of his methodological pronouncements.[29] And, at Teacher's College – where Lois Barclay Murphy pursued her Ph.D. – sociologist Dorothy Thomas was training a cadre of developmental researchers in the use of "time-sampling" techniques, in an effort to avoid "contaminating" observations of social behavior with interpretation.[30]

While Allport and the Murphys were located within institutional settings in which theirs were voices of dissent, it is also true that these same settings provided networks of support. As a rebellious minority within the larger body of academic psychology, the pool of internal supporters upon whom Allport and the Murphys could rely in attempting to reconstruct science was necessarily smaller than the one that existed in opposition to their goals. It would be wrong, however, to assume that the force of the claims that they pressed on their colleagues can be grasped simply in proportional terms. Where disciplinary arbiters saw firmly delineated professional borders, Allport and the Murphys saw permeable boundaries, across which they determinedly trafficked in their quest to reshape the practices of psychologists.

Allport and the Murphys saw themselves as sharing ideals with a community of intellectual critics, only some of whom were psychologists; this fluid world of iconoclasts offered individuals such as Allport and the Murphys confirming instances of their own stances and also provided corroborating resources on which to draw. Such sustaining associations served as vehicles through which they could continuously revalidate both their identities and their convictions. These associations also provided social materials from which new sets of relationships could be improvised as Allport and the Murphys experimented with new frameworks with which to approach the practice of psychology. By maintaining their memberships in a diverse set of professional coalitions, Allport and the Murphys were able to make choices *with* as well as against other members of the larger intellectual community, thus providing a certain amount of protection for themselves against disciplinary efforts to marginalize their achievements.

The form and content of the dissenting discourse developed by Allport and the Murphys is marked in various ways by the interlaced contexts of constraint and support from which it emerged. Thus it is useful to outline this background before moving into more detailed discussions of their work.

Starting Out: Professional Settings, Disciplinary Imperatives, and the Contours of Dissent

Gordon Allport

The complex maneuvers that preceded Harvard's offer to Allport of a post in psychology indicate the professionally inhospitable environment that he would encounter there during the 1930s. In 1928 E. G. Boring set out to fill the professorship in psychology at Harvard left empty by the resignation of William McDougall. Psychology, however, was still a subordinate part of the philosophy department at this time, as it was the philosophers who chaired the joint department.[31] The philosophers initially pushed for Gestalt theorist Wolfgang Köhler to replace the philosophically oriented McDougall, but Boring was determined to appoint a "pure" experimentalist. As Boring summarized the matter, the philosophers were looking for a person who demonstrated "breadth of interest, vision and imagination," whereas he wanted someone possessing "technical skill and knowledge within a given field."[32] Nevertheless, an offer was extended to Köhler, which he turned down. Although the issue of the McDougall professorship was resolved when Boring was promoted to it by President Abbot Lawrence Lowell (when Boring received a similar offer from Cornell), this still left a psychology position to be filled.[33] Boring renewed his campaign for a "pure" experimentalist and put forward as his choice Karl S. Lashley, a neuropsychologist at the University of Chicago, who had trained with John B. Watson and neurologist Shepard I. Franz.[34]

In his intellectual orientation, Lashley offered little that would be congenial to the philosophers, believing, as he did, that "the essence of behaviorism is the belief that the study of man will reveal nothing except what is adequately describable in the concepts of mechanics and chemistry." Pushing his viewpoint still further, Lashley averred that "the concepts of the physical sciences are the only ones which can serve as the basis for a science," and that "all psychological data, however obtained, shall be subjected to physical or physiological interpretation."[35] In brief, what Lashley's psychological research was meant to show was "that the statement, 'I am conscious' does not mean anything more than the statement that 'such and such physiological processes are going on within me.'"[36] Lashley brusquely objected to any attempt to "inject metaphysics into the science [of psychology]," asserting that "the developments of physics are independent of any theory of the ultimate nature of matter, and it is a bold metaphysician who ventures to take the physicist to task for ignoring things-in-themselves." Lashley claimed that "it is only by divorcing itself from metaphysics and values, and adopting the phenomenological method of science that psychology can escape the teleological and mystical obscurantism in which it is now involved."[37]

Lashley rejected the Harvard offer, however, and the philosophers, who knew Allport from his years at Harvard as a psychology student and as an instructor in social ethics, directed Boring to hire him.[38] As a social ethics instructor in 1924 Allport had pioneered what may have been the first course in personality psychology taught at an American university, but he had also kept his hand in psychology proper by assisting Boring and McDougall in the elementary psychology course. In 1926 Allport had resigned his instructorship in social ethics in order to accept a professorship in psychology at Dartmouth.[39]

The sensibility that Allport brought to the Harvard Psychology Department was in the humanistic tradition of William James and therefore was at odds with the neobehaviorist ethos that Boring was seeking to consolidate at the university. Boring had just published his monumental *History of Experimental Psychology*, a polemical book meant to propagandize for a definition of psychology that excluded all but "pure" experimental work.[40] One of the first doctoral exams at which the new assistant professor was asked to sit was that of B. F. Skinner. Later in life Skinner recalled that, during his oral examination: "I was embarrassed only once. Allport asked me, 'What are some objections to behaviorism?' and I could not think of a single one."[41] Skinner passed.

Over the course of the 1930s the neobehaviorist tenor of Harvard psychology was buttressed, first by Lashley's eventual acceptance of a Harvard professorship in 1935, and second through the attitudes of graduate students and young faculty such as S. S. Stevens, C. C. Pratt, and John Beebe-Center. Allport's one strong ally in psychology was the clinician Henry A. Murray, who was also pursuing studies in personality as director of the Harvard Psychological Clinic. Murray's hold on his position was weak, however, owing to his lack of publications and his rejection of reductionist methodology; at the time of his 1936–7 tenure review, an effort was mounted to deny Murray's promotion, bringing Lashley and Allport into open conflict.[42] In rising to Murray's defense Allport was, in essence, being forced to justify the validity of his own viewpoints.

What was at stake in this conflict was made clear by Lashley and Allport in letters to Harvard president James Bryant Conant. Lashley asserted that "our disagreement in evaluating the work presents, in a way, the conflict between the older humanistic and philosophical psychology and the attempt to evolve a more exact science through an objective and biological approach."[43] In response to this line of reasoning, Allport advised Conant that Harvard's "humanistic tradition in psychology" was under siege and in danger of destruction. "The critical standards of the 'exact sciences,'" Allport counseled, "admirable in their own right, are not catholic enough in outlook to serve as the norm for the newer science of the mind." Allport held that Murray's work provided "a much needed antidote to the prevailing barbarism of mental tests and statistical psychology" while never losing "sight of the human mind in its intricate entirety. This full-length view of the human mind has always been a distinctive characteristic of psychology at Harvard."[44] Lashley threatened to resign if Murray's tenure was granted, as did Allport, if Murray's tenure was denied.[45] In the end, Murray was indeed promoted, but, in all likelihood, the critical consideration was the fact that the Rockefeller Foundation – which financed the clinic's activities – indicated that its support of Murray was so strong that it was willing to virtually guarantee continued funding of his work for a lengthy future period.[46]

The conclusion of the Murray tenure crisis, however, did not signal the end of attempts by the neobehaviorist contingent at Harvard to dominate the department. In 1938 Boring and Allport came to loggerheads over Boring's attempt to promote his protégé, S. S. Stevens – a psychophysicist who studied audition – to a permanent position. Allport's opinion of Stevens, was, in short, that he was "more of a physicist than a psychologist."[47] Boring, however, saw Stevens as the future of "pure" psychology at Harvard, writing to him in 1935 that "operationism is getting identified in psychology with you and me and Harvard, and, although [Edward] Tolman may have anticipated us in psy. [*sic*], we are really doing the propaganda for it. And we are getting across."[48] Stevens fancied himself the progenitor of "'behavioristics,' which is a behavioristic psychology tuned up to keep pace with a fast-moving logical criticism."[49] Citing the precedent and authority of physicist Percy Bridgman, and philosophers Rudolf Carnap and Otto Neurath, Stevens declared that he brought tidings of a "revolution that will put an end to the possibility of revolutions," in the form of "a straightforward procedure for the definition and validity of concept," called "operationism" – that is a procedure in which "fundamental concepts" were to be defined by strict scrutiny of "the concrete operations by which the concept is determined." In Stevens's view, such a procedure, by insuring psychologists "against hazy, ambiguous and contradictory notions," would provide "the rigor of definition which silences useless controversy."[50]

In seconding the efforts of the Vienna Circle to "provide a secure foundation for the sciences," Stevens used operationism, "in the name of maximal rigor," as a platform from which to denounce statements about the "ultimate nature of reality," "subjectivity," "unique experiences," and any facts not "independent of the observer."[51] In expressing such sentiments, Stevens contested the validity of Allport's framework in a general way. Through remarks such as "we cannot, of course, put up with the psychology of a particular man; we must have a psychology of men, and be able to bring the 'mind' of any person into the scientific subject-matter," Stevens presented a direct refutation of Allport's efforts to bring individuality within the purview of "scientific" psychology.[52]

It was not only the principals who carried on these duels; seconds were brought in as well. When Boring coedited a volume with the heavy-handed title *Psychology: A Factual Textbook* – containing a prefatory comment certifying the book to be "free from the bias of metaphysical presuppositions or of psychological systems" – one of Allport's young coauthors, Philip Vernon, produced a dissenting book review.[53] Vernon held that the book's claim to be merely factual and metaphysics-free was "open to criticism," for "this claim is itself a metaphysical presupposition, and the book demonstrates only too clearly the disadvantages of a physical-scientific outlook in psychology."[54]

On the other side, Bills, whose mentor had been Lashley, launched a flat-out attack on Allport in his presidential address to the Midwestern Psychological Association in 1938. In Bills's opinion, work such as Allport's, which insisted on the scientific significance of individuality, was "an attack on quantitative psychology in its most sacred shrine. It calls for a duel to the death to decide whether personalistic psychology and other brands that eschew quantitative method shall be hurled from the ranks of scientific psychology or whether, instead, the quantitative criterion shall no longer be counted as an essential plank in the platform of psychology as science."[55] For those who might scoff at his tone of urgency, Bills bid them recall the chilling example of the 1913 Armory Show, asking rhetorically: "Could the artists who scoffed or chuckled at the 'Nude Descending the Staircase,' when it first appeared, have imagined that, a few years later, everyone would be taking it seriously?"[56] Stevens judged Bills's address as "excellent" and recommended it for "pointing out both the achievements and dangers of psychology's attitude toward method and principles."[57]

The aggressive, polemical stance that Allport would develop was, in great part, a consequence of the belligerent and hard-fought disputes that were part of the departmental politics at Harvard. Allport's willingness to engage his opponents publicly and his verve in framing the issues added force to his battle on behalf of what he saw as the Jamesian tradition at Harvard; and, in matters of style, it should be noted, James himself delighted in combative rhetoric. Beyond this end, however, Allport's forthright rejection of the status quo also contributed to his high visibility within the discipline of psychology. Much as the behaviorist John B. Watson's audacity twenty years before had signaled his resolution to incite a new intellectual movement, Allport's willingness to be controversial was accepted as a sign of leadership, as the various honors that came his way in the latter 1930s indicate. In turn, Allport's visibility strengthened the resolve of his adversaries to rebut his efforts.

If the Harvard milieu contained some of Allport's harshest critics, it was also the context in which he met the man whom he considered to be his most important mentor: Richard Clarke Cabot, an internationally prominent professor of clinical medicine who was also notorious as one of the medical profession's "most radical critics."[58] Cabot was a member of one of New England's most elite families, a man of affluence and influence who approached the practice of medicine with a sense of moral mission that fed his reformist zeal. Cabot introduced the practice of the clinical case conference in teaching medical students and later in life also helped to establish clinical training for members of the clergy.[59] Cabot's clinical appointment was in outpatient medicine, and his belief that the successful treatment of illness demanded an understanding of the social context of his patients' lives led to his advocacy of social work as an aspect of medicine; in 1905 Cabot financed and founded the Medical Social Service at Massachusetts General Hospital.[60] Cabot enraged his peers by publishing his findings on the high level of misdiagnosis at the hospital and was considered for censure by the Massachusetts Medical Society for his public statements that the high costs and poor levels of private medical care could be remediated only by reorganizing practitioners into medical cooperatives.[61] No doubt his colleagues in the medical school were relieved when, upon his return from serving overseas in World War I, Cabot accepted an appointment as chair of the social ethics department.

Cabot's interest in ethics had long extended beyond the world of medicine to society as a whole, and he authored a number of books aimed at the general public, such as *Social Service and the Art of Healing* (1909), *What Men Live By* (1914), and *Adventures on the Borderlands of Ethics* (1926), in which he explored issues of personal integrity and public morality. As chair of the social ethics department, Cabot believed that he could use his influence to combat "pure science" and "art for art's sake," and to initiate "nothing less than an ethically-based redirection and reinterpretation of the social sciences."[62] Cabot took over the helm of the department with gusto, introducing a new course on "Human Relations," which attracted an enrollment of 317 students for the 1920–1 academic year. In 1927 Cabot established an undergraduate field of concentration in sociology and social ethics administered by an interdisciplinary committee made up of such figures as A. M. Schlesinger from history, W. Y. Elliott from government, E. F. Gay from economics, E. A. Hooton from anthropology, and, from philosophy and psychology, Ralph Barton Perry and Allport.[63]

Allport took Cabot seriously as an intellectual critic and moral exemplar. In a letter to Cabot upon the publication of *Personality,* Allport recalled the disturbing effect that Cabot's scientific critique had initially had upon him: "In the first place, strange as it may seem, I am indebted to your frequently expressed skepticism regarding the 'science' of psychology. It was you who first made me see that as generally written it is a crude and arrogant discipline. For a time I felt always on the defensive toward your views – but gradually I came in part to share them." The result of Cabot's influence on his thinking, Allport related, was that he "began for myself to formulate an approach to psychology that seemed to me free from the worst of its historic blunders: its excessive empiricism, grotesque nativism, traffic in boggled ethics, superficiality, and undue abstractness (i.e., preoccupation with mind-in-general to the exclusion of mind-in-particular)."[64]

For Cabot the clinician and social ethicist, respect for individuals –

"minds-in-particular" – was ultimately a religious value. Writing with two of his colleagues in the *Harvard University Gazette* upon Cabot's death in 1939, Allport declared that Cabot's "respect for individuality was deep-rooted in his religious creed."[65] Cabot indicated that this religious spirit was manifested in the world by figures such as "Jesus Christ, Saint Augustine, St. Francis of Assisi, Tolstoi, Pasteur, Lincoln, Phillips Brooks, Jane Addams, Josiah Royce, and [his wife] Ella Lyman Cabot."[66] These secular saints, Cabot believed, were individuals whose ability to create and implement moral ideas showed that, "through growing, each in his individual way, man expresses his reverence for God. By helping one another to grow, man shows reverence for God's creatures." This is why, the elogistes noted, "bare knowledge, unapplied to human welfare, untested by the single case, and unenlivened by artistic feeling, never won [Cabot's] allegiance nor his admiration."[67]

In his letter to Cabot, Allport testified that the older man's influence pervaded the book. Allport credited Cabot's technique of "teaching through the use of case studies and biography" with influencing his own work: "From it I have drawn courage to state radically the objectives of a psychology of personality as I conceive them to be." Allport noted with gratitude both Cabot's championing of Allport's ideas against "an unduly conservative Dean" when he was "an inexperienced instructor under [Cabot's] supervision" and his financial patronage, which allowed Allport to take a semester's leave of absence to finish writing *Personality*. And, finally, Allport attributed his ability "to speak up boldly in this book, more boldly than I should have done otherwise," to both Cabot's "unremitting emphasis upon the value of individuality" and to the confidence he had displayed in Allport. It was "for these various reasons," Allport stated, that he had had Cabot "in mind probably more than any other single person" while drafting *Personality*.[68]

For Cabot's part, he wrote Allport that he was "looking forward to [the publication of *Personality*] with the greatest eagerness. It is sure to be one of the great books of the century."[69] Cabot invited Allport to serve as a trustee of the Ella Lyman Cabot Trust, Cabot's primary charitable endeavor, and Allport also carried through some of Cabot's other philanthropic projects upon his death.[70] In 1966 Harvard founded a new professorship, the Richard Clarke Cabot chair in social ethics, and appointed Allport as the first incumbent.[71]

The Social Ethics Department survived until 1931, when it was replaced by the Department of Sociology. This restructuring brought another ally into Allport's Harvard circle, in the person of the first chair of the new department, Pitirim A. Sorokin, a Russian emigré. In his student days, Sorokin had participated in the Russian Revolution as a member of the noncommunist left, and had been a member of the Constituent Assembly and secretary to Prime Minister Alexander Kerensky; he was banished from the Soviet Union by the Bolshevik regime in 1922.[72] When Sorokin arrived at Harvard he had already produced such works as *The Sociology of Revolution* (1925), *Social Mobility* (1927), and *Contemporary Sociological Theories* (1928) and demonstrated a wide-ranging interest in the consequences of such phenomena as disaster and revolution for interpersonal behavior, along with questions of social organization, disorganization, and reorganization.[73] Sorokin published the four volumes of his controversial synoptic study, *Social and Cultural Dynamics,* from 1937 to 1941 and would devote his efforts during the 1940s and 1950s to the study of altruism.[74] Allport dedicated the published version of a series of Yale lectures entitled *Becoming: Basic Considerations for a Psychology of Personality* to Sorokin, whom he characterized as a "colleague of powerful erudition and blazing conviction."[75]

During the 1930s, the interests of such associates as Sorokin in the Sociology Department and of Murray at the Harvard Psychological Clinic would serve as reminders that psychology could be viewed not only as an extension of classical physics but also as a *social* science with interdisciplinary moorings. Indeed, the strength of such alternative ties made possible the wrenching institutional reorganization that occurred in 1946, with the founding of the Department of Social Relations, an interdisciplinary experiment in which social and personality psychologists such as Allport left the department of psychology to join with like-minded scholars in anthropology, sociology, and clinical psychology. In the words of Clifford Geertz, who was a graduate student there in the 1950s, Harvard's Department of Social Relations was "social science in full cry; headier and more confident than before or since."[76] If the Harvard environment was one that presented Allport with numerous obstacles in moving his agenda forward, it was also one that provided strategic opportunities for negotiating around those obstacles.

Gardner Murphy

Gardner Murphy's twenty-one-year association with Columbia University began in 1919, when he entered the doctoral program in psychology after having served overseas during World War I in a mobile medical unit. Murphy had attended Yale as an undergraduate and possessed a master's degree in psychology from Harvard. Although Columbia University had, like Harvard, an elite reputation, its ambience was quite different. Harvard, situated in Cambridge on the banks of the Charles River, still recalled its genteel Brahmin heritage and retained an aura of exclusivity. Columbia, on the other hand, was an urban university in the midst of the nation's largest metropolis. One telling point of contrast lay in sheer numbers: during the interwar period Columbia produced more Ph.D.s in psychology than did any other American program.[77]

Columbia's program ranged more widely than did Harvard's, fostered by chairman Robert S. Woodworth's belief that there should be no "schools" in psychology; Woodworth himself exemplified this approach, for, although he was primarily concerned with experimental and physiological psychology, he also enjoyed teaching more distant topics, such as social and abnormal psychology.[78] The applied emphasis at Teachers College – a professional school whose prestige was burnished by Edward Thorndike's renown as an educational psychologist – and the numbers of part-time graduate students that it accommodated, also marked the Columbia milieu as different from that to be found at Harvard. As Heidbreder observed, the Columbia student "discovers immediately that psychology does not lead a sheltered life; that it rubs elbows with biology, statistics, education, commerce, industry, and the world of affairs."[79]

Heidbreder characterized Woodworth's "dynamic psychology" as "a modest, matter-of-fact, unaggressive system" that was "conservative in the literal sense of the term. Its purpose is not to found a new school, but to make plain what psychology has always been doing." To that end, reported Heidbreder, Columbia graduate students were made "aware of the immense importance . . . of curves of distribution, of individual differences, of the measurement of intelligence and other human capacities, of experimental procedures and statistical devices, and of the undercurrent of physiological thought."[80] As one student in psychology at Columbia during the 1920s and 1930s later recalled, the hallmark of the department was "objectivism implemented by the most rigorous experimental-statistical methodology."[81]

Woodworth's mimeographed experimental psychology textbook was known by students as the "Columbia Bible." As Andrew Winston observes, when the book was formally published in 1938 it became the "Bible" for a generation of North American psychologists as well, serving as "a major element of socialization for new recruits to the discipline."[82] In this book – and in revised editions of his best-selling introductory text, *Psychology* – Woodworth narrowed the definition of experiment "from a term that included any empirical research, to a term that included only studies that actively manipulate an independent variable."[83] In the third edition of his *Psychology,* which appeared in 1934, Woodworth explained that an experimenter is one who controls the conditions and "does not let things happen at random." To make sure that students understood the need to emulate natural science, Woodworth included a "description of Galileo's work on objects falling in a vacuum as the prototype of proper experiments."[84] In 1940 Woodworth informed his readers that psychologists had finally "decided they must follow the lead of physics, chemistry and physiology, and transform psychology into an experimental science."[85]

The nature of Murphy's relationship with the psychology department at

Columbia cannot be fully grasped without some understanding of the tension that he felt between his devotion to psychical research and his decision to identify himself professionally as a psychologist. While Murphy received his doctorate from Columbia in 1922, he avoided committing himself to a full-time instructorship there until 1925, and he did not become an assistant professor until 1929. Murphy's reluctance to launch a career as a psychologist stemmed from his hope that he could make a career for himself as a researcher of psychical phenomena. While pursuing a conventional Ph.D. in psychology, Murphy had simultaneously been working his way through a program designed – at his request – by the secretary of the Society for Psychical Research (SPR) in London. From 1919 to 1921 Murphy applied himself assiduously to his adjunct program in psychical studies, two hours a day for the first year, and then three hours a day the next. The task was not one requiring "'iron discipline,'" Murphy commented, "for I loved the material passionately . . . with all the ardent intensity of youth I had found what I believed in."[86]

In 1921 Murphy journeyed to Harvard to ask William McDougall's advice on seeking permanent employment with the SPR. McDougall informed Murphy, however, that an appropriate opportunity existed closer to hand, for Harvard had recently been named the beneficiary of funds that could support part-time psychical research. From 1922 to 1925 Murphy commuted weekly between New York and Boston, teaching part-time at Columbia while pursuing telepathy research and the investigation of mediums under the auspices of the Richard Hodgson Fund at Harvard. In 1923 Murphy attended the international SPR meeting in Warsaw as the delegate of the American Society for Psychical Research. On this trip he met René Warcollier, a French chemical engineer who had recently produced a book on his own telepathic researches, and with whom Murphy then carried out two years of experiments in transatlantic telepathy.[87] On a trip to Europe in 1929, both Murphys spent three weeks at Warcollier's home in Brittany as his guests. Gardner Murphy later recalled that they devoted much of their time to discussing questions regarding the subconscious "and the implications of Henri Bergson's philosophy for the problem of transcending time and space." After this 1929 visit, Murphy stated, "our contacts were continued through a steady flow of correspondence. Interest centered about the question of the nature of the telepathic image, the 'fragmentation' of the impulse, the role of emotion."[88]

Of this time, Murphy remarked: "I led a double life, keeping a toehold on respectable psychology while carrying on the work of a 'quack,' as psychologists saw the matter."[89] Most of the people around him, Murphy believed, "thought it [psychical research] was just a hobby of mine, an intellectual interest. It was of course, much more than that." Murphy looked on psychical research "as bearing on basic personal realities, and even cosmic realities, so that I invested very much more of myself in it than could possibly be evident." In 1924, however, Murphy began being pressured to accept a full-time instructorship at Columbia, an offer he at first declined.[90]

Murphy's professional stalemate was broken the next year by a dramatic turn of events. In 1925 Murphy became severely disabled when his eyes failed "utterly," a problem compounded by what he described as the "desperately distressing sequelae" he also experienced following a bout of influenza in that same year. At this point, Murphy felt compelled to give up his Hodgson-supported research.[91] Although he then joined the Columbia faculty on a full-time basis, as far as psychical research was concerned, Murphy explained: "In no way did I slacken interest, but life forced on me temporarily an orthodox path. I had saved no money. I could anticipate medical difficulties. I had no margin of safety anywhere."[92] Until 1935, Murphy's research would remain, in his words, "within the fold of complete academic orthodoxy."[93] When Murphy did return to an active role in psychical research in 1935, it was because of what he described as "a very extraordinary and revitalizing coincidence": the publication in early 1934 of J. B. Rhine's *Extra-sensory Perception,* and Murphy's parallel "quasi-miraculous restitution of health under Dr. Hay's treatment in November, 1934."[94] Rhine's work, at first blush, appeared to be the holy grail sought by psychical researchers: replicable experiments.

Immediately after undergoing a month-long rehabilitation program at Hay's sanitarium, Murphy traveled to Duke University to confer with Rhine and his colleagues. At this point Murphy reentered the mainstream of what Rhine had christened "parapsychology." Questions regarding how most appropriately to regard experimental practices were now doubly magnified for Murphy: in regard to psychology, and in regard to parapsychology. Where Allport could afford to call doctrinaire experimentalists on the carpet, Murphy was instead obligated to be more diplomatic, because the fate of parapsychological research seemed to hinge on the issue of experimental technique – primarily because those hostile to psychical research were able to insist upon it. As Paul Allison points out, the research done by parapsychologists in the 1930s and the 1940s was "almost a dialogue with their critics," a highly charged process, in which, as H. M. Collins and T. J. Pinch remark, there is a "ruthless exploitation of any weaknesses of protocol or security" by critics.[95]

In delineating the kinds of negotiations that parapsychologists must conduct in order to obtain legitimacy in the eyes of their scientific colleagues, Collins and Pinch indicate that one strategy that they can adopt is that of "metamorphosis": "'becoming scientists,'" by increasingly "incorporating into their work the complex experimental techniques available to physicists, biologists or psychologists."[96] Indeed, experimental psychologists demanded that those working in such contested areas as psychic research demonstrate adherence to ideals of the utmost rigor. On the one hand, this was a call that Murphy felt compelled to honor, in order to maintain his credibility as a member of the scientific community; on the other hand, one of the reasons that Murphy had been attracted to the study of psychical phenomena in the first place was that such activities seemed to contradict the mechanistic premises upon which a strict experimentalism was based.

As the "youngest of the experimental sciences," Murphy contended, psychology had "been content to model itself upon physics and biology and instead of challenging their tenets has felt that its own scientific status depended in large part upon acceptance of the standard world view."[97] Citing the example of Archimedes, Murphy speculated that, given the right place to stand, one could, however, move the scientific world: "The leverage for a replacement of 17th-century naive mechanism by other conceptions more characteristic of 20th-century scientific adventure is," he submitted, "research on extra-sensory perception."[98] Comparing the two paths between which he felt constrained to negotiate, Murphy confessed that "none of the experimental quantitative methods, so exquisite as science, ever had anything like the same appeal for me" as did the British investigations of psychic phenomena conducted from the latter nineteenth century to World War I.[99]

As a teacher of psychology at Columbia, Murphy observed that, "for the most part my job . . . was to inculcate respect for solid research method."[100] Because of his commitment to parapsychology, Murphy was in an awkward position in regard to the increasingly restrictive vision of experimentation being offered by influential figures such as Woodworth, whom Murphy greatly admired. Pursuing research in parapsychology already made one's scientific credentials suspect; for Murphy to protest, in the caustic manner of Allport, that the "objectivist" contingent in psychology had misconceived the nature of reality would be to make his position more vulnerable than it already was – it would be a rebellion on top of a rebellion. Responding in 1931 to the suggestion that scientists had become more friendly to psychical research than they had been in the past, Murphy replied resignedly that he doubted "whether a daily visitor to laboratories of psychology, with an opportunity to live in and breathe their intellectual atmosphere, would venture to suggest that the academic halls echo with gentler taunts than those which greeted [Henry] Sidgwick and [Frederic W. H.] Myers in 1882." Murphy instead found contemporary psychology to resemble "a closed fortress which casual arrows are not likely to penetrate" and that had new towers "constantly being built which are harder and harder to hit."[101]

Still, Murphy persisted, providing room in the Columbia University psychology laboratory for research into psychic phenomena – both his own and that of graduate and undergraduate students. One such student, Gaither

Pratt, was an assistant of Rhine's who came to study and experiment under Murphy from 1935 to 1937 while finishing his dissertation (an experience made possible by Murphy's commitment of $1,000 dollars of financial support from his own book royalties).[102] Harvard agreed to allocate an annual allotment of $1,200 dollars from the Hodgson Fund to Murphy for the support of a student working under his supervision. Columbia undergraduate Ernest Taves was the first recipient in 1937, and Murphy mentored Taves until the young man enlisted for service in World War II.[103] During this period Murphy also edited and oversaw the translation of Warcollier's research into an English text, *Experiments in Telepathy,* published in 1938 by Harper and Brothers.[104] Furthermore, although pleading overwork, Murphy agreed in 1939 to take on a reorganization of the *Journal of Parapsychology,* as coeditor with Bernard Riess.[105]

Murphy's protest against increasingly restrictive views of what constituted legitimate research methods, would, for the most part, reside between the lines during this period. Where Allport's dissent existed primarily on the page, Murphy's dissent existed more in the round, in his midwifing of various unconventional scientific enterprises.[106] During the 1930s Murphy lent his efforts to placing other possible Archimedean levers at diverse scientific borders: in his leadership of the avowedly activist Society for the Psychological Study of Social Issues (SPSSI); in his editorship of a provocative series of psychological texts for Harper's, and of the new journal *Sociometry* (which attempted to forge links between psychologists, sociologists, anthropologists, historians, and biologists); and in supporting Lois Barclay Murphy's engagement with developmental and depth psychology. Murphy's energetic modeling of the sophisticated tolerance that he expected other scientists to extend to him as a professional served to advance his own research commitments at the same time that it acted as an understated rebuke to those who insisted that scientific knowledge could only be securely won when pursued within ever more specialized precincts.

Later characterizing his approach to psychology during the 1930s, Murphy defined it as "broadly outreaching, a collector's interest in everything . . . it was comprehensive; it was eclectic." Murphy searched during these years for conceptions that were interdisciplinary in nature, and Columbia offered especially fertile ground in which to nurture such views, given the presence of such influential figures as Franz Boas, Ruth Benedict, and Margaret Mead in anthropology and of Robert Lynd in sociology; the latter, in particular, was a good friend of Murphy's. Murphy greatly admired the *Middletown* studies authored by Lynd and his wife, social historian Helen Merrell Lynd (who taught with Lois Barclay Murphy at Sarah Lawrence), and he adopted *Middletown* for use in teaching social psychology, creating a course that was "essentially a study of American urban life as seen through the Lynds's eyes, but with a rich utilization of all available psychological materials."[107] The relationships among this diverse circle of colleagues were further cemented by summers in New Hampshire, when the Murphys began vacationing with others who owned or rented homes in the Holderness area, such as the Lynds, Mead, and Lawrence K. Frank. The New Hampshire summers, in which family visits were intertwined with intellectual discussions of topics of common interest or of work in progress, constituted an alternative "institutional" environment in which both Murphys found support for the development of their research interests.[108]

Lois Barclay Murphy

As Gardner Murphy began expanding his efforts as an academic psychologist, Lois Barclay Murphy joined with him professionally, first as coauthor of their 1931 text *Experimental Social Psychology,* and soon after as a graduate student in psychology at Teachers College. At Teachers College, Woodworth's counterpart was his contemporary and close friend, Edward Thorndike, a man whom Woodworth hailed as a "sane positivist."[109] An iconic figure at Teachers College, Thorndike is perhaps remembered most famously in psychology for what Kurt Danziger has termed his "truly apostolic zeal" for science as measurement.[110]

Although Murphy's doctoral committee at Teachers College was generally supportive of her goals, she was able to stretch the quantitative commandments only so far in conducting her dissertation research. The most influential methodologist at Teachers College was Dorothy Thomas, whose prescriptions on research design set the department's standards. Thomas was a sociologist who had come to child study via undergraduate work with sociologist William Ogburn and economist Wesley Mitchell, a doctorate from the London School of Economics, and a stint in 1924 as a statistician at the Federal Reserve Bank. Thomas's published work ranged from her 1925 text *Social Aspects of the Business Cycle* to *The Child in America,* coauthored in 1928 with her husband, sociologist W. I. Thomas.[111] At Teachers College during the 1930s, Thomas's prescriptions on methods were accepted as definitive.

Thomas believed that, in experimenting with humans, "real objectivity" could be obtained only by using statistics.[112] Thomas herself remembered that when she "joined the staff of the Child Development Institute at Teachers College, in 1927, I was still somewhat distrustful of the subjective and the 'as-yet-unmeasured' as materials for scientific investigations." Thomas indicated that she "preferred to work exclusively with the objective, defined in almost mechanistic terms, and to count, measure, sample, fit curves, correlate, test for reliability, validity and the significance of quantitative differences, rather than to utilize descriptive materials or life histories, case records, and other types of personal documents."[113] Qualitative methods were irrelevant, in Thomas's view, "on the ground that they 'obviously [would] not yield data appropriate for statistical analysis.'"[114] Thomas and her associates were especially concerned with observer reliability, and experimented with techniques for recording overt behavior on a time-sampling basis, using five-second intervals. As one of her Teachers College students explained, "the main problem of method is in the control of the observer rather than of the situation."[115]

Not everyone was impressed with the efficacy of Thomas's quest for objectivity. One reviewer of a foundational Thomas text asked, "Is five seconds of no-overt activity a meaningful unit? Is it interchangeable with any other five second interval regardless of whether the individual concerned is planning his next move with building blocks or phantasying revenge? Is a slap interchangeable with a caress? They are both units of physical contact as used in this study. If they are interchangeable are they meaningful?" The reviewer thought not, arguing whether "units which take no account of individual motivation or specific situation" could be useful at all.[116] In the reviewer's opinion, the researchers had opted to make a "sacrifice of significance for objectivity."[117] Such critiques evidently made little impression, however, for a few years later one of the authors of the above study, Ruth Arrington, would castigate Murphy for publishing, in *Social Behavior and Child Personality,* a work displaying an "anti-objective emphasis," yielding, as it seemed to her, a "disappointing" amount of quantitative analysis.[118] That there was still a battle on over the values promulgated by restrictivist researchers can be seen in another review, in which Murphy's book was "heartily recommended not only to students of social behavior but to all investigators of child psychology as a refreshing antidote to the narrow delimited researches of recent years."[119]

After receiving her Ph.D., Murphy dealt with the objectivist strictures that permeated the social psychological study of children by, in large part, ignoring them. Unlike Allport, Murphy saw less ground to be gained by preaching to the recalcitrant; unlike her husband, she had no desire to try to work within the confines of a conventional psychological department. Murphy did not deliver papers at APA meetings – she did not, in fact, even join the APA – nor did she publish her work in mainstream psychological journals. Such actions would seem to have consigned her subsequent work to professional oblivion, assuming, that is, that the organizational mechanisms of academic psychology completely controlled the dissemination of research. Murphy was willing, however, to bet that she could advance her perspectives by targeting a different audience, one that would accept her premises and approaches: the progressive education movement. Murphy did join the Progressive Education Association and was Chair of the Research Committee of the Association of Nursery Educators. She published analytic work in the *Progressive Education*

Journal and the *Journal of Experimental Education* and placed popularly oriented articles in publications such as *Childhood Education,* which billed itself in 1940 as a magazine whose goal was "To Stimulate Thinking Rather than Advocate Fixed Practice."[120]

One reason that Murphy could afford to take an unconventional attitude toward professional advancement was that she was already in possession of a college teaching position at Sarah Lawrence both before and during her Ph.D. work. Originally hired to teach comparative religion when the college opened its doors in 1928 as an experiment in progressive education, Murphy shifted her assignment to psychology in 1935 as she neared completion of her doctoral degree.[121] At Sarah Lawrence, Murphy was surrounded by intellectually and emotionally supportive colleagues, such as Ruth Munroe and Eugene Lerner in psychology, and others such as Helen Lynd. Nor was financial support a problem, even given her outsider status in psychology: Ludwig Kast, president of the Macy Foundation, had financed her dissertation research, providing about $3,000 during the depression's nadir, and later, Lawrence K. Frank – a good friend of both the Murphys – would provide Rockefeller money and further Macy funds, most significantly more than $30,000 in 1936 to establish a nursery school laboratory at Sarah Lawrence.[122] Married to a psychologist whose own status was rising, Murphy was, of course, in no danger of cutting *herself* off from the field as a whole, even if she chose to absent herself from psychology's *institutional* mainstream. Given that she had carved out an alternate niche in which to pursue her interests, Murphy turned her back on the standard academic advancement network, trading away the possibilities it held out for more widespread recognition within academic psychology for the opportunity to work without the inhibiting effect of having disciplinary arbiters look over her shoulder.

Murphy's vision of herself as a researcher drew on the outlines of the career of a woman with whom she had worked as a young adult: psychologist Helen Thompson Woolley. Murphy's initial foray into psychology was as a member of the testing staff at the Psychological Laboratory of the Vocational Bureau of Education in Cincinnati, during the summers of 1921 and 1922, and then for a year from 1923 to 1924. Murphy's involvement with the clinic came about when her mother, in the course of initiating the local Parent-Teacher Association chapter, met Woolley, who was director of the clinic, and asked if there might be an opportunity for her daughter to volunteer.

Woolley – who, as Helen Thompson, had been one of John Dewey and James Angell's most brilliant Ph.D. students at the University of Chicago – was particularly well known among psychologists for her dissertation research on sex differences. This work, published in 1903 as *The Mental Traits of Sex: An Experimental Investigation of the Normal Mind in Men and Women,* undermined then-prevalent assumptions of biologically determined gender differences.[123] In her testing of University of Chicago undergraduates, Woolley had found more similarities than differences between men and women, and she argued forcefully that what differences did appear under testing were due to the effects of differential socialization.

In Cincinnati, however, Woolley had turned her attention to researching another set of differences: those that existed between children who left school at the age of fourteen to enter the labor force and those who remained in school. This change of emphasis was partly due to the fact that Woolley had moved to Cincinnati for the sake of her husband's career, and she lacked an academic post that would have allowed her to continue her earlier research; it was also due, however, to her strong commitment to social activism. In Cincinnati, Woolley was an active presence in reform circles, serving as chair of the Woman's Suffrage Committee during the ratification fight and assisting in the successful drafting and lobbying of child labor legislation. She was also a moral presence, as when she led "an exodus from a professional meeting in a leading hotel when the admission of a Negro member was questioned."[124]

At the clinic, Murphy was impressed by Woolley's insistent "emphasis on giving a live, vital picture of a child along with the technical information from the tests." Murphy remembered that Woolley "said emphatically, 'I don't want just an I.Q. – I want as complete a picture of the child as possible,'" one that took into account a child "being tired, scared, insecure or anxious." Although Murphy would not pursue a Ph.D. in psychology until a decade later, she marked these experiences in "trying to understand the individual child and the whole child [as] crucial in my subsequent approach all the rest of my life with children."[125] Murphy's identification with Woolley's perspective is especially interesting when one considers that, as distinguished as Woolley was, Murphy was simultaneously doing coursework at Vassar with the profession's most eminent woman psychologist, Margaret Floy Washburn, who at that time was president of the APA. Although Murphy credited Washburn – a student of E. B. Titchener's who became noted for her work on the question of mind and consciousness in animals – with "brilliant" lectures and a supportive stance, she also recalled sharply her "dogmatic" and dismissive comments in regard to both Freud and psychical research. Murphy judged Woolley, on the other hand, to have been "a very creative psychologist ahead of her time."[126] Like Allport, who held Cabot to be "in the vanguard of progressive thought," Murphy seems to have used Woolley as a gauge of what was both professionally possible and intellectually meaningful in terms of constructing a scientific identity.[127] In 1922 Woolley organized one of the first nursery schools in the country to serve as a laboratory for the study of child development, and her published case studies of individual children, such as "Agnes: A Dominant Personality in the Making," became classics. Murphy, too, would found a laboratory nursery school at Sarah Lawrence and publish case studies of her own.[128]

Murphy's work in the psychology of children was positioned within a vigorous community of progressive educators in New York. Especially important for Murphy was the Bank Street School, a progressive educational institution that had been founded in 1916 by Lucy Sprague Mitchell as the Bureau of Educational Experiments (BEE). Mitchell, who had been Dean of Women at the University of California at Berkeley and who was married to economist Wesley C. Mitchell, was active in the New York reform circles that included such figures as Lillian Wald, Pauline Goldmark, John Dewey, Thorstein Veblen, and Max Eastman.[129] In the later 1920s, when psychologist Barbara Biber joined Mitchell's venture, Bank Street would become a thriving center for the study of child life, especially in psychodynamic terms.[130] Biber had studied with Franz Boas while a student at Barnard and had worked for the Amalgamated Clothing Workers Union in New York; she earned a degree in psychology from the University of Chicago.[131] Biber's research focused on children's interactions with their environments and the consequences that these interactions held out for understanding their cognitive and emotional development.[132] Bank Street during the 1930s included a nursery school serving an average of sixty to seventy children a year, a Cooperative School for Student Teachers (a joint venture of eight experimental schools), and a research and publications division.[133] The institution was a prominent player in progressive circles, attracting the attention, for example, of First Lady Eleanor Roosevelt, who visited the school in the 1930s.[134]

Murphy received a fan letter from Biber in 1935, expressing the "hope for some means through which it might be possible to put our common attitude into joint activity." Biber had been impressed by an article of Murphy's in *Progressive Education,* and a report by a Bank Street colleague of a talk that Murphy had given at a conference.[135] Murphy joined the research collective at Bank Street and found it a welcoming and stimulating milieu; the laboratory school provided another venue for conducting research in addition to her own nursery school laboratory at Sarah Lawrence. One result of this joint research was the volume *Child Life in School: The Study of a Seven-Year-Old Group.* In their introduction, the authors stated that, although the price paid for cooperative studies may be great "in terms of time and effort spent in making a coherent, coordinated product out of the approach, the working habits, the quality of expression of four different people," the gains that resulted were substantial, "in terms of broadened viewpoint for each member of such a cooperative undertaking, and, we trust, in a more comprehensive, more varied and more reliable body of findings."[136] The cooperative worked from an interdisciplinary perspective, drawing on the work of psychologists

Kurt Lewin and Jean Piaget, anthropologist Margaret Mead, and sociologists Robert and Helen Lynd, among others.[137]

In regard to progressive education, it should not be overlooked that Sarah Lawrence was itself an experimental college, and undergraduate learning was a further research interest of Murphy's.[138] During the 1930s, the General Education Board of the Rockefeller Foundation awarded funds to Sarah Lawrence to produce a series of studies on their methods. Murphy played a major role in two of these: *Psychology for Individual Education* and *Emotional Factors in Learning.*[139] Both Sarah Lawrence and Bank Street offered supportive environments for women in which to work and to explore new avenues of investigation, and Murphy took full advantage of these opportunities.

Allport and the Murphys moved in intellectual and social worlds of considerable complexity, ranging from Gardner Murphy's intense identification with the mavericks of the Society for Psychical Research to the patronage provided to Gordon Allport by an iconoclastic Boston Brahmin, to Lois Barclay Murphy's relationship with the Bank Street research collective. Although participation in these various and diverse milieu did not invest them with disciplinary powers equal to those possessed by their more conventional peers, these experiences still provided substantial benefits: opportunities to collaborate on developing scientific frameworks at odds with prevailing conventions, emotional support, financial assistance, and sheltered environments in which to develop the confidence and skills necessary for arguing for new theoretical perspectives and forms of research.

Despite the clear difficulty, then, of combatting the resistant forces of disciplinary intransigence, Allport and the Murphys moved through the 1930s with a sense of assurance and conviction that their dissent could have an impact, moving the orthodox to see science in a new light. In *Personality,* for example, Allport frequently spoke with irreverence, arguing that science had become "an arbitrary creed," and that those who wished to "restrict the meaning of the sacred phrase, 'the scientific method,'" had lost themselves in "blind loyalty to an anachronistic ideal."[140] In many ways this encounter between upholders of the status quo and dissenters from it represented a clash between those intent on enforcing the letter of the law and those attempting to hold others to account for ignoring its spirit: the priesthood versus the prophets.

The analogy is one that – at least from the vantage point of Allport and the Murphys – is close. The dogmatism with which some scientists held forth on the search for "objectivity," for example, seemed to the Murphys to compare unfavorably with the adventurous biblical exegesis going on across the street from Columbia University at Union Theological Seminary (UTS). One of Gardner Murphy's students recalled that Murphy, in a 1933–4 graduate seminar on social psychology, "emphasized the point of view that 'scientific method' was a rigorous way of thinking and not necessarily the routine formalism of experimentation. He suggested that some of the best illustrations of this kind of rigor might be found among the scholars dedicated to the study of biblical history."[141]

When Allport offered chapel devotions during the 1930s, he frequently spoke on the theme of hubris, using texts from such Hebrew prophets as Isaiah and Micah. In one meditation from 1935, for example, Allport declared that the highest stage of wisdom "comes only to those who like the prophet in Ecclesiastes perceives the vanity of the search for wisdom, and who then are able to hold in perspective the achievements of the human intellect." Allport used the example of "Francis Bacon, the father of the Enlightenment" to illustrate "this highest stage of development" – although what Allport instructed his audience to note was the religious cast of Bacon's mission, citing the closing words of Bacon's *Advancement of Learning* before stating his own conclusion:

> "For so it may be said of my views, that they require an age, perhaps a whole age to prove, and numerous ages to execute. But as the greatest things are owing to their beginnings, it will be enough for me to have sown for posterity, and for the honor of the Immortal Being, whom I humbly entreat, through his Son, our Savior, favorably to accept these, and the like sacrifices of the human understanding, seasoned with religion, and offered up to his glory."
>
> These words are worth recalling to the scholars who in imitating Bacon's scientific attitude have added an intellectual arrogance to it, and in so doing have lost the essentially religious quality of enlightened thought.[142]

In a chapel reading the next year, speaking from Isaiah, Allport sounded a similar theme: "Students and teachers, and research investigators as well, often sit securely on some one branch of the tree of knowledge and feel entirely content. . . . We apply the epithet sophomoric to the mind that has a small fragment of knowledge and greatly exaggerates the importance of what it has."[143] Lois Barclay Murphy, too, found the Hebrew prophets of intellectual relevance in thinking through her own stance. Murphy later asserted that one of the most inspiring UTS experiences that she had had as a student was in Julius Bewer's class on Old Testament prophets. Murphy speculated that, "in an authoritarian period, when a highly formalized structure of thought and behavior is the accepted basis of life," change was yet possible, as had been seen "in the emergence of creative thinking in the Hebrew prophets." Murphy asserted that it was just such an impulse that had allowed Hosea, Isaiah, Jesus, and Paul to battle "the fetters of priestly authoritarianism" and to break the monopoly that was the result "of the apotheosis of the canon, the 'right' books of the authoritarian or priestly group."[144]

The prophetic image also applied to the feeling on the part of Allport and the Murphys that they were fighting against disciplinary arbiters who, in making a "ritual of method," were ensuring "that all the faithful crowd onto a carpet of prayer, and with their logical shears cut more and more inches off the rug, permitting fewer and fewer aspirants to enjoy status."[145] The impatience of these dissenting psychologists with those exhibiting an excessive devotion to scientific pieties is captured in a lampoon from the period, in the style of a mocking paean to Thorndike, who had famously declared that "whatever exists at all exists in some amount." The jest was, appropriately, rendered in the form of a prayerful creed:

> I believe in Psychology, the Study of Mental Life, the Source of Wisdom and Truth;
>
> And in the Experimental Method, its only Hope, our Pride, conceived of the Nineteenth Century, born of Physiology, suffered through Psychophysics, become introspective, objective, statistical. It descended into Behaviorism. Some day it may arise again from the depths. It ascended into Tests, and sitteth on the right hand of Teachers College, the Light and Hope of the Universe. From thence it presumes to judge the quick and the dull;
>
> I believe in the Endocrine Glands, the occurrence of Imageless Thought, the process of Redintegration, the maintenance of a Congenial Pace, the Probable Error of the Difference, and the I.Q. everlasting. Amen.[146]

That in which Allport and the Murphys did believe is the subject of the following chapters.

3

Defying the Law of Averages: Constructing a Science of Individuality

These unhappy times call for the building of plans that . . . put their faith once more in the forgotten man at the bottom of the economic pyramid.

Franklin D. Roosevelt (1932)[1]

By pressing the scientific claims of individuality, Allport and the Murphys signaled their dissatisfaction with the current state of scientific knowledge-seeking in two ways. The study of individuality was first of all justified as an end in itself, in order to right the lop-sided view of nature that they believed scientists had produced by ignoring particulars in favor of universals. But promoting the study of that which is "individual" as a scientific priority was more than a move to supplement already dominant views, for talk of individuality simultaneously represented other challenges as well: to intellectual practices that minimized diversity, broke apart wholes, feared the taint of the subjective, and banished qualities in favor of quantities. As a symbolic rallying point for those who wished to dissent from the status quo in the sciences, a commitment to foregrounding individuality signified a resistance to perpetuating images of scientific method held dear by science's elite.

The concept of universality possesses a status in the Western intellectual tradition that the idea of particularity lacks. The practices and rhetoric of the sciences incorporate a distaste for the investigation of singularities as singularities, subscribing to the belief that such a path offers little hope of obtaining secure and certain knowledge. As philosopher Jorge Gracia remarks, unlike scholarly attention to "universals," "discussions of the correlative notion of individuality are not abundant and, by comparison with the number and depth of treatments on universals, may even be considered scarce." Gracia observes that "the relative neglect of individuality in the philosophical literature is surprising, for surely individuality seems to be one of the most fundamental and brute facts of our experience" – a point that Allport and the Murphys repeatedly argued for during the 1930s.[2]

Gardner Murphy, as indicated in chapter 2, believed that the philosophical significance of singularity had been revealed in Gestalt theory, which demanded "nothing less than a new metaphysics" in which "the concept of unique characters, entities, or integers" would be "elevated to a position of genuine explanatory value." Scientists, Murphy held, could no longer assume, "as naive chemists of the early nineteenth century had assumed or as the almost equally naive physicists of the early twentieth century thought they had proved, that the physical world is made up of parts which are ultimately exactly alike."[3] Allport made himself unequivocally clear on the matter of metaphysics by brandishing a quotation from Goethe on the opening page of *Personality:* "Die Natur scheint Alles auf Individualität angelegt zu haben [Nature seems to have planned everything with a view toward individuality]."[4] Allport also made plain that he believed something had gone seriously awry in science since Goethe's time, for while "the man in the street is never in danger of forgetting that individuality is the supreme characteristic of human nature . . . with the scientist the case is different." If individuality appeared to be "self-evident" to the laity, that was hardly true for professionals, in Allport's view: the "sciences find the very existence of the individual somewhat of an embarrassment and are disturbed by his intrusion into their domains."[5] The widespread adoption of mass methods of research – used in the search for generalizations and universals – had been responsible, Allport charged, for the "ridiculous" results that had convinced "so many educated people that psychology is a sappy science."[6]

Historians of psychology such as Kurt Danziger have indeed found the "individual" disappearing in the practices of the era's academic psychologists, who, "in order to make universalistic knowledge claims . . . took to presenting their data as the attributes of collective rather than individual subjects." To the extent that an individual was represented in the devising of statistical regularities, it was as a point within a population distribution; as Danziger notes, individuals were now distinguished by their "deviation from a statistical norm established for the population with which they had been aggregated."[7] But radically empiricist psychologists resisted these trends, contending that individuality was a legitimate feature of experienced reality and thus required study, not dismissal. If the phenomenon of individuality seemed to slip through conceptual nets spun of general laws, then psychologists such as Allport and the Murphys recommended that their colleagues work at developing idiographic research practices equal to the task. Allport offered promissory notes for new methods by shepherding a number of projects by Harvard graduate students on the question of studying unique events and entities through the journal presses by the end of the 1930s.[8] Lois Barclay

Murphy set for herself the question of devising ways to elicit the inner world of preschool children, and of ascertaining how the outside world was internalized and transformed by individual minds. Gardner Murphy served as a sounding board for these goals, and he continued to investigate such novel spheres as parapsychology and sociometry, while beginning to draft chapters of a major treatise on personality that would appear in the latter part of the 1940s.[9]

Changing Cultural Conceptions: "Individualism" versus "Individuality"

Viewed from the intellectual framework of these psychologists, the scientific status of individuality was seen as a political as well as a philosophical problem. At the turn of the century William James had described the "practical consequences" of "the pluralistic or individualistic philosophy [as] the well-known democratic respect for the sacredness of individuality."[10] James's views retained this meaning well into the twentieth century, as in Horace Kallen's assertion that James held that "in both the life of man and the life of nature, individuality remains the irreducible surd."[11] When Lois Barclay Murphy used Abraham Lincoln as an archetypal image of the fusion of democracy and individuality in a radio address, she, too, was speaking in this Jamesian idiom. In observing that "all of our averages, our stereotyped ideas of how tall and how wide and how well-adjusted a child" should be prevented Americans from accepting "our future Abe Lincoln with his individuality and uniqueness as a boy," she was rebutting both the practice of collapsing idiosyncrasies within statistical generalizations and the potential political implications of this process, declaring that "our democracy will lose the strength that" such values "alone can give it."[12]

A generation before, John Dewey had observed that psychology could hardly fail to be a "political" science, given that its charge as a discipline was to study the individual; and indeed, when Allport addressed the American Psychological Association as its president in 1939, he drew Dewey's remarks to his audience's attention.[13] Dewey had argued that every science, "in its final standpoint and working aims is controlled by conditions lying outside itself – conditions that subsist in the practical life of the time." Given that such was the case, it was therefore true that "the way in which the individual is conceived, the value that is attributed to him, the things in his make-up that arouse interest, are not due at the outset to psychology." Differing estimates, for example, "of the worth and place of individuality" that occurred in an autocracy, an aristocracy, or a democracy would be embodied within that society's psychological science.[14] From a similar standpoint Allport asserted that, contrary to "the ethics of democracy," which directed that "respect and value are to be ascribed to each individual person . . . mass methods of research do little to engender understanding or respect for the individual person." Under the heading "Squaring Science with Democracy," Allport argued that "a shift of interest to personal documents, a substitution of the clinical for the mass point of view, cannot fail to bring psychological science closer in line with the ethics of democracy."[15] In short, in Allport's view, "the aspiration of democracy is to foster the integrity of each individual," and if psychological science failed to do so it was therefore shirking its democratic obligations.[16]

When historians visit the depression years, they often report back that "intellectuals insisted that society should engulf the individual, and for them communitarian values gained precedence over individual values." They find that intellectuals became fond of the phrase "the people," as in Carl Sandburg's epic poem, *The People, Yes.* The intense interest of the era's intellectuals in exploring the "culture concept" is also noted – the word "culture" no longer designating the "high-water marks" of a civilization (its singular Shakespeares and Beethovens), but instead referring to the *shared* patterns of a *community's* way of life.[17]

The dispersal of the individual within amorphous collectivities was indeed a characteristic trend within many areas of American life. During the 1930s, followers of laissez-faire economic doctrines, for example, rendered individuals faceless before the "invisible hand" that swept into one force their separate actions, while, for those intellectuals adopting the rhetoric of the Communist party and of scientific socialism, individuals were addressed only insofar as they were submerged within the "masses." George Gallup's new techniques for public opinion polling made the "average American" a ubiquitous part of the national scene. Poet W. H. Auden, who had taken up residence in the United States, responded to the creation of a world populated by statistical constructs with an ode to "The Unknown Citizen," published in the *New Yorker* in 1940. The poem was preceded by the phrase, "To Social Security Account Number 067-01-9818 This Marble Monument Is Erected by the State":

> Our investigators into Public Opinion are content
> That he held the proper opinions for the time of year;
> When there was peace, he was for peace; when there was war, he went.
> He was married and added five children to the population,
> Which, our eugenist says, was the right number for a parent of his
>   generation,
> And our teachers report that he never interfered with their education.
> Was he free? Was he happy? The question is absurd;
> Had anything been wrong, we should certainly have heard.[18]

That social science elites were rendering singularity irrelevant through the solvent of generalization had not escaped public notice.

Along these same lines, film director Frank Capra characterized his "individual-against-the-corrupt-system" trilogy – *Mr. Deeds Goes to Town* (1936), *Mr. Smith Goes to Washington* (1939), and *Meet John Doe* (1941) – as examples of "the rebellious cry of the individual against being trampled to an ort by massiveness – mass production, mass thought, mass education, mass politics, mass wealth, mass conformity."[19] James Thurber, writing in the "little guy" idiom in the preface to his 1933 collection of stories, *My Life and Hard Times,* confessed that they were the memoirs of a man whose only accomplishment of excellence was an "expertness in hitting empty ginger ale bottles with small rocks at a distance of thirty paces." With apologetic irony, Thurber explained that "such a writer's time is not Walter Lippman's time, or Stuart Chase's time, or Professor Einstein's time," but merely a record of "his own personal time, circumscribed by the short boundaries of his pain and his embarrassment, in which what happens to his digestion, the rear axle of his car, and the confused flow of his relationships with six or eight persons and two or three buildings is of greater importance than what goes on in the nation or in the universe."[20]

To focus on the working out of intellectual commitments to various forms of collectivity during the depression years would be to miss, therefore, the simultaneous emergence of a powerful concern with the place of the individual in modern society. Alan Brinkley, in his ground-breaking book, *Voices of Protest: Huey Long, Father Coughlin and the Great Depression,* demonstrates that Long and Coughlin were "manifestations of one of the most powerful impulses of the Great Depression, and of many decades of American life before it: the urge to defend the autonomy of the individual and the independence of the community against encroachments from the modern industrial state." As Brinkley argues, the followers of Long and Coughlin "yearned for no shining collective future" but instead called for "a society in which the individual retained control of his own life and livelihood; in which power resided in visible, accessible institutions; in which wealth was equitably (if not necessarily equally) shared."[21]

In pledging himself to rescue "the forgotten man at the bottom of the economic pyramid" Roosevelt was also responding to the general public's discontent with the deindividuating consequences of America's political and economic institutions. Much of Roosevelt's political success, in fact, stemmed from his ability to establish a seemingly personal bond with a wide array of citizens, giving them the feeling, in William Leuchtenberg's words, that "they could confide in him directly."[22] Roosevelt himself said that he aimed "not merely to make Government a mechanical implement, but to give it the vibrant personal character that is the very embodiment of human charity."[23] In his "Fireside Chats," the president used radio addresses to make himself an intimate presence in millions of American living rooms; such exercises were an expression of his belief that democratic government "must make men and women whose devotion it seeks, feel that it really cares for the security of every individual."[24] Indeed, a public presence was given to innumerable "forgotten" individuals through a variety of New Deal programs – as when those receiving a Works Progress Administration (WPA) or Farm Security Administration (FSA) paycheck recorded the oral histories of former slaves, or photographed sharecroppers, or told the stories of eccentric characters in local guidebooks, or painted scenes from a town's history on post office walls.

Like the politicians of this era, advertisers were quick to cater to the desire of individuals for personal recognition in advertising campaigns that sought to tie corporate products to a "re-personalization" of American life.[25] Roland Marchand observes that consumers "hungered to be addressed as individuals, in personal tones," and commercial personalities such as Betty Crocker crowded the radio airwaves in "the simulation of one-to-one personal conversations between the personalized emissaries of the corporations and individual consumers."[26] Products themselves eventually became personalized, as in a Lucky Strikes cigarette ad promising that "when all else fails, I'm your best Friend."[27]

If artists, politicians, and the manufacturers of consumer goods all recognized the complex and contradictory longings that were contained within the concept of individuality, it is not surprising to find these same concerns expressed within the psychological science of the time, in ways that sometimes found the two discourses converging. For example, when Allport applauded a colleague, who, "in stressing the person . . . with remarkable directness calls psychologists back from their esoteric bush-beating and confronts them with their forgotten man," he played upon the words in which Roosevelt responded to public discontent.[28] Similarly, he echoed Auden's claim that the "unknown citizen's" voice could not be heard by social scientists when he conjectured that if a subject in an experiment were to protest "that it is evident to him that he had a rich and vivid experience that was not fully represented" by the experimental results that he would be "firmly assured that what is vividly self-evident to him is no longer of interest to the scientist."[29]

Intellectuals such as Allport and the Murphys responded affirmatively to both sides of the prevailing debate on the validity of communitarian and personal values, seeking to find ways in which individuality and contextuality could be studied in juxtaposition. It is important to recognize that the "individual" with which they were concerned is not the "individual" of laissez-faire capitalism, an atomic free-agent ricocheting off other human atoms in a dizzying dance in which markets call the tune. Allport, for example, remarked in a talk that the free-market assumptions of Adam Smith had "resulted in employing 'hands' not persons," and thence in "child labor, long hours, sweat shop[s]."[30] These scientists were concerned instead with the dynamics of *individuality,* a view that assumed that the human beings of which any system is comprised are to be taken as richly singular, rather than abstractly uniform.

By this usage, "individuality" signified a belief that the integrity of the individual is a social good. In the view of Allport and the Murphys, this principle would be realized in an egalitarian democracy, but not in a society dominated by industrial capitalism, which they regarded as having devolved into an aristocratic system in which the privileges of personal freedom were unevenly distributed. Indeed, in a 1933 manifesto by progressive education activists, Dewey bluntly made precisely this distinction between the presumably antithetical precepts underlying "individualism" and "individuality," remarking that criticism of individualism did not similarly entail a "depreciation of the value of individuality." Dewey pointed out that "the form which the historic individualism of the eighteenth and nineteenth centuries took is now adverse to the realization of individuality in and for *all.* It favors and supports legal and economic institutions which encourage an exaggerated and one-sided development of egoistic individuality in a privileged few, while militating against a full and fair opportunity for a normal individuality in the many."[31] As many of those working the *via media* political terrain between revolutionary socialism and laissez-faire liberalism during this period found, rethinking democracy meant rethinking, as Dewey argued, "the worth and place of individuality."[32]

Variations on this theme were legion. Harold Ickes, Roosevelt's Secretary of the Interior, excoriated the nation's industrialists for their espousal of "rugged individualism," a doctrine he declared to be "founded upon the antisocial, unchristian theory of 'dog eat dog,' 'may the devil take the hindmost.'" In practice, rugged individualism meant "regimentation in mill, mine and factory so that a few may grow rich and powerful at the expense of the many;" and, as Ickes explained, the many were "as fully entitled as are the wolves to live their own lives, preserve their liberties and seek happiness."[33] What was needed, Ickes demanded, was the recognition that "the soul of the American system" inhered not in "rugged individualism," but in an "individualism" that, as bequeathed to the nation by its founders, was "consistent with the principle of the greatest good for the greatest number, [where each] could give free rein to their own initiatives, develop their own individualities, and live such lives as suited them."[34]

Speaking from an anarchist perspective, Emma Goldman similarly contended that "individuality is not to be confused with the various ideas and concepts of Individualism, much less with that 'rugged individualism' which is only a masked attempt to repress and defeat the individual and his individuality." Such "so-called Individualism" had "meant all the 'individualism' for the masters," and was, for Goldman, nothing more than "social and economic laissez faire," a creed that had as its "highest wisdom . . . 'the devil take the hindmost.'"[35] Harvard philosopher Ralph Barton Perry, commenting on the ambiguity adhering to the meaning of individualism, stated that it "is often identified with a man's inconsiderate or ruthless assertion of his own individuality; but individualism as one of the ingredients of a democratic community is essentially a respect and relish for the individuality of others. The first is the easy way, and the second is hard."[36]

Dewey was not alone, then, in arguing that "the spiritual factor of our tradition . . . is obscured and crowded out. Instead of the development of individualities which it prophetically set forth, there is a perversion of the whole ideal of individualism to conform to the practices of a pecuniary culture." Dewey's answer, however, was not simply that an earlier tradition must be recovered, for such an approach would treat "individualism as if it were something static, having a uniform content," and thus ignore the fact that "the mental and moral structure of individuals, the pattern of their desires and purposes, change with every great change in social constitution." The task at hand, therefore, was that of "forming a new psychological and moral type," in response to "the great pressure now brought to bear to effect conformity and standardization of American opinion."[37]

Lois Barclay Murphy agreed that a structural shift was under way, judging "that the scientific disciplines are being reformulated. Cultural pressures deeper than themselves are redirecting them toward a fuller understanding of the individual."[38] Allport sounded a comparable theme in his claim that "one benign effect" of "a decade of depression, war, and misery" had been to bring out "upon the center of our cultural stage the struggles of the common man, the picture of his daily life, his courage, his homely values." The cultural stage to which Allport pointed was that of the media; there it could plainly be seen, Allport indicated, that "the layman has become interested in the *personal document;* and so too has the social scientist, caught up in the general cultural tide."[39]

"Caught up in the general cultural tide," social scientists such as Allport and the Murphys could feel assured that they were participating in a larger movement in which the meaning of the individual was being refigured. At the same time, for a significant portion of their academic colleagues, the fact that individuality was of interest in the common public world only made the whole matter more suspect as a scientific topic.

The Scientific Status of the Single Case

In a historical survey of trends in psychology, Allport observed that his contemporaries were studying personality more frequently than had been the case in the past, "but, paradoxically enough, the individual is studied less often. Most psychologists do not regard clinical cases, individual life histories, or single historical events as appropriate material for professional publication," and his survey demonstrated that there had been "a general decline of concern for the single case."[40] This decline was not simply a matter of benign neglect. As one psychologist declared in regard to the scientific status of individuality, it could be flatly dispensed with, for "the psychology of the concrete particular individual is not a science. It is an art, to be practiced by the gifted few."[41] During the 1930s, Allport would act as the chief polemicist for the reality of individuality within psychology, a task he took up with decided zest. His book, *Personality: A Psychological Interpretation,* was hailed as the successful attempt of the "leading American investigator of the problem of personality structure" to consolidate his position, and as "a critical book . . . that defines agreements and differences."[42] Indeed, one reviewer predicted that "the system which it espouses will be bitterly contested."[43] Allport kept up this campaign in a monograph published a few years later, *The Use of Personal Documents in Psychological Science.*

In the presentation of his claims, Allport met his opponents head-on. Acknowledging that "the prevailing bias in psychological and in social science is *nomothetic* [based on general laws]," and that "it is exceedingly difficult – especially in America – to make the idiographic phase of knowledge seem enticing – or, indeed, even plausible – to many scientific workers," Allport directly attacked the logic of the "scientific method."[44] The source of this implausibility regarding the scientific status of the single case, Allport argued, could be traced to scientists' acceptance of a "narrowly conventional view of what scientific method must be."[45] In fact, Allport charged that scientists obey dicta such as "Scientia non est individuorum," because, as currently practiced, "science is an arbitrary creed." Scientists, according to Allport, placed a priority on "the discovery of regularities and uniformities *characteristic of a whole class* of objects." As a result of this choice, they had restricted themselves to a search for "generalized truth, with occurrences that are common to events of one class. A 'class,' to be sure is a question-begging concept, for it in turn is an abstraction designed to cover common occurrences. So it turns out that the 'order in nature' which the scientist seeks is after all quite a circular matter."[46]

Allport contended that the search for generalized truth was due to socially imposed practices, and that the "typical procedure the scientist feels compelled by convention to follow" stemmed from "a certain *professional* attitude toward nature." It was "the fact that this attitude is only one of many kinds of attitude of which he is capable," Allport suggested, that "demonstrates at the outset a certain arbitrariness in his method of study."[47] By holding to convention, "absorbed by the shadow of Method rather than by the individual objects upon which the shadow lies," psychologists had, Allport claimed, transformed "what is merely an artifice of method into a doctrine of reality."[48] All that they had gained by such procedures was an "entirely mythical" concept called "the generalized mind."[49] By changing convention, therefore, scientific method could be recast in form to avoid reifying what were methodological preferences.

It might indeed be true, Allport concluded, that, in "blind loyalty to an anachronistic ideal," some of his colleagues would repudiate the "new movement within psychological science" that was attempting "to depict and account for the manifest individuality of mind."[50] To repudiate this movement, however, would be to reject the claim that individualities were as much facts of nature as were generalities.[51] In Allport's indictment, scientists who banished idiographic knowledge from science's precincts "pretend to deal with Nature, but are oblivious to the fact that Nature, as Goethe said, seems to have planned everything with a view to individuality."[52] Allport argued instead that what needed to be understood was that "acquaintance with particulars is the beginning of all knowledge – scientific or otherwise."[53] As long as scientists insisted on "stripping the person of all his troublesome particularities," the subsequent distortion would result in nothing but the destruction of an individual's "essential nature."[54] In speaking of an epistemology that apprehended particulars, Allport was furthering lessons he had learned from postgraduate study in Germany, as well as developing features of James's philosophy, especially his radically empiricist concept of a pluralistic universe.

In addition to the United States Allport claimed Germany as his "intellectual home."[55] Allport highly esteemed the work of German psychologist William Stern, and he counted himself as one of the many students who "borrowed rays from his 'star.'"[56] Stern, who was director of the Psychological Institute at the University of Hamburg when Allport came to know him as a student, had made a name for himself as a pioneer in many fields of psychology, including intelligence testing, the psychology of testimony, and child psychology. What Allport found of greatest significance, however, was the synthesis that Stern was forging between his philosophical doctrine of "critical personalism" and his psychological studies.[57]

Stern's theoretical system was based on the premise that "the defining property of 'person' is *concrete purposive activity*." The world was replete with "essences which are because they have effects; which are wholes, through the fact that they exemplify in themselves a significant manifold of parts; which are bearers of a teleological causality, in that the meaning of the totality determines the realization of its subordinate part-purposes; which are concrete and individual, in that they alone give significance and sense to all abstraction and generalization." The category of person was thus a capacious one and could be applied to "the human, the sub-human, the super-human, to the organic and the inorganic, to individual and societal forms."[58] In Allport's view, Stern had divined "prophetically, that individuality would be the problem of the Twentieth Century."[59]

In bringing Stern's metaphysics to bear on the study of human behavior. Allport placed the idea of personhood front and center in his psychology. In his discussions of personalistic doctrine, Allport emphasized that the spontaneous as well as reactive behavior of persons was taken into account, a subject, he claimed, that was "generally neglected by psychology at large." Allport pointed especially to the prominent attention that personalistic perspectives awarded to such phenomena as "intention, attempt, effort, and that distinctive human ability for *conscious planning*."[60] Allport saw personalistic psychology as thus aiding in the attempt to further the "modern revolt" in which "the individual person as a many-sided unity must serve as the center of gravity for each and every investigation and formulation of theory undertaken by psychology." In elaborating on the relationship of Stern's views to his own aims, Allport instructed readers of *Personality* to note well that "the goal is not merely to free the study of personality from over-rigid conceptual barriers drawn by general psychology, but to demolish and reconstruct the entire edifice of general psychology from the ground up."[61]

Allport's sympathy for Stern's views on individuals as many-sided unities paralleled his affinity for James's emphasis on the diversity inherent in a pluralistic universe. In the ancient question of the "one and the many" James came down squarely on the side of "the many": "The world we live in exists diffused and distributed, in the form of an indefinitely numerous lot of *eaches*, coherent in all sorts of ways and degrees."[62] This is why, as James put the matter, "the whole notion of *the* truth is an abstraction from the fact of truths in the plural, a mere useful summarizing phrase like *the* Latin Language or *the* Law."[63] If it was true that, in a pluralistic universe, "individuality outruns all classification," then the investigator of nature could take one of two paths: either rule idiographic knowledge out of court, or adopt an epistemology adequate to the task.[64] The place to start, Allport suggested, was with instances of personal individuality, and he directed psychologists to abandon their scientific prejudices against single instances and subjectivity, and to take up the study of personal documents. Allport wagered that, "as compared with the generalized canons of his science," the researcher will find "the case document more absorbing, more enlightening, and fundamentally more real."[65]

There are two reasons, Allport posited, that particular cases serve us more instructively as guides to reality than do generalizations. The first reason is that the advantage stems from our inherent perceptual orientation: Allport suggested that "we normally fixate the concrete, and find it more natural to contemplate than the nomothetic abstractions for which our attention has to be trained." The second reason that the study of concretes is valuable is that it is a *qualitatively* different process from abstract analysis, for the investigator takes on a different relationship to what is being studied: "By allowing himself to participate in autobiography or in any gripping drama of human life, the psychological and social scientist recovers time after time a fresh sense of the reality and vitality of his field of study."[66] Allport drily noted that "nearly all of the classical productions in psychological science are but the personal records of their authors. . . . Helmholtz in his incomparable *Physiologische Optik* wrote little more than the autobiography of one pair of eyes; and, even later in the laboratory era, Ebbinghaus' extraordinary experiments were almost exclusively a record of his own mnemonic abilities and of his enviable patience." Such estimable authors did not, of course, class these productions as personal documents, for, as Allport remarked, "it was assumed that the introspective deliverances of the philosopher-psychologist were necessarily infallible and that they were sufficient and typical for mankind at large."[67] Such psychological titans had been so impressed with the reality of their personal experiences – and so wedded to the concepts of scientific respectability – that they had had to suppress the individuality that had given the experience its force.

James, of course, advanced a number of arguments as to the benefits of attending to the concrete. In *Varieties of Religious Experience,* James remarked that, as "compared with [the] world of living individualized feelings, the world of generalized objects which the intellect contemplates is without solidity or life." Appropriating the analogy of "stereoscopic or kinetoscopic pictures seen outside the instrument," James pointed out that "the third dimension, the movement, the vital element" was missing.[68] It should not be surprising, then, that Allport determined that "the first great book in psychology to rest its case entirely upon the use and interpretation of personal documents" was this one from James's hand.[69] For Allport, the *Varieties* fulfilled the promise that James's previous psychological work had left unredeemed; Allport argued that, even though the early James "had defined psychology as the science of *finite individual minds,*" he had nevertheless "allowed himself to follow the prevailing current of emphasis upon *mind in general.*" Allport found that the James of the *Varieties,* however, "seems to want to see what finite individual minds are really like at their more complex levels of integration."[70] Allport especially noted that "kernels of pluralism are present in the method he adopts. An Hegelian absolutist would be unlikely to trouble himself with a finite life history, but as a pluralist James believes that no one sees further into a generalization than his own knowledge of particulars extends." It was in this work, Allport remarked, that "James felt that he might put his developing presuppositions of radical empiricism to test by drawing upon the self-reported religious experiences of thoroughly religious people describing their most acutely religious moments."[71]

The radically empiricist acceptance of personal experience as a "psychological datum" was a potential scientific minefield. In the eyes of its detractors, knowledge of individuality was damned by more than the fact that it required the study of single cases: even worse was the fatal taint of "subjectivity" that seemed to pervade the methods needed to apprehend individuality. Many scientific workers were critical of the fact that the analysis of subjective data openly required interpretation, thus introducing an irreducible amount of arbitrariness into scientific research. Allport acknowledged that, in using subjective data, "there seems no way of insuring that only compulsory interpretations will be drawn." He responded, however, with a *tu quoque,* pointing out that "psychologists notoriously interpret even experimental facts according to different theories (of learning, of motivation, of the unconscious)." Allport argued that, since "even the most objective of data in psychology (such as the excursions of a needle on a kymograph caused by a reflex action) can be ordered to diverse explanatory systems" it seemed doubtful "that personal documents actually are any worse off at the hands of psychologists than are other forms of raw data."[72] In Allport's view, it was "the versatility of the human mind in contemplating its own infinite complexity" that led to conflicting interpretations, not "the subjectivity in personal documents."[73] Allport concluded, therefore, that charges by the proponents of methodological rigor that "the use of subjective data derived from single cases is not 'scientific' is both question-begging and provincial."[74] All that scientists had done in banishing subjective data and evidence of the "individual" from their researches, Allport cautioned, was to deprive themselves of a primary "touchstone of reality."[75]

Allport insisted that the point of admitting the plural existence of irreducible individualities in the world was not to throw up one's hands in despair; rather, in the case of psychology, it was to find "some level of analysis that does the least possible violence to the structure of personality."[76] In contrast to the reductionist practices of the experimentalist contingent, Allport urged that "the only reasonable thing to do if one wishes to study a phenomenon is to put a specimen before one's eyes and look at it repeatedly until its essential features sink indelibly into one's mind." Where the experimentalists put the emphasis on *doing* – on using laboratory apparatus to manipulate their subjects – Allport instead emphasized *apprehending,* which is a more passive stance, allowing as it does for a more active role on the part of research participants, either explicitly (as when individuals create autobiographical material) or implicitly (as in the dynamic relationships structuring an object's "essentiality"). "Later," Allport directed, "dissection and ablation may be used to gain acquaintance with details. But unless the fundamental interrelations are first grasped, analysis is likely to be aimless."[77]

To gain knowledge of "finite, individual minds," Allport counseled his colleagues that they would have to trust the minds of those they studied. Instead of dissolving the individualities of "minds-in-particular" into the abstraction of "mind-in-general," Allport approvingly cited the example of James, who, in the *Varieties,* had found his materials "not in the haunts of special erudition, but along the beaten highway."[78] It had been as a consequence of "listening receptively" to the "eloquence" of the voices that he found there, Allport indicated, that James had "fashioned a masterpiece of descriptive science."[79] To this end, Allport exhorted experimentalists to again hear the voices that their laboratory protocols now silenced: "If we want to know how people feel: what they experience and what they remember, what their emotions and motives are like, and the reasons for acting as they do – why not ask them? This is the simple logic of the introspectionist's position that commends itself to many in spite of the scorching displeasure of behaviorists and objectivists."[80] It was not introspectionism that currently gave most vivid expression to subjective experience, however; as Allport remarked: "The personal document is, after all, a natural outgrowth of a clinical situation in which the patient is always encouraged to speak for himself."[81] James, too, had emphasized this point, when he contended that the "physician's attitude," and "the clinical conceptions, though they may be vaguer than the analytic ones, are certainly more adequate, give the concreter picture of the way the whole mind works, and are of far more urgent practical importance."[82]

The clinical perspective is one with which both Allport and Lois Barclay Murphy had direct experience. In regard to Allport's own exposure to clinical work, it should be noted that, as an undergraduate, he did casework in various social welfare projects; that his father was a physician; that his wife, Ada Lufkin Gould Allport, after earning a master's degree, worked in the field of clinical psychology; and that his mentor, Richard Clarke Cabot, was also a clinician.[83] In addition, a close friend and intellectual ally of the Murphys was also a colleague of Allport's: physician Henry A. Murray, director of the Harvard Psychological Clinic.[84] For her part, Murphy had been exposed to psychology in Cincinnati in the clinical setting of the Psychological Laboratory of the Vocational Bureau of Education. Murphy's research during the 1930s was funded by the Josiah Macy Foundation, whose programs were charged with helping "to develop more and more in medicine, in its research, education and ministry of healing, the spirit which sees the center of all its efforts in the patient as an individuality." Foundation president Dr. Ludwig Kast, in a review of the years 1930–6, stated that proper diagnosis and treatment entailed "not only a clear conception of the patient as a total organism but also a concern for the patient as an individual personality often in need of sympathetic insight which the family physician of old was able to offer."[85] During the 1930s, both Murphys also began to study more closely psychoanalytic approaches to personality theory.

Material and intellectual support from individuals and perspectives related to the field of clinical study offered Allport and the Murphys a crucial base from which they could pursue scientific commitments at odds with the generalizing imperatives of their colleagues. If the legitimacy of "subjective" knowledge was suspect within orthodox experimental circles, inferential inquiry was part and parcel of scientific practices modeled on clinical approaches.[86]

The Subjective Side of Science

Murphy first approached the study of personality in her dissertation research, published as *Social Behavior and Child Personality: An Exploratory Study of Some Roots of Sympathy.* Although Murphy's primary interest in this work was in displaying the ways in which personality and cultural values were entwined, her focus never wavered from presenting her findings as they were expressed by particular individuals, embedded within highly specific contexts. In the course of this and related research, Murphy argued for the inextricability of the subjective point of view in the framing of psychological knowledge.

As the text of *Social Behavior and Child Personality* unfolds, individual children's personalities emerge in conjunction with Murphy's synthetic analysis of their sympathetic behaviors. Rather than adopting a heavily quantitative presentation, as was then becoming accepted practice, Murphy instead presented much of her research in protocol form.[87] Instead of noting on a tally sheet, for example, that physical contact between one child and another had occurred, an observer wrote the following description of a behavior episode that transpired on October 27, 1933:

> Peter accidentally jumped on Lila's finger, as he jumped on the inclined board, with Lila lying down.
> Lila cried.
> June, Winifred, and Jude stared curiously.
> Peter and Lila played on the inclined board, then Peter lay down and said to Lila, "Come step on my finger."
> Lila did so, not hard.
> Peter said, "Come on."
> Lila did so, hard.
> Peter said, "Now."[88]

Murphy insisted on gathering and presenting her research in this manner, she explained, because "there are overtones of response in the raw records which are lost in generalizations," and she wanted her audience to "hear the children talk and see them run . . . to see the children as individuals." Murphy insisted that her readers perceive that "each child brings to a situation his own characteristic patterns of verbal, emotional, and manipulatory responses, which color his responses in characteristic ways."[89] In his review of Murphy's book, Allport emphasized this aspect of Murphy's research, stressing that "each case, in the last analysis is unique. Nancy was sympathetic, but always with a peculiar anxious helpfulness all her own; Winifred's sympathy was a blend of affectionate fearfulness; Douglass had an aggressive way of jumping to the physical defense of others."[90] In his APA presidential address, Allport marked Murphy's study as indicative of a new vein of psychological work shedding "light upon the process of understanding the contexts and imperatives that determine behavior from the subject's point of view."[91]

In her determination to "come as close as we can to this world of the child," Murphy and the three part-time assistants she trained observed thirty-nine children during their two-hour free play periods over two school years, recording some 5,000 episodes of their behavior and conversations for a total of 432 hours.[92] Even though Murphy believed that she had honored her commitment to study "the relation between social behavior and the context in which it occurs," she was frustrated that the professional oversight of her dissertation had not allowed her to investigate "the whole world of inner feeling." "We must recognize," Murphy stated, "that the world which the child perceives is not the world which the adult perceives as he watches a group of children, and it is after all only the adult's record which we have here."[93] In her subsequent research Murphy would couple her efforts at naturalistic field observation with a new subjective psychological tool: projective testing.

In her initial sympathy research, Murphy had, in addition to her observational study, devised some complementary experimental conditions (which she preferred to call "framed situations"). Some of these framed situations involved showing a child a series of pictures – of a woman crying, a child crying, an animal attacking another animal, and so forth – which might elicit sympathetic responses. Murphy was intrigued by the fact that in these framed situations the children often seemed to be reading their own concerns and anxieties into the pictures, rather than responding directly to the pictured situations. In discussing this unexpected result with Lawrence K. Frank in 1934, Frank suggested to Murphy that what she had was the beginnings of a projective technique.[94]

The next year Murphy attended a seminar given by a Viennese refugee psychoanalyst, Erik Erikson, in which his discussion of a play session with a disturbed child had moved her "deeply." Murphy began thinking more seriously about using children's play with toys as a method for studying the subjective life of normal children.[95] During the mid- to late thirties, in fact, Murphy was becoming acquainted with a number of psychologists, clinicians, and psychoanalysts fleeing Europe in the wake of Hitler's rise to power, such as Peter Blos, Erich Fromm, Heinz Hartmann, Abraham Kardiner, Bruno Klopfer, Marianne and Ernst Kris, John Levy, Bela Mittelman, David Rapaport, Friz Redl, and René Spitz.[96]

While Murphy eagerly entered into the emergent psychoanalytic world that was forming in New York as a consequence of the sudden influx of exiled European intellectuals – a community whose scientific presuppositions contested the assumptions of objectivist American social scientists in a number of ways – she felt most personally responsive to the work of Anna Hartoch and her husband, Ernst Schachtel, who were pioneers in Rorschach research. Hermann Rorschach had been instrumental in transforming an investigative format based on responses to inkblots into a clinical technique designed to foster free association and the projection of unconscious patterns of perception and fantasy.[97] In the Rorschach situation, an individual was to look at ten inkblot pictures and suggest what they might be. The point, in Klopfer's words, was that the ambiguous nature of the inkblots prevented individuals from falling back on stereotyped ways of responding, therefore leaving no "avenue of escape from the necessity of solving the task in one's own way." Klopfer's assumption was that each subject, in "reacting to the Rorschach situation, is as much left to his own resources as when he faces a serious and decisive life situation where conventional patterns of behavior seem to be out of place, and one must willy-nilly, be himself."[98]

Murphy was profoundly impressed by Hartoch's discerning interpretation of Murphy's own Rorschach responses, and she immediately began studying Rorschach's *Psychodiagnostik,* a book she found so "enthralling" that she insisted that her husband read it aloud with her in the evening.[99] As a Rorschach analyst, Murphy considered Hartoch a "genius" and came to view her as an intellectual beacon, much as she had Woolley during her early college years.[100] Struck by Hartoch's "imaginative regard for every trace of individuality" exhibited in a person's Rorschach responses, Murphy worked closely with her as she learned to use projective methods to study the children who were her research subjects.[101]

As a member of the Rorschach community, Murphy was invited to offer her views on the question, "Shall the Rorschach Method be Standardized?" at a 1939 roundtable discussion. The question was not one that had arisen from within the circle of Rorschach researchers but had been instead imposed on them from a familiar outside source. What the roundtable participants were to address was the recent demand by "orthodox experimentalists" that the Rorschach method be purged of "all 'subjective' elements" – with the ideal in view of reducing "the scoring and interpretation to a seemingly foolproof, mechanical and, therefore, 'objective' procedure."[102]

Murphy dissented from the "orthodox experimentalist" position. Arguing that the Rorschach needed to be recognized "as a major guidepost in the emergence of a somewhat new scientific trend in psychology," Murphy believed "that shortcuts and rigid stereotyping of procedure, of interpretation or even analysis of the material obtained from projective methods in general and, especially, the Rorschach test, will interfere with growth." Analysis of the structure of perception and the content of fantasy by means such as the Rorschach, Murphy contended, "is going to appeal to a variety of creative scientists who will inevitably bring somewhat different approaches and different insight to their work with projective methods of studying personality."[103] Indeed, among some champions of the study of individuality even the Rorschach inkblots were seen as being too structured. Stern, for example, was dissatisfied with inkblots as a projective tool because "the outlines of the patterns are sharp and unmistakable . . . and wholly symmetrical. This narrowly restricts freedom of imagination. The perception of symmetrical forms, insects, butterflies and bats, anatomical crosscuts and pairs of people standing face-to-face, tends to be overemphasized."[104] Stern instead favored the use of "cloud pictures" of "irregular spots of different shades of black and gray, more or less blurring into one another," arguing that this method left it "to the subject to single out the items to be interpreted, i.e., to determine what parts of the picture seem to represent the outlines of single figures."[105]

For Murphy, the Rorschach was but one of an increasing array of projective methods that she and other psychologists were developing and refining; the technique to which she was most deeply committed was one employing Miniature Life Toys (consisting of commercially available items such as cribs, beds, toilets, refrigerators, cooking bowls, dining tables, boy and girl dolls, automobiles, and so forth). Murphy formally introduced her new line of research to the child development profession in a 1941 monograph she coauthored with Eugene Lerner, her Sarah Lawrence collaborator. The monograph appeared under the auspices of the Society for Research in Child Development, an audience composed of many of the objectivist-oriented researchers against whom she had rebelled as a graduate student at Teachers College. Murphy opened the monograph by appealing to her readers to acknowledge the one-sidedness of the body of work that they had amassed, that is, the great quantities of data deriving from "'controlled' stimuli (with more emphasis on what the experimenter *wants* it to mean than on what it *might* mean to the subject." Murphy argued that work "on the other side of the ledger" was necessary "before we can really begin to balance our books in human psychology." Readers of the work at hand would therefore find that the contributors all displayed "a marked readiness for overstaying one's time in descriptions of selective responsiveness."[106]

Murphy rebuked the child development community for its tendency to favor "loosely or hurriedly obtained descriptions" that could not "possibly support sturdy theories." Murphy asserted that, "in under-investigating the nature and varieties of internal sets as such, scientists have surely hurried matters rather than carefully 'described' personality and behavior." In place of focusing their efforts on establishing ever-more precise observer reliabilities, Murphy suggested that researchers should commit themselves instead to a "patient quest for valid horizons." Recourse to projective techniques, Murphy stated, "directly impl[ies] the assumption that we do not even know how to elaborate the obvious – let alone construct systematic theories through second-line or third-line interpretations."[107] If Murphy, when she published her dissertation research four years earlier, had had to be more circumspect in criticizing those who disapproved of her dissent from neobehaviorist mores, this time she had tens of thousands of Macy Foundation dollars behind her, as well as the sanction of Frank, who wrote a preface for the monograph.

In his preface, Frank declared that "it has become necessary to question the common assumption in experimental psychology and psychometrics that the stimulus-situation presented to a subject will always mean what the experimenter-tester expects it to mean to his subject." To understand the personality process, Frank explained, something like what Murphy was devising was needed: "some procedure that will show how the individual, *qua* individual, sees the world and feels toward people through that selective awareness and characteristic way of organizing experience which is more or less uniquely his own."[108] If psychologists were ever going to rise to the challenge of accounting for individuality, Frank indicated, they would no longer be able to work from within a nineteenth-century scientific framework that dictated that "individual deviations" should be "regarded as a more or less unfortunate aberration of nature which ideally was conceived as operating through regularly recurring uniformities upon which scientific generalizations might be built."[109]

In a piece of his own on projective methods, Frank urged psychologists to redesign their studies so that a "stimulus-situation" no longer meant "what the experimenter has arbitrarily decided it should mean (as in most psychological experiments using standardized stimuli in order to be 'objective')," but was chosen instead so that it was the individual research subject "who gives it, or imposes upon it, his private, idiosyncratic meaning and organization." Once psychologists had learned to treat discrete individual events as legitimate aspects of scientific study, they would see that "individual deviations," rather than being "a sort of unavoidable but embarrassing failure of nature to live up to our expectations," represented instead the failure of scientists to acknowledge a primary feature of experienced reality.[110]

Rather than dwelling on statistical generalizations, therefore, Murphy felt at liberty to offer a "small collection of intensive case studies." Murphy stated that, "for the moment at least, in this monograph, rigidity of procedure is definitely secondary to insight, and traditional patterns of control are followed only in so far as they are helpful to the development of insight into child personality." Although aiming for "orderliness in our quest for insight through projective clues," Murphy emphasized, "we are *not* interested in orderliness as such, if and as long as orderliness would stand for mere compulsiveness – in basic ignorance of what we are to be orderly about."[111] The matter at hand, as Murphy presented it, was detailing the "first steps in a process of clarifying new scientific procedures," so that others would be able to join Murphy and her associates in advancing this alternative agenda. The researchers had, accordingly, put their "major effort . . . into the development of 'projective methods.'" The play materials and play methods that Murphy and her collaborators had designed sought "to inquire into the ways in which 'individual personality organizes experience' – into the individual child's '*private world* of meanings, significances, patterns and feelings.'"[112] The different projective situations were "assumed to provide the child with various 'plastic fields' upon which to project 'his way of seeing life' – by having to 'organize' them and 'react affectively' to them."[113]

Murphy's strategy of closely observing individual children playing with miniature life toys differed in a number of ways from conventional therapeutic technique. Where children would typically be scheduled for observation as if for a test – like the Stanford-Binet – it was instead Murphy's practice to take "children on a thoroughly casual basis, saying 'yes' whenever possible when they ran up asking: 'Can I go play with the toys?' and never taking a child who was unwilling to go." The termination of the play situation was equally relaxed, ending whenever the child announced that they were through playing.[114] Here, in essence, the testing situation receded in prominence as a distinctive event, and the idea of an "experiment" from the researcher's point of view became submerged instead into the ebb and flow of the everyday school world of the child – becoming but another field event to be observed.

The ability to keep the conditions of the projective situation this informal and familiar were unusual, and highlighted Murphy's belief that methods of producing scientific data were inextricably entertwined with the kind of data one produced. In the case of Murphy's work, the fact that it was conducted within the environmental setting of a progressive education classroom is significant. In fact, this point was made in another research report, one recounting Murphy's work with somewhat older children at the Bank Street School.

In the Bank Street research, described in *Child Life in School,* the authors stated that their "purpose was to describe a level of maturity in terms of the relational patterns between the individual as a personality and the needs, the possibilities, the demands which confront him at any period of growth." In the formal school situation, Murphy and her collaborators contended, observers would fail to encounter enough situations "in which personality finds sufficiently free and varied expression" from which to work. The authors claimed that the progressive school, on the other hand, since it offered "the opportunity for creative expression, the relative freedom of social relations among the children, [and] the relaxation in the relation between the teacher and the child," therefore provided "a distinctly rich situation for the study of individual personality. It includes much more of what is all of life to an individual than does the formal school – more activity, more variety, more feeling, more interest, more expression of feeling and interest." In consequence, therefore, the "basic life patterns of individuals" would be exhibited more readily, because they were channeled less restrictively.[115] In contrast to the objectivist strictures of experimentalists, Murphy and her coauthors argued that only at a site where a *minimum* of controlled conditions existed – here, the progressive education schoolroom – would valid scientific data be produced.

In Murphy's research, then, it was not only the individual children who were being studied, but also the *research situation* itself. In presenting their case studies, Murphy and her collaborators felt confident that their readers would find it "obvious" that "the experiments could not have the same subjective or emotional value to each child" – even though each child encountered the same procedures and the same materials in the testing situation. The authors stressed that the researcher should not assume – as was customary – that each participant was responding to the same stimulus, asserting that, "for one child, the experimenter's room was a haven of refuge from the usual school situation where 'you get bossed all the time.' Another child was afraid of missing the roof playtime, which meant more to him than anything else. The situations were satisfying in varying degrees to different children, depending upon the emotional and activity needs of each child." The authors stated that "this, of course, is an important finding in itself, since the differential value of different material-social situations to individual children is as revealing as are responses to particular items of material."[116]

In candid defiance of conventional scientific rhetoric emphasizing adherence to the canons of experimental rigor, the Bank Street researchers remarked that, "although interpretation was necessarily limited by the data of the study and the insight of the investigators, we were inclined to give it a fairly free hand, within what seemed to us to be the legitimate bounds of scientific inquiry." That Murphy and her coauthors relied on their own judgment of what the legitimate bounds of scientific inquiry should be depended upon the fact that they possessed the means to pursue their investigations outside the highly structured world of the "university laboratory," and that they chose to publish their results through a mainstream press, rather than in the more tightly controlled realm of disciplinary journals. The oppositional stance that the authors of *Child Life* struck can be seen as well in the way they vouched for the soundness of their research conclusions. Rather than relying upon the expected tokens of scientific propriety – statistical manipulations – they instead referred to specific personal characteristics: "the safeguards against erroneous conclusions," they stated, "are broad background for interpretation and skillful sensitivity in observation."[117]

This effort from within science to recast its practices entailed a new vocabulary of its own, as in the recurring appearance of the word "insight" in the work of those such as Allport and the Murphys. In passages where their psychological colleagues stressed "rigor" or "objectivity" as the ideal toward which research activities should be directed, these radically empiricist scientists sometimes chose to speak instead of the value of "insight." In one previously cited instance, Lois Barclay Murphy stated that, "in this monograph, rigidity of procedure is definitely secondary to *insight,* and traditional patterns of control are followed only in so far as they are helpful in the development of *insight* into child personality."[118]

To speak of "insight" in a considered way was to openly introduce elements of interpretation and subjectivity into the practice of scientific knowledge-seeking. As a term, "insight" denotes a sense of personal vision, of the achievement of intuitive discernment by means of an understanding sympathy. Such a description did not fit the customary characterization scientists offered of their own research procedures, one that instead emphasized the generation of "indisputable facts" by means of detached and impersonal methods. Such a description did correspond, however, with religious understandings of how knowledge was attained.[119]

Philosopher Josiah Royce, for example, in *The Sources of Religious Insight,* introduced his discussion by explaining that "by insight, whatever the object of insight may be, one means some kind of knowledge." That kind of knowledge was, Royce asserted, of "a special sort and degree," for it "unites a certain breadth of range, a certain wealth of acquaintance together with a certain unity and coherence of grasp, and with a certain closeness of intimacy whereby the one who has insight is brought into near touch with the objects of his insight." Royce rendered equivalent the insightful knowledge that can be gained by a painter, an author, or a scientist: "a man may get some sort of *sight* of as many things as you please. But if we have insight, we view some connected whole of things, be this whole a landscape as an artist sees it, or as a wanderer surveys it from a mountain top, or be this whole an organic process as a student of the sciences of life aims to comprehend it, or a human character as an appreciative biographer tries to portray it."[120] As evidenced by this inventory, it was not surprising then, that Royce maintained that "insight may belong to the most various sorts of people," and was knowledge of the most democratic kind: "Many very learned people have attained almost no insight into anything," Royce offered, while "many very unlearned people have won a great deal of insight into the matters that intimately concern them." Unlike the conventional image of scientific learning, then, Royce argued that knowledge obtained through "insight is no peculiar possession of the students of any technical specialty or of any one calling."[121]

In emphasizing the singular over the aggregate, in engaging in scientific scrutiny of subjectivity, and in speaking in terms of "insight," scientists such as Allport and the Murphys directly challenged prevailing assumptions that the unique had no place in science. Whether primarily arguing for the place of individuality in nature, as did Allport, or in crafting techniques to explore it, as did Murphy, these scientists brought their research to bear on the nature of scientific practices in general, as well as on the validity of the results derived from those practices. This dual challenge was not lost on contemporaries, as in psychologist Arthur Bills's statement that "the disturbing fact is that any persistent and searching inquiry into the criteria of *psychology as science* leads inevitably to a discussion of the essential principles of science itself." In holding that science "is necessarily abstract and general," and its proper sphere restricted to general laws and quantitative differences, Bills was speaking for many orthodox social scientists.[122]

Because radically empiricist psychologists, in their insistence on discussing the nature of individuality, were speaking of phenomena believed by those in the mainstream to be incapable of "being reduced to laws or principles of uniformity," a defender of the status quo such as Bills could only conclude that their work smacked of "rank heresy."[123] A defender of heretics, such as Frank, could counter by arguing that the arbiters of orthodoxy had failed to keep pace with a changing scientific vision that had already transformed other research fields. It was not, proponents of dissent would claim, the instances of individuality encountered in the natural world by scientists that represented "impediments to the scientific truths we seek," but rather the thinking of those who saw these instances as impediments.[124]

Much of this debate came down to differential understandings of what was meant by scientific law and order – and to whether or not discontinuity could be considered a scientifically respectable principle. The logic of allowing individuality a legitimate place within scientific discourse entailed that discontinuity deserved one as well. Allport and the Murphys indeed championed discontinuity in their discussion of such concepts as "emergence," and this effort placed them at odds with peers who demanded a stricter understanding of the principle of law than could be encompassed by arguments allowing for singularities.

The Question of "Law" and "Order"

The "science" for which such psychologists as Bills were claiming to speak was the standardized version that had become familiar during the interwar years through the pronouncements of scientific publicists – from within the professional elite and in journalism – and through textbook accounts defining scientific method. Still, the standardized stencil constructed out of such accounts existed cheek by jowl with scientific activities over which the stencil made a poor fit, and it is from such "nonstandard" cloth that the work of Allport and the Murphys was cut: the "new physics" of Einstein, Bohr, and Heisenberg; the cultural relativism of anthropologists; Gestalt theory in psychology; and, in biology, the doctrine of "emergent evolution."

The phrase "emergent evolution" came to denote the view that "evolution is and has been a process in which there 'emerge' new qualities which cannot be explained in terms of or reduced to their antecedents."[125] An "emergent," therefore, was not the same thing as a "resultant." As Donald Worster notes, "the most important point about emergence was that it made for unpredictable outcomes in nature. When A and B were mixed in the same pot, in other words, the result might be a fresh synthesis rather than a mere mechanical or additive mixture: not AB but C."[126] Emergent evolutionists rejected the idea that nature could be best understood as simply matter and motion. Historian of biology Sharon Kingsland observes that proponents of emergence instead emphasized "that at each level of organization the system as a whole acquired new features because the constituent parts themselves acquired new properties, new modes of action, by becoming part of the system."[127] Kingsland argues that, in the United States, emergent evolution "became an intellectual tool in the struggle for authority between biology and the physicochemical sciences, a struggle that in the 1920s was pushing physiology toward a more reductionist method."[128]

During this period, Allport and the Murphys were also interested in the implications of the concept of "emergence," both for the authority struggle underway within their own discipline, but also for how it might illuminate the path toward constituting new forms of political authority. Individuality, uniqueness, and discontinuity were all hallmarks of emergent evolution, as they were of the democratic vision to which Allport and the Murphys were committed. The doctrine of "emergence" sanctioned unpredictability, a concept that traditionally had been held in as much disrepute in science as in political theories of governance.

In his introductory psychology textbook, Gardner Murphy associated the Gestaltists' protest against atomism with the rise of field theories in physics and chemistry, and, in biology, with "*the theory of 'emergence'* [which] defines the properties of wholes in terms which go beyond the physics and chemistry of tissues and calls for quantitative and qualitative study of the uniqueness of dynamic patterns."[129] Murphy also claimed that, "as the theory of emergence has shown, it does not follow that the ingredients of which experience is made up must themselves be *like* the qualities which they engender."[130] In the 1935 version of this text, Murphy used a passage on emergent evolution to illustrate the idea of nature as a creative force, stating that "the doctrine of 'emergent evolution' teaches that nature is constantly producing genuinely new living forms. A true creative process is at work. Not, of course, that nature is making something out of nothing, but that she hits upon combinations of old parts which give a new quality which could in no way be guessed from a knowledge of the quantities of the parts."[131] In speaking of a "true creative process" at work in nature, Murphy was alluding to Henri Bergson's treatise, *Creative Evolution.* In this 1907 work, Bergson rebelled "against the fixities and rigidities which both logicians and materialists had ascribed to reality," arguing instead that reality inhered "in movement and change."[132] Bergson's ideas enjoyed a celebrated, if brief, vogue, reaching such individuals as Vassar student Lois Barclay Murphy, who wrote in a 1920 letter to her aunt: "I am reading Bergson's *L'Évolution Créatrice* in French. I know it so well in English that I find it quite easy."[133]

Bergson's ideas found an especially appreciative audience in those readers who had encountered related ideas in such works of James's as his *Essays in Radical Empiricism, A Pluralistic Universe,* and *Some Problems of Philosophy.* Indeed, James was generous in his praise for the younger philosopher, writing to Bergson that "I feel that at bottom we are fighting the same fight, you a commander, I in the ranks. The position we are rescuing is 'tychism' and a really growing world."[134] As Peter Hare points out, novelty and growth are central features of James's metaphysics.[135] James remarked, for example, in *Some Problems of Philosophy,* that "phenomena come and go. There are novelties; there are losses. The world seems, on the concrete and proximal level at least, really to grow."[136] What James relates in *A Pluralistic Universe,* Richard Bernstein contends, is that "reality in its concreteness is continuous, changing, and *active.*" Our experience of reality, according to James, is not that of "a closed system; it is ontologically open. 'Chance' is not a name for our ignorance, a sign of a defect in our understanding of the chain of causes. The appearance of novelty is rooted in the very character of a continuously growing and developing reality."[137] For those acquainted with the outlines of

James's metaphysics, therefore, the doctrine of "emergent evolution" played upon familiar themes.

Allport saw merit in James's "tychism," stating that "tychism, or the admission of chance, may be distasteful to the scientist, but by admitting the occurrence of chance he becomes a chastened and more open-minded observer of the world around him." Allport was also alert to the prominence that the principle of emergence was receiving among his colleagues.[138] In the late 1920s, for example, in a paper in which he critiqued the scientific conventions of mechanism and reductionism, Allport warned that "the natural sciences (including psychology) might almost be said to have been committing slow suicide with the dissecting knife of analysis." But Allport discerned reason for optimism, in that "quite recently, the natural sciences have commenced auto-resuscitation. The unity of the organism is affirmed, the principle of emergence has appeared. A new era of synthesis has dawned."[139]

Allport was drawing here on biologist H. S. Jennings's 1927 article in *Science,* in which Jennings declared that "no longer can [Emergent Evolution's] exposition be described as a voice or two crying in the wilderness. . . . Philosophical congresses discuss it, eminent zoologists discant upon it; still more significant, it has acquired a name that identifies it."[140] Allport applauded Jennings's assertion that "no longer can the biologist be bullied into suppressing observed results because they are not discovered nor expected from work on the non-living parts of nature."[141] Indeed, in place of a "timorous subservience to the inorganic," Jennings averred, "the doctrine of emergent evolution is the Declaration of Independence for biological science."[142] Ten years later, Allport similarly proposed that "the principle of functional autonomy is a declaration of independence for the psychology of personality."[143] Functional autonomy was Allport's name for his theory of motivation, in which he proposed that adult motives be understood as "infinitely varied, and as self-sustaining contemporary systems growing out of antecedent systems, but functionally independent of them."[144]

Allport put forth "emergence" as an important causal principle, one which would remedy the overreliance of American social scientists on "Hume's theory of causation and the long tradition of Anglo-Saxon empiricism which has always exalted the principle of 'frequency of connections.'"[145] Allport was thinking of circumstances in which "*latent trends,* not yet operating, [would] in time ripen and lead to certain consequences"; such "tendencies" were always, Allport argued, singular conditions, "not (as in statistics) an expression of past frequencies."[146] As an example of such a phenomenon, Allport contended that one hundred years before, the German poet Heinrich Heine had "made a prediction which today seems to be a remarkably exact forecasting of Nazism." Heine's insight, in Allport's estimation, derived from his ability to pick out a confluence of disparate trends – "hero worship and mercilessness (derived from the ancient myths), self-righteous rationality

(Kant), supremacy of the Idea (Fichte), inventiveness (natural science), the restraining power of Christianity, plus certain political considerations" – which were "*einmalig,* occurring never before in history."[147]

Allport claimed that "Heine based his prediction upon no experiment, no statistics, no reliance on the frequency of occurrence, on no standard terminology of constructs." What Heine did employ, Allport suggested, was the "whole dynamic context" in coming to his "understanding of the total field of events."[148] Statistical weighting was not at issue here, for the accuracy of the prediction hung not on discerning the "*present intensity*" of latent trends, but instead upon "the *perceiving* of *relations,*" and ascertaining "their mutual interdependence *in the future.*"[149] Correct causal attributions depended on a complex process of relational thinking which was the result of an "intuitive synthesis of many cross currents, imagining the consequences of the interaction of these currents; and predicting perhaps an entirely new emergent phenomenon that does not have any existence at the time the prediction is made."[150] Allport held that "the epistemology involved with this line of reasoning has more in common with the *intellectual agens* posited by [Gottfried] Leibnitz, [Alfred North] Whitehead, and [Wolfgang] Köhler than it has with the passive intellect posited by [David] Hume, [John Stuart] Mill, and [Edwin] Guthrie." Successful prediction, Allport charged, "often involves the forecasting of emergent changes (novel forms of behavior) on the basis of now latent trends which have shown little frequency and not actual intensity."[151]

My point is not that Allport and the Murphys derived their theoretical emphasis on variety, individuality, and antireductionism from a study of the work of emergent evolutionists, but that they viewed such intellectual activity as offering support and validation for their own perspectives, and thus used these ideas as resources in their own debates with their neobehaviorist colleagues. The pursuit of ideas about emergence, as with the development of Gestalt psychology and field theory, cut across disciplines, and seemed to blur the demarcation between the physical and the animate worlds. Frank, for example, in a 1935 essay in *Philosophy of Science* entitled "Structure, Function and Growth," applied terms associated with physical science to the organic world, stating that, "in so far as we envisage the interaction of space-time configurations (organisms) and the 'fields' of the environing situation, we may contemplate the 'emergence' of new structure-functions."[152]

The new physics, Frank argued, had already shown that it was necessary for those in the biological sciences to alter their understanding of cause and effect. Instead of "seeking so-called causes," Frank suggested that researchers must learn "to think of events taking place within an organic configuration with a past history so that organic development and functioning, breakdowns, etc., are to be viewed as different ways in which the organic configuration reveals its make-up, the on-going energy transformations, and the realization of its unique life career."[153] Such commentators, in shifting discussion of scientific phenomena away from "cause and effect" and toward something as yet undefined, inevitably ran up against conventional understandings of "universal law." While scientists such as Allport and the Murphys did not want to give up the idea of perceiving order in nature, they were nevertheless dissatisfied with conceptions of "law" that, by definitional fiat, excluded such phenomena as individuality and discontinuity from first-order scientific status. Scientific laws, in Allport's view, acted like a Procrustean bed, shearing off inconvenient particularities in order to create mythical entities. In the process, the characteristic essences of objects were lost from view. Allport believed that modern scientists had committed a grave error in founding causal explanation on the universality of events rather than on their essentiality. Unfortunately for the investigation of nature, Allport commented, "the die-hard nomothetist feels that in sheer loyalty to science he *must* search for nothing but common and basic variables, however great the resulting distortion of the individual structure."[154] If one viewed "structuration" as a causal phenomenon – as did Allport – then such a decision presented grave consequences for the validity of scientific knowledge.

Allport argued that, in their haste to build models divorced from the particular, and derived instead from generalizations, psychologists had no way to work back to specific, concrete cases, for they took their *models* as constituting reality. Allport contended that those psychologists attending to a fictitious "mind-in-general" at the expense of the reality of "minds-in-particular" could only hope to learn what "is common to an imbecile and an Aristotle."[155] The model constructed by generalizing the human mind from these two examples would approximate neither of the sources from which the model was made. From Allport's perspective, such a "mind-in-general" as was generated by a model was no mind that existed in nature; any law that described it described a mythical entity, and was therefore an invalid law.[156]

Allport noted that it was always necessary to particularize generalized models when they were applied back to the concrete world. Allport asserted that "the application of knowledge is always to the single case," as in the science of engineering, which is applied "only in building particular conduits or bridges."[157] It was the same in the social sphere, for "in the human realm we have to particularize our nomothetic knowledge before it is of any value, and it must be particularized through its modification in the light of concrete existing circumstances." Allport's point was that "general laws of human behavior known to us are altered and sometimes negated by the idiographic knowledge available to us concerning the personality we are studying."[158] In making such claims, Allport was working against the legacy of the nineteenth-century movement in statistics that "took individuals to be, if not random, at least too variable and inconsistent to serve as the basis of the moral sciences," and in which the "average man was invented as a tool of social physics, and was designed to facilitate the recognition of laws analogous to those of celestial mechanics."[159]

Allport was convinced that the time had come to revamp the idea of "law," and he argued that whether or not a science treating individuals would be able to result in laws depended "upon the conception of law." Allport conceded that, in the conventional usage of the idea of law – that is, as "a statement of an invariable association common to an entire class of objects" – it would be the case that, "as Stern says, 'individuality is the asymptote of the science that seeks laws.'"[160] Allport contended that "a more liberal interpretation of the nature of law" was needed, and offered as his preference equating the idea of law with "*any uniformity that is observed in the natural order.*"[161] The case was, as Allport summed up the matter, that "somewhere in the interstices of its nomothetic laws psychology has lost the human person as we know him in everyday life. To rescue him and to reinstate him as a psychological datum in his own right is the avowed purpose of the psychology of personality."[162] Ultimately, of course, those who defined such phenomena as individuality, singularity, subjectivity, or discontinuity as belonging beyond the boundaries of good and true science would fail to find revisionist philosophical arguments regarding the status of law to be a compelling enough reason to change their minds on the matter. Orthodox empiricists remained resistant to the insistence of their radically empiricist colleagues that the inability to incorporate larger swaths of experiential reality was leading to faulty knowledge of the natural and social worlds.

Gardner Murphy gave witness to this continuing state of tension in a passage on James in the 1949 revision of his 1929 text, *Historical Introduction to Modern Psychology.*[163] At the end of the chapter Murphy noted that, for his contemporaries, "concerned with being up-to-the-minute . . . concerned with the winning of prestige by performing beautiful little experiments like those of the biologists," James "remains 'back there,' framed in time." Along with James's work, a concern with human experience had been lost to the view of scientific authorities, although not, Murphy suggested, to "those thoughtful Americans who are not professional psychologists: artists, physicians, men of affairs; the writers and the readers of novels, essays, plays; those who roam widely and speculate freely upon the infinite fullness, complexity, subtlety of human experience."[164]

Murphy observed that, "in forgetting James one may feel that one's scientific conscience may come to rest," although he questioned "whether, in forgetting the vistas, the feet are necessarily planted in a direction that leads to major discoveries." Murphy closed his discussion by asserting that it would yet "be worth while, sometime, for the general historian of science to tell us just what happens when science forgoes the concern with the teeming richness and immediacy of personal experience."[165]

4

The Pursuit of "Impure" Science: Constructing a Science of Social Life

Pay no attention to the man behind the curtain!

The Wizard of Oz (1939)

In elaborating the "social" half of the-individual-in-social-context construct, Allport and the Murphys sanctioned crossing beyond the confines of scientific propriety into the realm of "impure" science – "impure" that is, by the definition of scientists who sought to cleanse science of the taint of the social in their quest to obtain "pure" knowledge. As students of the "common public world of nature," Allport and the Murphys, on the other hand, assumed that their endeavor was indeed a social enterprise, and that science was, as a result, shot through with moral and political values. As researchers, they believed that neither they nor their colleagues could – or should – sidestep confronting questions regarding the place of scientific elites in a democracy by hiding behind a veil of "objectivity."[1]

As Allport and the Murphys conceived it, one of their duties as social psychologists was to insist that the antidemocratic social structures of American life be identified and forthrightly examined – so as to discharge their public obligations as scientists working in a democracy, but also in order to chart the terrain in which they themselves, as scientists, labored. They fully expected that "impure" science, done properly, would be regarded as "dangerous" and "subversive" by those in positions of economic power, as Gardner Murphy maintained in his 1938 speech as chairman of the Society for the Psychological Study of Social Issues (SPSSI). Murphy's warning to psychologists that they must take responsibility to see that their "factual discoveries" were not placed in the service of the "exploitation of some by others" signaled the existence of a dissenting viewpoint that questioned the democratic pretensions of scientific practice.[2]

In large part, the moral universe these dissenting psychologists championed had been constructed during their parents' lifetimes and was articulated most forcefully by Hull House founder Jane Addams, who remained a vital presence in the progressive left until her death in 1935.[3] In such classic texts as *Democracy and Social Ethics,* Addams contended that "the path of social morality" was not to be found "by travelling a sequestered byway, but by mixing on the thronged and common road where all must turn out for one another, and at least see the size of one another's burdens."[4] It was only by choosing the common road that the "temper if not the practice of the democratic spirit" could be achieved. The case that Addams put in regard to infusing the national polity with the temper of the "democratic spirit" is one that Allport and the Murphys, a generation later, would carry into their activities as members of the academic scientific community.[5]

Viewing scientific life in Addams's terms could lead one to suspect that inhabitants of the laboratory had situated themselves at some distance from the "path of social morality." How could an enterprise whose practitioners deliberately sought to remove themselves from social "contamination" – traveling "sequestered byways," in Addams's language – be expected to result in "the temper if not the practice of the democratic spirit"? The relevance of Addams's social ethics in evaluating the moral bearings of science can be seen in the remarks of Lois Murphy's father Wade Crawford Barclay – a staunch supporter of Addams during his Chicago ministry – who suggested in 1936 that scientific knowledge was being used to "serve the private gain of the few" rather than the "higher interests and welfare of all."[6] Barclay's Methodist Federation for Social Service (MFSS) ally Harry Ward expressed similar sentiments, arguing that industrial relations, imperialism, and the question of who owned science and invention were joined together. As early as 1918, in *The Gospel for a Working World,* Ward had asked: "Are the yellow and the black races always to be hewers of wood and drawers of water and makers of profit for the white people? Are the nations which own the capital and machinery, the science and the invention, to have control of the undeveloped resources of the earth and the labor of the backward peoples, taking the first fruits for themselves?"[7]

Such views on the democratic accountability of science rested on the belief that the social aspects of scientific activity were undeniable. Barclay's bishop, for example, Francis McConnell – a key figure in the activist wing of Chicago's Progressive movement – expressed this point in 1932, by arguing that "the scientific spirit of a time is not an infallible, incorruptible guide to pure truth. The scientific spirit, grievous as this may sound to some scientists, is largely determined by the general social temper of a time." Echoing the perspective of William James, McConnell contended that, whether "we are dealing with the will-to-believe or with the will-not-to-believe, we are dealing with will; and will is subject to general social influences, which a thinker can no more escape than he can escape breathing the air that surrounds him."[8] McConnell asserted that "the given" included the fact that scientists brought to their investigations "assumptions and moods" that could "determine not only the interpretation of the facts but the discovery of the facts themselves." The mind of the scientist, McConnell insisted, was not a "bare sheet upon which impressions are produced without any response or activity by the sheet itself" but was instead "a living agent, with peculiarities of its own. It moves more easily in some directions than in others."[9]

Like these analyses by proponents of a radicalized Social Gospel, the dissent lodged by Allport and the Murphys extended beyond a critique of the *uses* to which scientific expertise was put to the idea that scientific investigation was itself irretrievably enmeshed in the social world. If science was to be placed into a proper relationship within a democratic social structure, attention would need to be paid to the entire spectrum of scientific life, not simply to the application of the results of technological research (although this, they believed, was also an essential scientific responsibility). Where Allport and the Murphys had used "individuality" as a destabilizing theme in their quest to shake scientists loose from their abstractionist and antisubjectivist moorings, their investigation of the "social" presented an equally disruptive threat to scientists' poses as detached and politically neutral seekers of knowledge.

As the 1930s progressed, public expressions of doubt regarding the social and ethical values of pure and applied science increased. In 1933, for example, Henry Wallace, Franklin D. Roosevelt's secretary of agriculture, observed to the American Association for the Advancement of Science (AAAS) that "previous to 1933 more than three-fourths of the engineers and scientists believed implicitly in the orthodox economic and social point of view. Even to-day, I suspect that more than half of the engineers and scientists feel that the good old days will soon be back when a respectable engineer or scientist can be an orthodox stand-patter without having the slightest qualm of conscience."[10] Within science as well, voices could be heard urging social responsibility on their comrades. One sociologist claimed that the American scientist "lacks moral courage, has no integrated social philosophy, has tremendous self-complacency and egoistic smugness, feels no social obligation or communal responsibility, is provincial-minded and so highly specialized that he is almost psychopathic."[11] Those who had been traveling the sequestered byways of the turn of the century were, during the depression years, still keeping their distance from those passing along the "thronged and common road where all must turn out for one another, and at least see the size of one another's burdens." According to the idioms of the 1930s, however, they were now found cloistered inside "ivory towers."

Assaults on Science's "Ivory Tower"

During the 1920s and 1930s, advocates for "pure" science, such as the physicist Robert Millikan – who saw no reason for nonscientists (such as the Congress) to subject their work to scrutiny – were ultimately unsuccessful in convincing Americans that such special privileges for science did not threaten democratic political arrangements.[12] Although the scientific leadership of the interwar years actively promoted science's virtues, by the end of the 1920s, as Ronald Tobey argues, "scientists had failed to create a public consensus on the ideology of national science."[13] In the United States, the demands of scientists for autonomy due to the presumed inability of the lay public to comprehend their work conflicted with the democratic ideal of participation by all.

Science's image would continue to be problematic in the 1930s. The scientific establishment felt itself to be under attack by a sullen public who blamed scientific "progress" for throwing people out of work. By the decade's end, economist Wesley C. Mitchell, president of the AAAS, was complaining that "there is a widespread disposition to hold science responsible for the ills men are bringing upon themselves – for technological unemployment, for the rise of autocracies, for the heightened horror of war."[14] In this address, entitled "The Public Relations of Science," Mitchell maintained that scientists needed to clarify the social function of science to the lay public, by showing that the democratic way of life and the scientific way of thinking had grown up together, each nourishing the other.[15] "We are," he concluded, "on the defensive."[16]

What made Mitchell's remarks even more striking was the fact that the scientific leadership had been aggressively working to win the public relations battle for the better part of the decade. Robert Rydell has described the considerable lengths to which teams of corporations and scientists went during the depression to use the two world's fairs – Chicago's 1933 "Century of Progress" and New York City's 1939 "World of Tomorrow" exhibitions – to restore science and industry to the public's favor. Scientific leaders regarded the expositions "as unsurpassed opportunities for embarking on a 'cultural offensive' that would create a scientific culture with pragmatic values and thus provide Americans with faith in corporate leadership and in scientific expertise."[17] Scientific worthies such as chief engineer and vice-president of American Telephone and Telegraph Frank Jewett suggested building a "Temple of Pure Science" for the Chicago exposition, in order to offer "'a quiet unconscious schooling to the thoughtful people' who visited the fair."[18] By the opening day of the Chicago exposition, the official guidebook had distilled the received scientific wisdom into a terse epigram: "Science Finds – Industry Applies – Man Conforms." This stern QED proof was even given tangible existence, in the form of a heroic statue named "Science Advancing Mankind," in which two human figures were physically propelled into the future by a robot pushing them from behind.[19] Nor were such national exhibitions the only sites at which the scientific establishment waged its advertising campaigns. Similar promotional characterizations permeated the remarks of science writers for daily newspapers; as Paul Carter notes, given the "leader-follower point of view pervading the writings of the science publicists, it is small wonder that in some quarters their outpourings were taken not as enlightening instruction but as arrogant propaganda."[20]

Mitchell was right in sensing that science's image problems were connected in some way to the vigorous public debates being waged during the depression era over the meaning of democracy. Mitchell was wrong, however, in thinking that things could be put right by simply stating that science and democracy were two sides of the same coin. The trouble was rooted deeper than this, in the suspicion of a number of thinkers that claims of "objectivity" – a characteristic that the majority of scientists believed to be synonymous with "scientific" – could be deployed in ways that subverted the process of democratic decision making.[21]

A 1930 piece in the *New Republic* offered a striking deconstruction of the ethos of scientific objectivity. Taking as its point of departure a visiting British scholar's address at a conference of agricultural economists, the article related that the delegate had asked his colleagues to examine the work of researchers in the light of the following questions: "What ends are they seeking? Why are they seeking them? How do the ends fit each other? What kind of rural life, what kind of society, do they want to bring about? What is their program[?]" The anonymous author conceded that, "to the lay mind, the burden of the paper may not appear startling." Yet, in challenging his colleagues' assertions that they dealt only in objective facts in pursuit of total impartiality, the impertinent lecturer had, the *New Republic* explained, "in their minds bordered on heresy."[22]

Objective scientists, the article argued, acted as if they "had received higher orders to investigate phenomena, looking neither to left nor to right. The experts were a sort of Light Brigade, charging nobly into the valley of classifications and measurable differences." Dismissing the science = democracy equation favored by the scientific leadership, the *New Republic* piece opted for a rather different analogy: with "an expert for everything, and every expert in his place," scientific souls were brought "much of the same shelter enjoyed by the orthodox believer in an authoritarian church." But despite such attempts at protecting themselves from the mundane world, scientists, the article concluded, "cannot escape the stream of life. You cannot successfully departmentalize such things as knowledge, policy and values. Knowledge is not knowledge unless its composite relations and arrangements are dictated by some sense of value. Otherwise the scientist would not know where to look for his facts or how to array them." The claims of scientists that their work was the embodiment of impartiality was flatly rejected, with the comment that "at the very least, they must bestow their blessing upon what is."[23] In the course of a mere three years the *New Republic's* image of the falsely impartial scientist blessing "what is" would reach the pages of *Science,* in the guise of Wallace's "orthodox stand-patters."

Many of the terms of the assault on the "ivory tower" existence of scientists played off the language of manifestos issued by insurgents in the progressive education movement, who were protesting the detached postures adopted by many academics. Boyd Bode, for example, one of the leading progressive education activists, argued that the "common man," since coming "into his own," had "become more sensitive to the conflict between the aristocratic way of life and the way of life where, as a practical man, he lives and moves and has his being." Bode suggested that this sensitivity "makes him more skeptical, even if he cannot match theory with theory; and he has lost his awe sufficiently to engage, on occasion, in ridicule." As evidence of this loss of respect, Bode reminded his readers of the liberal use being made by ordinary folk of such dismissive epithets as "highbrow" and "ivory tower" – terms, Bode suggested, that were serving as "irreverent way[s] of saying that scholarship and culture tend to become entangled with a mythological realm of trans-experiential reality, which is too remote and too sublime to be checked and tested by ordinary experience. When or insofar as this happens, higher education becomes a kind of cult in which human values evaporate into a set of abstractions."[24] The affinity within the progressive education movement for including scientists under the heading of intellectual "aristocrats" can be seen in the pages of the house organ of radical progressive educators, *The Social Frontier.* Here, in the 1934 premier issue, Benjamin Ginzburg carried the assault to the laboratory door.[25]

Ginzburg opened by noting that "bitter times" had come to scientists, who had seen "the accusing finger of the outside world point[ing] to them as in large part responsible for the great crash." "Outside the ranks of professional scientists," Ginzburg informed his readers, "the tragedy of modern science is being followed vicariously but with close attention by the intelligent and reflective layman."[26] The "tragedy," to Ginzburg's mind, was that scientists, in refusing to acknowledge the material forces underwriting their production, had relinquished control of their destinies as intellectuals and as citizens. Questioning the validity of the modern image of the scientific hero, Ginzburg located the start of the troubles in the nineteenth century, for "vigorous as the Victorian scientists were in the quest for facts, theories, and laws in the realm of animate and inanimate matter they were at best timid and at worst blind with respect to the social facts of their day." Ginzburg's scorn even extended to that great scientific evangel himself, Thomas Huxley, who, Ginzburg stated, "could easily demolish Gladstone and the miracle of the Gadarene swine, but when it came to such questions as property or even birth control, his reasoning was as insipid intellectually as it was backward-looking morally."[27]

Ginzberg thus found that "the great social respect for science was not based entirely on a pure admiration for disinterested truth," but on "its great practical powers [which] made an intimate appeal to the instincts responsible for the growth and triumph of capitalism." Science had become "the brain of modern capitalistic industry."[28] Ginzburg ended his indictment by maintaining that it was no longer possible for scientists to "hide from themselves the knowledge that the times are out of joint and that they have a responsibility in helping to set them right." To the extent that scientists refused to acknowledge that they were "linked by all the subtle ties of habit and immediate self-interest to the decaying old order," Ginzburg predicted that they would "turn away from the sight of their fellow men, and as long as the sinews of research hold out . . . seek to peer ever more intensely into their telescopes and at their voltmeters, exploring the secrets of the outermost spaces of the cosmos and of the infinitesimal recesses of the electron and nucleus."[29]

Gardner Murphy adopted the same reproving rhetorical turns, in speeches aimed both at his scientific colleagues and the lay public. As the outgoing chairman of SPSSI, Murphy used his platform to highlight the idea of a citizenry whose scales were falling from their eyes, reminding his audience of "existing types of attitude research which suggest that the shams and hypocrisies of society are being more and more clearly discerned by college students and even by grade school children." Such social research, Murphy argued, would remain, however, of "slight importance until it becomes so competent and so comprehensive as to lead into social action," at which point, Murphy predicted, "more and more vigorous steps are certain to be taken to prevent the continuation of our studies and the publicizing of our results."[30] Sharing Ginzburg's economic analysis, Murphy offered his listeners an abbreviated historical tour of the relations of science and power:

> If science makes any difference to society, society will be concerned about it. The Pharaohs of Egypt made social changes and used the expert advice of architects and stone cutters, irrigators, and money lenders. The Elector of Saxony used his primitive knowledge of economics to predict the outcome of the storm against Luther. Napoleon systematically cultivated physics and chemistry because they assisted his engineering and artillery officers, and when firmly in the saddle tried to develop economics, sociology, and statistics, in the service of his empire. It is as sure as destiny that psychology will become the tool of

> the community, just as every useful discipline has been the tool of the community. The society of the coming centuries will be a society constructed in the light of a technical psychology which does not yet exist.[31]

Murphy argued that such a "technical psychology" would act as "dynamite," and that those "dominant in society will try to protect themselves against the explosion." This fact, he offered, is "a sufficient answer to the ivory-tower remark that we should stick to science and let public practice alone. The answer is that public practice will not and cannot let psychology alone."[32]

Such commentary was not intended simply for the internal consideration of the scientific community. Murphy used much the same language in a public lecture he delivered under the auspices of the Graduate School of the United States Department of Agriculture, which was later published as "The Growth of Our Social Attitudes" in *Understanding Ourselves: A Survey of Psychology Today.* Linking the study of psychology to its social context, Murphy remarked that "it seems to me that in a democracy one of the most important problems of a social psychologist is the study of social attitudes, the way in which they grow, the factors which control them and the degree to which they actually serve the best interests of individuals and communities."[33] Murphy therefore recommended the study of "the interrelations of persons and individuals' hates, prejudices, and ideals." The central point he wished "to stress [was] that the psychologist can no longer maintain his isolation in his ivory tower, studying the color zones of the retina and the spinal reflexes of the frog."[34]

Two years later, in a companion piece to his presidential address to the APA, Allport touched on similar themes. Noting the "growth of operationism and the upswing of physiological and animal investigations," Allport interpreted these trends as the expressions of a large group of psychologists who, in "shunning the floodlight of 'social significance,' seek . . . the salvation of psychology by adhering more and more tenaciously to rigorous methodologies derived from the natural sciences and from modern logic." Opposition to this stance, Allport asserted, could be seen in the "demand of a consideration of human values and for social research," of which the formation of SPSSI was itself one indication.[35] "The year 1938 presents in fairly dramatic form," Allport claimed, "two alternative courses for the future. Briefly, these courses might be designated as 'psychology for science's sake' and 'psychology for society's sake.'"[36]

Allport urged his audience to acknowledge that the choices they made, like the discipline in which they practiced, "cannot help but be politically conditioned." On this point, Allport, like Murphy, turned to the past for illustration, paraphrasing an argument first made by John Dewey. Allport stated that, in regard to the political nature of psychological research, Dewey had pointed to "the fact that doctrines of the fixedness of human nature flourish in an aristocracy and perish in a democracy. The privileges of the elite in ancient Greece, and the doctrines of the Church in medieval times, provided the setting for psychological theories of their day."[37] In the modern world, Allport bluntly claimed, "theories of statehood play a major role."[38] The lesson of democracy, Allport lectured his audience, was that the scientific community should keep "alive diversified investigation and a diversified sense of importance," and demonstrate a willingness to ensure "that the vast horizons of our science shall not prematurely close down, neither through bigotry, nor surrender to authoritarianism, nor through our failure to pay our way in the civilization that is sustaining us."[39] Implicit in his address was the caution that the sources of bigotry, authoritarianism, and a disinterest in honoring one's debts as a citizen were being nurtured by proponents of the view that science must be always "empirical, mechanistic, quantitative, nomothetic, analytic, and operational."[40] Speaking elsewhere of "psychology in the near future" to a group of undergraduates, Allport was even more forthright in his warning about the "darker aspect of the movement I am calling operational immaculacy," as a result of which science was becoming "a cult all by itself, prettily but dangerously isolated from the mainstream of existence." Allport urged his audience to consider the promise of a science built on "contextual adequacy," pursued by psychologists "who know the vicissitudes of life, of industry, of strife, community living, of suffering, and of compassion." The ultimate fate of a psychology that did not incorporate the ethics of democracy, Allport argued, could be seen in Germany, which had at one time been "our model of freedom in the psychological world," and whose psychologists were now performing "slave labor for an enslaved science."[41]

Fellow members of SPSSI also drew connections between the nature of science and the nature of society, as Columbia Teachers College psychologist George Hartmann made clear in his convention speech in 1938: "Science itself is a social institution that can exist only when certain field conditions are present. The great tragedy (as well as the great discovery) of our age is the recognition that political and economic powers – presumably the subject-matter of the social sciences – are among the great determiners of the content, procedures, and conclusions of science as such." Hartmann observed that such influences might be less immediately discernible in the United States than in Russia or Germany, but he declared that "the emancipated members of this Society are not unaware that forces outside themselves often participate in crucial ways" in the practice of scientific research.[42] In his SPSSI address of that same year, Gardner Murphy stated that outside forces would attempt to censor any research threatening its prerogatives, and that psychologists must therefore take responsibility for making their "work directly available to the public."[43] Statements such as these questioned more than whether or not it was right for scientists to retreat into an "ivory tower" outside the public sphere in order to practice "science for science's sake": they also rejected the claim that the "ivory tower" existed in a zone bereft of specific social, political, and economic relationships. As an earlier generation of psychologists had used introspective techniques on themselves in order to try and describe sensory processes, a number of social psychologists during the 1930s were applying sociopolitical analyses to their work in order to speculate about "scientific-individuals-in-social-context."

During this period the Murphys assumed a pacesetting role in promoting and defining social psychology, as Allport had been doing in the field of personality psychology. Over the course of the decade the Murphys worked on several fronts in an attempt to bring social psychology a higher visibility within the profession. Gardner Murphy, for example, made significant expenditures of time and effort in sponsoring a cadre of doctoral candidates in social psychology, including such future leaders as Theodore Newcomb, Eugene Hartley, Muzafer Sherif, and Kenneth B. Clark. This commitment was augmented by his editorial decisions as an editor at Harper's to publish such works as Sherif's *Psychology of Social Norms.*[44] Lois Murphy, in turn, undertook a second graduate degree, and her dissertation, published as *Social Behavior and Child Personality* in 1937, offered a model of new-style social psychology. Together the Murphys compiled the 1931 text *Experimental Social Psychology,* which they brought out in a revised edition in 1937. In between these two editions they also coauthored a theoretical brief on behalf of their views in the 1935 *Handbook of Social Psychology,* as well as various other efforts on their own.

In their earnest attempt to survey the patchwork realm of social psychology in *Experimental Social Psychology,* the Murphys painted the field with a broad brush; the book might most accurately have been entitled *Empirical Social Psychology,* for the major intent of the text was not to establish experimental norms as much as it was to gauge what social psychology might look like from beyond the armchair. At any rate, the experimental framework that the Murphys were using was one that paid homage to the field of cultural anthropology. "It must be recognized," they asserted, "that nearly all the experimental work in social psychology, such as makes up the subject matter of this book, has value and is definitely meaningful only in relation to the particular culture in which the investigation was carried on."[45] In many respects, their efforts to gain a sense of where matters currently stood recalled Gardner Murphy's immersion in a self-imposed course of study on psychical phenomena a decade earlier, preparatory to entering into an intellectual domain in flux so as to make the greatest impact possible. As the authors stated in the introduction, "We should, if challenged, be entirely willing to maintain that all psychology is social psychology. This would, however, be like marking out squatters' claims in a land which we have no capital to develop."[46] The book was divided into three parts, the first of which assayed basic principles such as the question of nature and nurture; the second part focused on the development of social behavior in children; and the third covered issues such as response in group situations and attempts to measure personality and social attitudes. When the even larger revised version appeared six years later, psychologist Leonard Doob acknowledged the disciplinary debt owed the Murphys due to their survey efforts, stating that "the guild should be grateful that [the authors] are willing to sacrifice a portion of their own research time for the scientific and social good."[47]

The distance traveled by social psychology during the 1930s can be grasped by the change in tone between Mark May's review of the 1931 text and Doob's review of the 1937 revised volume. Both May and Doob, located as they were at Yale's interdisciplinary Institute of Human Relations (IHR), were active in the movement to ensure that psychology attained status as a "hard" science, and represented the arbiters of rigor who made up part of the audience whom the Murphys hoped to reach (Mark May had, in fact, been Gardner Murphy's first choice as coauthor of *Experimental Social Psychology*).[48] Of the earlier volume, May wrote that "perhaps the most outstanding impression which this book leaves with the reader is the spotty nature of the experimental data in social psychology." May's disdainful evaluation of the still-emerging subfield was that "it is literally strewn with odds and ends of experiments, with here and there a larger piece of work extending over a period of years." Social psychology's inferior breeding could be found, May suggested, in the "striking fact . . . that so few experiments are repeated by other investigators. That which is so common in the physical sciences rarely happens in the social sciences. The result is an increasing accumulation of unverified data." May remarked that, given this circumstance, "the authors of this book have done well to call attention to the gaps in the data as well as to their inconclusive nature."[49]

By the time of the revised volume, Doob was noting that "this new orientation, which has become a popular meeting place for so many social scientists under the banner of 'culture and personality,'" was actually indicative of two trends. The perspective reflected in the volume served, first of all, "the need which social psychologists have felt during the depression for relating their work to a live, social context," an impulse that had found institutional expression in the founding of SPSSI. Doob characterized the second trend as "the growing, rather desperate concern within all sciences, especially the so-called social ones, for breaking down the artificial boundary lines between related disciplines" – a viewpoint being given tangible reality through the bestowal of Rockefeller millions for interdisciplinary institutes such as the one that then existed at Yale's IHR. Where, in the first edition, Doob judged that the "authors apparently felt conscientiously obliged to confine themselves within the limits of that awe-inspiring adjective, 'experimental,'" in the second edition he applauded them for having "torn themselves away from this dogmatic restraint" so as to draw more explicitly on the neighboring disciplines of sociology and anthropology.[50]

Promoting the establishment of "social" psychology as a working subfield within the academy offered institutional opportunities for researchers such as the Murphys to conduct their work with less disciplinary elbowing from censorious peers plying more established areas, and the trends Doob cited aided in the mainstreaming of the new subfield. But the tangible benefits of subfield entrepreneurship should not overshadow the fact that the idea of "social psychology" could encompass more than a limited subarea within the larger discipline: *all* of psychology could be potentially thought of as "social psychology." Indeed, the impetus for grounding social psychology in these larger terms predated the rise of either depression-era sensibilities or foundation officials' philanthropic inclinations; as Allport argued in 1929, the field of social psychology would remain a superficial enterprise "unless there is a revival of interest in social philosophy."[51] By the time of the Murphys' revised volume there would be, as Doob's review suggests, a much greater number and variety of players taking their places on the social psychology field. But at the opening of the decade, as nurtured by those dissenting from the status quo, social psychology was viewed as a practical point of entry for making the social philosophy within which American science was situated more explicit.

And a Little Child Shall Lead Them: Children's Lives and the Revelation of the Social Order

To a great extent the Murphys' approach to social psychology in the 1930s was through the study of the behavior of children. The *Experimental Social Psychology* texts, for example, present substantial sections on the "genetic study of social behavior," or what would now be called "developmental social psychology." Choosing children as research subjects in social psychology ensured that the critiques of the larger culture implicit in the Murphys' work commenced from the vantage point of the next generation of Americans. This activist agenda was crystallized in the subtitle for the revised edition of *Experimental Social Psychology:* "An Interpretation of Research upon the Socialization of the Individual." It is, however, in such texts as *Social Behavior and Child Personality* and its companion volume, *Child Life in School,* that Lois Barclay Murphy's version of what Allport termed "psychology for society's sake" could be seen most clearly.

In *Social Behavior and Child Personality,* Murphy took as her subject the study of sympathy – a choice much at odds with a profession that, when it studied emotions at all, preferred to examine more hostile versions, such as "frustration" and "aggression." *Social Behavior and Child Personality* was in fact as much a meditation on "some roots of democracy" as it was an exploration of "some roots of sympathy," for, like her childhood hero Jane Addams, Murphy was concerned with how the youngest travelers on the "path of social morality" acquired the "resultant sympathy" that Addams proposed was the "foundation and guarantee of Democracy."[52]

In focusing her topic on the question of the "cultural and personality factors that are tied up with different sorts of sympathetic behavior," Murphy used the behaviors of children as a guide to the alternative world of compassion and cooperation that existed, like the tenets of her father's radical Protestantism, side by side with the attitudes promoted by the capitalistic ethos, such as frustration and aggression.[53] Intent on tracing more than the developmental trajectories of toddlers, Murphy was also, by design, studying the "common heritage of language and logic, of economic and political institutions, which is shared" by every group "in this civilization, however diverse its attitudes toward the tradition which it must use and manipulate."[54]

One of Murphy's goals, then, in investigating the emergence of positive social responses, as it was phrased in *Child Life in School,* was to move social psychologists further along "the road to mitigating one of the serious social disadvantages of our times, namely, the relative indifference of individuals to events and circumstances which they do not recognize as affecting their lives immediately, and the affective dissociation between what they may know to be true and what they feel impelled by the force of emotional identification to do something about."[55] In their early years, the authors stated, children take "unto [themselves] the ideals and values of the adults who are important" in their lives, each of them thus becoming "a partner in our cultural ideology." For these progressive educators, the point at which to intervene in changing a socioeconomic system from which they dissented lay not in anticipating the workings of the historical dialectic in the revolt of the masses, but in seeing how, in teaching "children the specifics of adapting to a given social structure," we are "really creating that social structure ourselves."[56]

Murphy's attitude toward this social structure can be seen in her deadpan recounting, in the guise of an "anthropological observer," of American culture patterns in *Social Behavior and Child Personality:*

> The organization of these tasks [jobs] and the provision of materials for labor is undertaken by individuals whose primary contribution is that of planning, and who retain for themselves all that is left of the

> rewards of production after the labor of their workers is paid for. These leaders in production compete with others to produce more things at less expense, and to sell more things with greater margin; usually the labor of those who do the work is paid for at a very cheap rate. Many workers, in fact, do not have adequate food, clothes, nor housing to keep themselves comfortable according to the standards of civilization. So many of the tasks are performed with the help of machines that become rapidly out of date, that a great deal of changing in tasks occurs; in fact, new machines are constantly produced which can do the work of men, so that many men lose the opportunity for work and for gaining a livelihood. On the other hand, some of the leaders make a large excess over what is needed to live on, which is handed on to their children, who are not, therefore, under such pressure to work, since their living is already provided.[57]

It is these kinds of socioeconomic arrangements, Murphy maintained, that "create conflicts and barriers to cooperation and sympathy." "The fundamental nature of this society," Murphy found, is "its competitive individualism. Here lie the roots of both the drive for power and the obligation to pity; the conflicts between economic, intellectual, sexual, and age groups, and the cohesive forces that hold groups together."[58]

Thus it is clear that, although her aim was to study the early emergence of sympathetic behaviors in children, it was not only the youngsters and their actions that were the subject of Murphy's inquiry but also the kind of culture of which they and their actions were representative. In studying the developing sense of self and its social expressions, Murphy was seeking to highlight cultural structures of meaning as much as she was trying to understand how individual children build for themselves "a position in [their] social environment."[59]

In describing the process in which she believed cultural ideologies and personal modes of meaning-making coalesced, Murphy favored as a descriptive term the verb "interiorization."[60] In the process of interiorizing – a kind of "incorporation" or "taking in" – Murphy inferred that, in socialization, forces are generated not only from the environment but also from within individuals. By framing her research strategies so as to try to grasp children's "interiorizing" processes, Murphy allowed herself to pursue two goals simultaneously: capturing the means by which individuals "become partners in our cultural ideology" and identifying what that ideology itself might be, as interpreted by its youngest members. Projective methods, in Murphy's view, could open up more than subjective experience; they offered the possibility "of tapping emerging conceptions of the world at an age when verbalization is not yet satisfactory." Using projective methods in longitudinal studies thus possessed the potential to delineate the increasingly great incorporation of cultural imperatives by its youngest members. The advantage of using projective techniques with children so young, Murphy speculated, was that they could highlight "the emergence of such concepts before they have become obscured by a realization on the part of the child of the social reaction they elicit, before he knows whether they are approved or disapproved, and before he has a chance to develop concepts conforming more with the approved pattern of the group in which he is growing up."[61]

In the article she coauthored on projective techniques with Ruth Horowitz, Murphy offered suggestive examples of interiorization that she felt should be of interest to social psychologists. For example, Horowitz and Murphy remarked that "the clearest indication of the accretion of culture values" in their studies was elicited from children in response to the presentation of a page containing pictures of twenty-four different flags. Children of varying ages were asked to rank the flags in order of preference, in response to the question "Which is the best looking flag?" Among younger children the flag of choice was that of Siam, which had a white elephant on a red background. Older children, however, "tended to choose the American flag, or flags containing elements similar to those that make up the American flag, stars, stripes, and the three colors," and, for the oldest children, "the American flag was almost the invariable first choice, and all the other rankings were given to flags bearing some resemblance to the flag of the United States."[62] In a similar attempt to find out at what level children become aware of economic differences, Horowitz and Murphy devised "a series of pictures making use of the common symbols for such differences" – housing, clothing, neighborhoods, furniture, and the like.[63] The two psychologists also ran a series of pictures attempting to elicit racial attitudes. In probing for children's cognizance of such inflammatory constellations of values – regarding nationalism, class location, and racism – Horowitz and Murphy demonstrated that the analytical tools of social psychology could be used to critique the status quo.

As Murphy began laying down her lines of approach to the question of what was being "interiorized" and when, she was working her way as well toward conceptualizing what she took to be the twinned process by which personal worlds and social worlds merged in the mutual construction of identities and social structures. To reach this goal, Murphy believed that it was necessary to shrug off restrictionist definitions of experimentation, for the elaborate mechanical-statistical models adopted by her peers allowed for neither a "depth" psychology nor a "depth" sociology (neither of which was presumed necessary in considering, for example, the behavior of rats or pigeons, which were increasingly becoming the favored "subjects" of psychological research).[64] Murphy's impatience with conventional models of experimentation can be seen in her preference for describing her research interventions as "framed situations" rather than as "experiments" (although she used this term as well). Murphy's "framed situations" were intended to retain the characteristics of sociality even as they partook of the image of experimental science. In conventional terms, the division that experimentalists posited existed between the "social" (that which is "outside" the laboratory) and the "scientific" (that which is "inside" the laboratory) validated the assertion that science itself lay outside society: the procedural dichotomy confirmed in practice the principle expressed in the philosophical dichotomy. The Murphys offered a competing view in an essay entitled "The Influence of Social Situations upon the Behavior of Children." The article belied its bland title, offering a provocative vision of how researchers should reformulate the questions they asked about social behavior.

The Murphys began by taking up the question of where the causal forces marking human life reside. Rejecting the metaphor of individuals as "putty" or "clay" molded by social forces, the Murphys maintained that "an organic system has vastly more definite organized resistance to external forces than putty or clay possesses." Rather than thinking of an "organic system" as "a bundle of discrete response units" they suggested that their readers conceptualize this system instead as "a way of behaving." The human organism, they offered, is "shaped by its culture and by individual habits of long standing," and thus manifested "many extraordinarily stable and dependable ways of maintaining its personal and social structure." The causes of behavior lay neither primarily "in" the organism nor "in" the situation. To move to the next step in understanding the nature of the "human organic system" it was necessary, they held, to avoid the logical mistakes implied in two alternatives: the "organism error" and the "situation error."[65]

The "organism error" was committed by thinking that the following statement is true: "Behavior traits are fixed attributes of organisms; they maintain a degree of stability comparable to that shown by a finger-print or a well-defined birthmark." To speak of "traits" without "reference to stimulus situations" when comparing individuals, they counseled, was to make an error in causal attribution: the Murphys contended that "as far as we know, there are not traits which can be defined *solely* in terms of characteristic ways of responding."[66] Elaborating on this point, they stated that "we still speak a great deal of 'characteristic' behavior traits which reside in organisms, though we have never defined the range of social situations in which these forms of behavior have been observed. 'Mary is aggressive;' 'John is introverted.' This is like saying that 'Mary attacks' without telling what she attacks, or that 'John turns away' without telling what he turns away from."[67] Alternatively, to say that "any form of behavior is the *result* of a social situation is a gross case of a 'situation error' which is in itself no less serious than the 'organism error.'" "It is utterly misleading," the Murphys concluded, "to speak of traits as if they were located *in* the child. They exist only in the relation between the child and his social surroundings." To understand causation, it was necessary to reconstruct the *relations* between phenomena, rather than to isolate single variables.[68]

The task of social psychologists, as the Murphys wished to frame it, was to "press deeply into so-called subjective aspects of the response of individuals to social situations" as they looked for the "psychological conditions" and the "mechanisms" that were together "responsible for variations" in human behavior. Their championing of a subjective approach, they admitted, "may seem to undermine the criteria of scientific work which have been made only after laborious experimentation with objective methods." In place of work "couched solely in statistical terms" the Murphys instead advised that "it may be well, however, to see our scientific problem and its methods not only in relation to the methods and research criteria of physical and biological sciences but in relation to criteria evolved in other sciences which deal with responses of persons, such as historical criticism, biography, comparative philology, cultural anthropology, etc."[69] The test of validity, then, should rest with the "*convergence of evidence*" as opposed to mere statistical verification.[70]

To take the individual and the social together, the Murphys argued, was to investigate "complete stimulus-response arcs" instead of "logically discrete items whose psychological integrity had been violated."[71] As an example of this point, the two drew on Lois Murphy's then work in progress in her sympathy study. One of the findings of her research was that the most sympathetic children were also the most aggressive. The Murphys noted that, for researchers who fragmented behavior into discrete units, it would seem "both psychologically and statistically queer to have positive correlations of approximately .50 between 'aggressiveness' and 'sympathy,'" given that "notions of 'good' and 'bad' and the behavioral dichotomy which they assume are so deeply ingrained in our attitudes toward behavior." How could such a seemingly inconsistent finding make sense? The Murphys theorized that the issue, when interpreted "from the point of view of the child," would be "between social responsiveness and inhibition, and it is only natural that if you respond at all to another person you will fight him when he grabs your shovel and help him up when he gets a bad fall."[72]

Psychologists, the Murphys contended, failed to appreciate the situational nature of behavior, because their work was "based on false assumptions regarding the *desirability* of . . . 'consistency,' which have been carried over from absolutistic ethics and conceptions of the 'stable character.'" Working from the presumption, for example, that "the trait of sympathy or helpfulness . . . is a consistently desirable one," investigators "might from an *a priori* standpoint question the healthiness of the development of an individual who did not show social responses in situations that called for them."[73] However, anyone "who observes a group of children for a considerable length of time" – those who, in other words, attended to "complete behavior arcs" – "appreciates the fact that a child who responds to every bump or fall, in a group of fifteen or twenty children, is having no opportunity for that constructive activity which is an important part of his own development."[74] In terms of that development, it was probably the child "who ignores the minor discomforts of his associates and [who] responds with resourceful cooperation to the more crucial situations" who would be a more capable actor in his or her environment. "The truly logical and scientific approach," the Murphys protested, "will eliminate the more general or superficial implications of consistency."[75]

Work such as this by the Murphys constituted in part their push during the 1930s to define "what [social psychology] is and what it may hope to become."[76] In studying the thinking and behavior of children as social partners in the culture's ideology, they were attempting to outline the patterns of American culture at large, and these patterns as they had been internalized by psychological researchers. In later decades, as developmental psychology became more clearly differentiated as a subfield of its own within psychology – indeed, as all subfields increasingly came to be sharply set off from each other – the lessons for *social* psychology that the Murphys were drawing from the study of *children's* lives would be overlooked.

Individuality and Contextuality: Differences of Emphasis between Allport and the Murphys

Although their attempts to understand the convergence of the individual and the social overlapped in a number of ways during the 1930s, clearly Allport preferred to focus questions in terms of individuals, whereas the Murphys embraced a more sociologically oriented starting point. In moving toward conceiving of individuals in a fluid relation with their environments, the Murphys were stretching the individual component of the individual-in-social-context construct as far as it would go without the concept of "individuality" losing coherence. This choice of emphasis is one that brought them into some tension with Allport's viewpoint, which favored the individuality side of the equation.

In his notes on the proceedings of the 1935 Rockefeller conference on Gestalt Psychology at Princeton, for example, Allport classed the Murphys (along with Lawrence K. Frank and George Stoddard) as "situationists."[77] While, in reviewing Lois Murphy's book *Social Behavior and Child Personality,* Allport judged "the research underlying this investigation virtually a model of its kind, many-sided, sound, and comprehensive," he firmly reproved Murphy for indulging in what he characterized as "unsupported sociological enthusiasm."[78] Allport was uncomfortable with the large role Murphy assigned to culture as an explanatory construct. Allport remarked that Murphy, "like other enthusiastic recruits to 'culture and personality'. . . starts, it seems to me, with a bias in favor of culture, and hence inclines to overstate her case." In support of his criticism Allport pointed out that this "predilection for cultural explanation leads the author to speak frequently of children 'mirroring' or 'reflecting' cultural patterns. Children may indeed do so, but evidence rather than assertion is needed to establish the case."[79]

Allport suggested that a more appropriate way of speaking of the cultural was to be found within Murphy's text: "The study [is] remarkably revealing of the specific ways in which children come to *learn from their elders* (and this, I submit, is what psychologists should always do when they deal with cultural causation)." An example of such a more specific and focused argument, could be seen in "the observation that the rudimentary sense of property rights seemed to advance when children who 'had it first' were defended by the teacher." Allport likewise found significant "the author's account of the powerful socializing effect of eulogistic and dyslogistic verbal symbols, 'good,' 'bad,' 'nice,' 'not nice.'" It is "in such tangible, specifiable instances as these," Allport argued, that "we find culture reduced to manageable proportions. Any broader allegations concerning cultural determinism seem crude and question-begging."[80]

Allport, too, could be called to task for his own theoretical privileging of the individual side of the "individual-in-social-context" construct; in Gardner Murphy's address as chairman of SPSSI, he noted that "much of Gordon Allport's superb plea for the study of the uniqueness of the individual would be more convincing" if it spoke in languages beyond that "of Emerson and Spranger."[81] In truth, however, the Murphys and Allport shared more ground than they sometimes surmised. The Murphys, for example, in their theoretical piece on the situational nature of social behavior, were at pains to point out that "a number of writers confuse the conception of consistency from the point of view of the *norm* with the conception of consistency from the point of view of the *organism*."[82] Allport similarly stressed that, when "we make rational reconstructions of our findings (constructs), we do so from the point of view of our presuppositions and communicate them to other scientists sharing these same presuppositions." During this process of appealing to norms, Allport asserted, "the pattern of trends peculiar to the organism, what the organism is trying by himself to do, is almost completely lost to sight."[83] While supporting different emphases, all three nevertheless refused to let subjective consciousness, will, and activity evaporate before the searing air of neobehaviorist methodology.

Furthermore, in the preface to *Personality,* Allport voiced a position that closely paralleled Lois Murphy's on the role of culture. Anticipating that he could be criticized for not devoting more space to a discussion of "the close relationship between personality and culture," Allport contended that "such criticism can arise only from a misunderstanding of my purpose. I do not deny that personality is fashioned to a large extent through the impact of culture upon the individual. But the interest of psychology is not in the factors *shaping* personality, rather in personality *itself* as a developing structure." Allport's perspective was that culture became "relevant only when it has become *interiorized* within the person as a set of personal ideals, attitudes, and traits."[84] But issuing a prefatory caveat was hardly a sufficient response to owning up to a consideration of the cultural aspects of the individuality question; as expressed in the bulk of their writings during this period, the differential commitments of Allport and the Murphys are clear.

Part of the disagreement between Allport and the Murphys stemmed from a difference in the extent to which they saw the use of sociology by psychologists as problematic. During this period Allport was particularly wary of sociological perspectives regarding the individual, a point he made bluntly in *Personality,* stating that "it is characteristic of all sociological definitions of personality that they deny to it the attribute of self-sufficiency. In one way or another personality is always considered a reflection of, or dependent upon, the social ground. One succinct and fairly typical statement says that personality is 'the subjective side of culture.'"[85] The problem with this view, to Allport's mind, was that, in addition to overlooking biological considerations, social psychology as practiced by sociologists considered "the subjectification of the customs and social traditions within a single human life as the whole story. It is obviously a one-sided view."[86] But there was more to Allport's sociological disaffection than this. The fact that the pronouncements of objectivist sociologists dovetailed with the views of his psychological opponents in denying the scientific validity of "individuality" alarmed him. Allport's concern for the scientific status of individuality was too great for him not to worry when otherwise staunch allies such as the Murphys seemed to him to disappear on occasion within a fog of "social context."

While differing theoretical concerns certainly underlay such disagreements, there was another dimension to this debate that was perhaps greater than any other: Allport focused his research on adults, whereas Murphy focused hers on children. But in this difference the two shared more than might be evident at first, for the emphases of both speak to the continuing relevance of the type of social philosophy advocated by Addams. In the Progressivism that Addams shared with John Dewey, for example, the focus fell on children as the nation's citizens-in-training, and on identifying developmentally appropriate experiences that would make democratic ethics a living reality for students. It was this kind of theoretical stance that Murphy's father brought to his innovations in religious education, in which he advocated allowing "the nature of the pupil" to determine "the materials and methods of nurture."[87] Lois Murphy's later professional interest in the ways in which society's imperatives impinge on young personalities as they fight to express themselves in a world they enter on unequal terms spoke to this strain of Addamsian progressivism. The remaking of society was presumed to depend upon the remaking of childhood experience.

But Addams's thinking was also one of the main tributaries of social ethics, an intellectual tradition that maintained a departmental foothold at Harvard until the 1930s. Richard Clarke Cabot, chair of the social ethics department and Allport's mentor, placed Addams with the class of "saints and heroes" that included Jesus Christ, St. Francis of Assisi, Louis Pasteur, and Abraham Lincoln.[88] Joining such earlier examples of the genre as Addams's *Democracy and Social Ethics* were works of Cabot's such as his 1926 *Adventures in the Borderlands of Ethics.* In this aspect of his writing, Cabot's educational focus centered on the ethical choices facing adults, such as those that might arise when "practicing any profession or carrying on any occupation." Allport's interest in adult personality – in the thoughts and acts of "experienced" humans who were potentially cognizant of the issues at stake in their own personal embeddedness within society – shared in this strain of Addamsian social philosophy in the same way that Lois Murphy's interest in children carried forward another facet of Addams's theoretical concerns.

During the 1930s, psychologists such as Allport and the Murphys had yet to find a satisfactory way to juxtapose the concepts of individuality and sociocultural contexts with ease. Ultimately, the analytical tools of "individuality" and "social context" could go only so far in the attempt to capture the dynamic processes denoted by the individual-in-social-context construct. Lois Murphy spoke to this difficulty in the concluding pages of *Social Behavior,* arguing that the concept of a "field" would be needed to replace such formulations as "child-and-situation." In this regard, Murphy made reference to the perspective of German Gestalt psychologist Kurt Lewin and his "topological" approach, which, she argued, moved researchers past "the phenomena which the behavioristic approach sifted out, past the mechanisms of identification and projection which psychiatry showed us, even beyond the consideration of the child himself as a total personality, to the recognition that the child-and-situation are one total field, the activities in which are influenced by all the extrinsic and intrinsic features of those parts that we [Murphy] called 'child' and 'stimulus-situation.'"[89] In rethinking what researchers called "child" and "stimulus-situation" into something conceptually new, these psychologists found the "field" concept to be a generative intellectual resource, for it afforded new insights into questions of individuality and contextuality that the idea of the "social" alone could not. For these radically empiricist scientists, pursuing the analogy of the "field" would ultimately mean taking natural history seriously as a scientific model.

How, then, would these psychologists treat the "scientist" and the "stimulus-situation"? Placing scientists within the field of social life made such a question one of intense concern for Allport and the Murphys, as for others who were advancing democratic critiques of American society.

"There Is Also a War between the Fascist and Democrat in Each One of Us"

The consequences of blurring the distinctions between "subjects" and "objects" and between "causes" and "effects" were of potential significance beyond the halls of academe: the ability to make such demarcations theoretically served as one sign of the superiority of scientific rationality to the workings of the untutored mind. Indeed, the ability to assume the proper pose of "pure" mental detachment necessary for such evaluations was presumed to be the prerequisite that allowed scientific thinkers to transcend the "social field" impinging on everyday thought.[90]

This image of "pure" science of course had its vigorous defenders, but the advocates of "impurity" had strong voices as well, voices that were amplified by the intense interest during this period in the "culture concept." Speaking of *all* science, not just so-called social science, Hartmann flatly declared science to be a cultural entity: "If the study of a culture cannot be pursued without an examination of the premier values that the culture embodies, then I think it follows that physical science and technology are affected by, if not saturated with, the special values of the persons who transmit and develop them."[91] All sciences, Hartmann maintained, were value laden for "the reason that the creature (science) must in some way reflect the creator (the scientist)."[92] Allport made the same point when he declared that, in Germany, "science without moral guidance has turned out to be sheer savagery."[93] The question was, within what social field were American scientists working, and what was the potential for Americans to turn to fascism? This was one of the central questions that Allport and the Murphys pursued in their social psychology during this period.

The ethical values imparted to Allport and the Murphys as youngsters were based on the confrontation of activists of their parents' generation with the rise of industrial capitalism and what these activists believed to be the immoral and undemocratic consequences of the increasing concentration of wealth and power in the hands of "captains of industry" who exploited the working classes. As adults themselves, Allport and the Murphys would find that such lessons would help orient them to the sociopolitical world of the 1920s and 1930s to some extent, but that there nevertheless existed troubling new developments of which even the perspective of a radical progressivism seemed unable to adequately respond. The emergence of fascism as a political philosophy and as an increasingly powerful political force was the most troubling case in point, spurring these legatees of the Progressive ethos to try to reforge their own political frameworks, so as to match the scope of new threats to the achievement of social justice.

It has been too little appreciated that progressives during the 1930s, as Kuznick observes, "viewed the democratic victory over fascism, not the socialist victory over capitalism, as the salient political struggle of the day." Were fascism to triumph, the question of social democracy would be rendered irrelevant. Kuznick rightly points out that Columbia University – where Franz Boas had "redoubled his efforts against racism and Nordic supremacism in the 1920s" – had become "a hotbed of antifascist sentiment in the late 1930s, much of which coalesced around support for the Spanish loyalists."[94] Indeed, SPSSI members – Gardner Murphy and Gordon Allport prominent among them – played leading roles in organizing aid to the Spanish loyalists, under the auspices of the Psychologists' Committee of the Medical Bureau to Aid Spanish Democracy.[95] Efforts to assist the antifascist forces in Spain were, however, but one aspect of fighting fascism, for the concern of intellectuals such as Allport and the Murphys that fascism was gaining a foothold in the United States was as great an issue for them as was the advance of fascism abroad.

Lawrence Levine's survey of films during the Great Depression is an effective cautionary to historians who find the soul of the 1930s to be bound up in the struggles between collectivists and capitalists. Levine notes that, during the early New Deal, it was common for magazine readers to come "across articles entitled 'Roosevelt – Dictator?' 'Fascism and the New Deal,' 'America Drifts toward Fascism,' 'The Great Fascist Plot,' 'Is America Ripe for Fascism?' 'Must America Go Fascist?' 'Will America Go Fascist?' 'Need the New Deal Be Fascist?'" Sinclair Lewis's book *It [Fascism] Can't Happen Here* predicted that indeed it could; the Works Progress Administration (WPA) Theater project produced a dramatic version of the story, opening the play in twenty-two separate productions in eighteen cities on the night of October 27, 1936. In a similar manner, as Levine relates, the film industry, through its story lines, sought to come "to the defense of the traditional American democratic system in the face of rising authoritarianism." Levine adds, however, that the "streaks of pessimism and doubt" contained in the scripts left many productions enveloped in "a quiet but pervasive sense of despair concerning the future of both the individual and democracy."[96]

Allport, too, asked "What Is Fascism?" in a one-page piece that may have been given as a radio address, circa 1939. Fascism, he stated, was the result of the "weariness and despair" of those who "are unable to piece together a rational solution to their difficulties under the controlling guidance of the democratic outlook upon life." The followers of fascism, Allport suggested, were invigorated with hope in their "voluntary and systematic retreat from reason," but this hope blinded "them to its degenerative effects."[97] In a 1940 chapel address at Harvard, Allport emphasized that German fascism was "not a question of metaphorical evil," but of "undisguised violent, physical torture, suffering worse than savages can inflict."[98] Allport held out the hope that "most men do evil with only one hand earnestly and with the other hand make gestures that will mitigate the evil they do." Speaking to his fear of American fascism, Allport advised his listeners that "it is not only the Germans who are fighting this war in themselves. There is a civil war in each one of us . . . a war between the fascist and democrat in each one of us."[99]

In fighting this internal war, individuals could go one of two ways: the "easy way" – the road to fascism – was "to dispense with reason and give way to the simple myths and prejudices of totalitarian philosophy." Elaborating on a text from Micah, Allport stated that "the hard way" – the way of democracy – entailed the courage to bear the "indignation of the Lord," and to face up to the obligation to "try to repair our own iniquities and stupidities, and think out a juster social order that will be buil[t] as Christ instructed upon the premise of the infi[ni]te value of each single human being."[100] One aspect of that "juster social order" would be, as Allport wrote to a colleague in 1936, the defeat of "the mischievous Immigration Registration Act now pending in Washington. It is *pure Fascism.*"[101]

Intellectuals such as Allport and the Murphys believed that the political crisis of their time could not be reduced to a choice between capitalism and economic collectivism: to observe the growing appeal of fascist philosophies of the state was to see a less clearly delineated picture of the path toward social progress. Indeed, Allport believed that, as obstacles to the realization of some form of socialism in the United States lessened, the real "active oppos[i]tion" would come not from "Capitalism but by Fascism." The opposite of socialism was not capitalism, Allport contended, but fascism: both fascism and socialism were philosophies with ethical goals, whereas capitalism was merely a set of practices.[102]

In trying to throw light on the course ahead, Allport and the Murphys turned to the study of social "attitudes." In a thinkpiece in Carl Murchison's 1935 *Handbook of Social Psychology,* Allport defined attitudes as frames of orientation that "determine for each individual what he will see and hear, what he will think and what he will do." Citing James, Allport argued that it is our attitudinal frameworks that "'engender meaning upon the world;' they draw lines about and segregate an otherwise chaotic environment; they are our methods for finding our way about in an ambiguous universe."[103] During the 1930s, this "ambiguous universe" was explicitly viewed as being as much a problem of politics as it was of psychology.

For the scientist interested in fostering social change, then, the study of attitudes was of the utmost significance, for, as Allport explained, they "are often as rigid as habits." Paraphrasing James again, Allport illustrated his point by drawing a political moral: attitudes were "a fly-wheel and a conservative agent in society; they save the privileged classes from the envious uprisings of the poor; and cause the poor in spite of their bitter experience to think and to vote in ways that are inimical to their own interests." Society's conservatives, in Allport's view, were individuals who were "aware of, but confused by, the injustices of society," and who protected themselves "from personal discomfort by the reflection that 'you cannot change human natures.'"[104] The conservative thinkers of this 1935 piece on attitudes greatly resembled the fascist sympathizers whom, as mentioned earlier, Allport would describe several years later.

Allport's concern that domestic fascism might take root can be seen as well in the attention he gave to the "media of public control," most significantly in *The Psychology of Radio,* a book he wrote in 1935 with Hadley Cantril, a student of Allport's at Dartmouth who had followed him to Harvard. The book is essentially an investigation of the rhetorical conventions of radio communication, and the implications that this new medium of persuasion held out for the polity: references to Mussolini and Hitler in the opening section indicated the political issues at stake; references to insurgent political figures Huey Long and Father Coughlin signaled their immediate significance. In discussing Long and Coughlin, the authors noted that "a sound argument is always less important for the demagogue than are weighted words."[105]

Radio, Cantril and Allport argued, was particularly significant among the "media of public control" in "forming opinion and in guiding action." This point was buttressed by the cases "of Huey Long, of Mussolini, [and] of Hitler," whose "listeners had ready-made attitudes toward these leaders that needed only to be intensified and directed through vocal appeal." But in the realm of attitude formation radio possessed a new power, one the authors find in the example of Father Coughlin, who "was not a well-identified leader before he used the medium of broadcasting. His principles were not known nor were they widely accepted." And yet, "were it not for Father Coughlin's feat in creating exclusively on the basis of radio appeal an immensely significant political crowd, one could scarcely believe that the radio had such potentialities for crowd-building." Coughlin's example offered psychologists and citizens a lesson in the "creation as well as [the] shaping" of political attitudes literally right out of the air.[106]

By itself, Cantril and Allport stated, radio "is as democratic, as universal, and as free as the ether." Under capitalism, however, "it is an altogether elementary psychological fact that dissenting opinions and germinal attitudes favoring radical change in the American way will not readily be encouraged by an instrument controlled by vested interests." The authors advocated that, "in order best to serve the American public, radio should be removed from the dictatorship of private profits, and at the same time be kept free from narrow political domination."[107]

Cantril and Allport's study of the effects of the pervasive presence of radio was intended to throw into relief the cultural and economic context in which "radio as a social institution" was embedded. They asserted that, "although human nature may be everywhere *potentially* the same, the ways in which it actually develops are limited by the constraints of each particular social system." Such constraints were difficult to discern as a matter of everyday routine, for they "become second nature to the individual. He seldom questions them, or, indeed even recognizes their existence and he therefore takes for granted the great majority of the influences that surround him in everyday life." What radio listeners were taking for granted was the corporate control of the airwaves, a fact that Cantril and Allport held possessed grave implications beyond the question of private profits being appropriated from a public resource. As the authors explained, "the problem of the rights and responsibilities of broadcasting companies is a delicate one, for it involves the two explosive issues of censorship and propaganda."[108] Cantril and Allport defined censorship as the "process of blocking the expression of opinions, and thereby of arbitrarily selecting the listener's mental content for him," and propaganda as the "systematic attempt to develop through the use of suggestion certain of the listener's attitudes and beliefs in such a way that some special interest is favored." On the issue of propaganda Cantril and Allport made forcefully clear their assumption that advertising was propaganda, and that "the socio-economic framework within which radio operates always creates a temptation for its managers to exert censorship along some lines and to facilitate propaganda along others."[109]

It was the exploration of issues such as these – the ways in which the social psychology of radio perpetuated the conservative nature of the status quo – that helped shape the research mission of SPSSI. Gardner Murphy spoke explicitly to the theme of entrenched conservatism in his exhortatory address to SPSSI members in 1939. Murphy predicted that, as the findings of social psychologists began, "within a decade or two, to make a real difference to society," there would be "resistance to the application of our findings, and devious means used to keep our work from the service of the public, just as new patents are daily held out of use because they might conflict with established interests." Murphy argued that social psychologists, if they were to see their findings put to use, would have to "do an infinitely better job than we are now doing in the study of the basic psychology of *resistance* to social change, and in systematic research into the ways of making people aware of their own needs and interests."[110]

A key difference between the social critique mounted by intellectuals such as Allport and the Murphys and the more programmatic thrust of those who were committed to following some variant of "scientific socialism" was that Allport and the Murphys judged that the forces holding back social change were more complex than promised by an unfolding historical materialism.[111] In terms of affecting social change, many Marxist-oriented intellectuals focused on the working-class as the engine of revolution; in contrast, Murphy in the 1930s focused on the members of the middle class *as barriers to change.* Murphy argued that "one of the outstanding tasks" ahead of psychologists would be the "patient and comprehensive analysis of middle class psychology, to enable members of this class to understand fully their own situation, and to forestall the otherwise inevitable fascist trends." Murphy noted that "when progressive or radical doctrines are naively applied, as in America almost constantly during the last twenty years, the result has been the alienation of the middle classes, the farmers, and indeed many of the working class, with consequent brutal repression of labor's demands and the spread of fascist sentiment." The great mistake made by "contemporary American progressives and radicals," Murphy cautioned, was their bias toward thinking "in terms of economic analysis, not in terms of psychology, failing to see the huge research problems that have got to be solved before the public can be reached."[112]

Murphy sternly reminded his fellow social reconstructionists that "a fragment of research which it is hardly wise to ignore is the Gallup material on the reaction of the middle classes to CIO [Congress of Industrial Organizations] tactics." Those who considered themselves revolutionary Davids versus the establishment's Goliath must not leap too soon to the conclusion that their blows had hit their mark: in Murphy's analysis, "the very understandable but very short-sighted activities of a radical leadership which has not yet become research-minded" had too often "boomeranged." Murphy offered the bitter illustration of the fate of the Spanish loyalists as evidence for his claim. Large numbers and excellent morale had brought early successes to the loyalist forces, Murphy recounted, and "the near-impossible was done again and again. The loyalist officers, however, were not well enough trained to foresee all the consequences, and the result was that they found it physically impossible to hold the positions which they had taken, against the air and tank attacks which followed."[113] Murphy drew a parallel between the loyalists and "almost all American progressives," who had often made "the mistake of putting pins in the map to show their progress, forgetting in their excitement that the forces of reaction must be studied and intelligently handled if such gains are not to be swept away." As Murphy pointedly put the matter, "Most Davids forget that most Goliaths are going to be able to return the first blow."[114]

Thus, Murphy urged his audience, it was "the reaction which follows one's own forward step [that] is precisely the thing that needs to be fully understood and blueprinted." Not only action, but reaction, must be studied. Murphy admonished his colleagues that "blueprinting of the future is not the same thing as autistic thinking regarding the desirability of a new society; it is realistic mapping of issues in such a way that the mountains in the way can be crossed."[115]

Scouting those mountains was to be the intent of an SPSSI yearbook, "Resistance to Social Change," whose editorial board was to consist of Gardner Murphy, his students Rensis Likert and Eugene Horowitz, Hadley Cantril, and others. The yearbook was to focus on the power of existing social and economic arrangements to frustrate efforts at restructuring society, and it was to cover a wide range of topics. Cultural perspectives were to be represented by case studies of the "Chinese resistance to change compared with the westernization of Japan," and temporal perspectives – as in an examination of "reforms rejected in one period and accepted in another, e.g., industrial unionism and New Deal measures" – were to be covered as well. In seeking to "describe the human relationships which make change difficult – the many lines of defense of political institutions against change," such practices as "the two-party system, the legislative process, executive checks, judicial checks especially as they operate in the field of administrative law," in addition to the "monopolistic and conservative practices of special organizations like the American Medical Association and the American Federation of Labor" were offered as fruitful targets of investigation. The "psychological nature of social order and social stability" was to be looked at by examining "institutional patterns and social organization as a barrier to change," and by preparing psychological analyses "of social patterns summarizing studies of conforming behavior, emotional allegiance to symbols, pluralistic ignorance, [and] institutional fictions especially the belief in an unchangeable social structure."[116] The proposed yearbook never appeared, however, most likely because of the entry of the United States into World War II, and the subsequent mobilization of scientists for government duty.[117]

In seeking to examine the "war between the fascist and the democrat in each one of us," Allport and the Murphys were attempting to delineate the forces at play in American society, not simply to describe the nature of American life, but to gather knowledge that could be put to use in changing that society. Their study of the larger social context was intended to illuminate what Lois Barclay Murphy's father characterized as "the great unyielding forces: natural conservativism, and inertia to social change; the determination of privilege and power at whatever cost to hold on to what it has; and impersonal, corporate power, unfeeling and well-nigh irresistible."[118] These activist psychologists saw themselves as laying the groundwork for a "science for society's sake."

In making the case for "impure" science, Allport and the Murphys assumed that science was a value-laden enterprise, in which it was proper and right to "pay attention to the man behind the curtain," whether that figure was preparing a radio broadcast, or socializing pre-schoolers, or engaged in scientific research. Here they corroborated the assertion of McConnell that the mind of the scientist was not a "bare sheet upon which impressions are produced without any response or activity by the sheet itself" but was instead "a living agent, with peculiarities of its own. It moves more easily in some directions than in others."[119] It was because of this belief that they encouraged their colleagues to move beyond the cloistered walls of the ivory tower, and to step out instead upon the "thronged and common road."

5

Natural History and Psychological Habitats

> To get the feeling of what it is like to be a creature of the sea requires the active exercise of the imagination and the temporary abandonment of many human concepts and human yardsticks . . . Time measured by the clock or the calendar means nothing if you are a shore bird or a fish, but the succession of light and darkness and the ebb and flow of the tides mean the difference between the time to eat and the time to fast, between the time an enemy can find you easily and the time you are relatively safe. We cannot get the full flavor of marine life – cannot project ourselves vicariously into it – unless we make these adjustments in our thinking.
>
> Rachel L. Carson (1941)[1]

In their concern with characterizing individuality and social context during the 1930s, Allport and the Murphys shared much with the techniques and goals of naturalists. Practices and values inherent in aspects of the American natural history tradition legitimated dissent from the mechanistic and reductionist values promoted by neobehaviorists, and also offered a scientific framework that carried powerful associations to such politically charged art forms in the 1930s as documentary film and literature. In drawing on the authority of natural history, and in appropriating some of its key constructs and practices, Allport and the Murphys were able to identify conditions under which it would be possible to display and investigate aspects of scientific "reality" that their orthodox colleagues asserted were incapable of "scientific" treatment.

Researchers such as Allport and the Murphys adapted ways of knowing from natural history not only to pursue new areas of investigation within psychology but also to argue for a transformed relationship between scientists and the objects that they studied, one that deemphasized mechanical manipulation and instead privileged observation and intuitive discernment. Such demands for new sources of scientific understanding had received prominent attention by nineteenth-century American romantics, as in Henry

David Thoreau's lament that "science in many departments of natural history does not pretend to go beyond the shell; i.e., it does not get to animated nature at all."[2] William James's radically empiricist beliefs were in agreement with this view, for in his emphasis on grasping a "thicker" experiential reality than that provided by "classical" science James also sought a more dynamic sense of nature. Radically empiricist scientists such as Allport and the Murphys felt a strong affinity for this tradition of protest. Lois Barclay Murphy's proposal to "study the soil" in which behavior grows, and her desire to capture the "whole world of inner feeling" that lingered in the "overtones of response in raw records," spoke to her commitment to thinking her way "beyond the shell" at which Thoreau believed too many investigators stopped.[3] Indeed, Murphy's words articulated an ethic shared across disciplines, as exemplified in Rachel Carson's first book project, in which she urged her readers to adopt an attitude that would allow us to "project ourselves vicariously into nature."

Historians of biology have recently begun to appreciate, in Gregg Mitman's words, the existence of a "borderland between the biological and social sciences through the study of interrelationships between and among individual organisms and their environment." Mitman suggests that this borderland is characterized most succinctly by the idea of "ecology," at least for the first half of the twentieth century, before professional ecologists had incorporated "sophisticated quantitative techniques and theories from the physical sciences, hoping to make ecology a 'hard science.'"[4] Although I believe that Mitman's point is a sound one, I choose here to modify his claim slightly, by using instead the concept of "natural history" as a marker for the orientation displayed in the work of Allport and the Murphys. One reason for this choice is the fact that, when these psychologists referred to "fieldwork" as a scientific practice, they typically had in mind not only the activities of biologists but also those of astronomers and geologists; natural history, in this respect, is a more inclusive term than is ecology. Furthermore, especially for the earlier decades of the twentieth century, the term "natural history" was more prevalent in common parlance and indicates a certain frame of mind that the scientists that I examine here shared with the wider public.

The term is relevant to this discussion for another reason as well: natural history techniques could be deployed in that scientific space that was presumed to be antithetical to the haunts of the naturalist: the laboratory. Historian of biology Sharon Kingsland has pointed out that the attitude of experimental biologists Charles Manning Child and Charles Judson Herrick in this era was "essentially that of the natural historian, refined by the experimental method that shaped the new biology emerging at the turn of the century." Investigators working from such a perspective found it crucial to know the life history of the specimens they were examining under the microscope.

Kingsland finds that while such natural history orientations were "challenged by the reductionist physicochemical biology and behaviorist psychology that gained in importance in the 1920s," they were "never wholly eclipsed by this approach."[5]

In fact, the 1930s found such laboratory biologists as Barbara McClintock and Ernest Everett Just furthering natural history perspectives under the microscope. Evelyn Fox Keller has argued that McClintock, in determining the effects of genetic crosses, eschewed such tools as biochemical assays in favor of "techniques more familiar to the naturalist – she observed the markings and patterns of colorations on the leaves and kernels of the corn plant, and the configurations of the chromosomes as they appeared under the microscope."[6] Scott Gilbert makes a similar observation about the practices of cell biologist Ernest Everett Just, who, in Gilbert's words, "sought the mystery of life, not its mastery," adopting an "attitude [that] was more akin to the naturalist than to the experimental embryologist of the 1930s." A prominent feature of Just's approach was his insistence on making detailed microscopic observations in ways that would avoid altering the integrity of the organisms he was studying.[7] In his 1939 text, *The Biology of the Cell Surface,* Just in fact maintained that "description is the method of most use in biology. It plays a larger role here than in physics not because experimental biology is a younger science than physics, as one often avers, but because of the heterogeneity and complexity of the composite life-unit. The more heterogeneous and complex an object, its parts undergoing multitudinous kaleidoscopic changes, the more important description becomes."[8] Drawing on an approach to nature that owed more to *Naturphilosophie* than to the Newtonian mechanical worldview, Just, like Allport, even opened his work with a quote from Goethe: "Natur hat weder Kern / Noch Schale, / Alles ist sie mit einemmale [Nature has neither kernel nor shell, but is both all at once]."

In speaking of the uses made of "natural history" by Allport and the Murphys, I do not mean to imply that there existed some clearly delineated paradigm that these individuals reproduced in their work. My argument is that each of these psychologists appropriated those aspects that, within a diverse set of traditions, furthered their attempts to make concrete their understandings of reality. Broadly speaking, "natural history," as I will use it here, encompasses field studies and descriptive methods of organisms as a whole, within their environmental contexts. Observational methods are one fundamental aspect of this natural history model; two other elements are the affirmation of singularity and of temporality in nature. Speaking to this point, Stephen Jay Gould has contended that sciences such as biology, paleontology, geology, and astronomy "cherish detail and find their greatest satisfaction in the resolution of a particularly complex and singular event like the Cretaceous mass extinction." These sciences require a historical perspective because of the relevance of temporal sequence, where "everything interesting happens only once in its meaningful details."[9] The anthropologist Alfred Kroeber made this same assertion in 1936, remarking that "a dateless and placeless finding in human history, natural history, palaeontology, geology, or astronomy would make no sense."[10] Such was the logic behind Allport's statement that although "individuality is never twice repeated, it represents nevertheless order in nature."[11]

Another salient characteristic of a "natural history orientation," as suggested by biologist Daniel Lehrman, is the attempt by researchers to frame their questions "in terms that are isomorphic with the experience of the subject, rather than with the operations of the observer." Lehrman claims that whereas "the behaviorist orientation expresses detachment, lack of interest in the nature of the animal, and stress on control and/or prediction of behavior as the aim of professional effort," in a naturalist-oriented approach investigators instead feel free to express "a feeling of continuity" with their subjects, are interested "in all data that illuminate that inner organization of the animal from which its behavior emerges, and stress an understanding and appreciation of the animal as the goal of research."[12] As Allport argued at the end of the 1930s, "unless we can first comprehend our subject's value-context, we are unable to know the significance of his behavior as he performs it, for the simple reason that the behavior we perceive is instantly ordered to *our* own presuppositions without any regard to what *his* presuppositions may be." Psychologists, Allport urged, needed to find ways to frame their work in terms of their subjects' contexts, and to study behavior "in the light of the subject's imperatives."[13]

In dissenting from the push to institutionalize quantitative method and laboratory experimentalism as the sole markers of "legitimate" science, the model of natural history was explicitly invoked by psychologists such as Allport and the Murphys. Seeking to rebut "the feeling [that] has grown that experimental and quantitative method are the hallmark of science," Gardner Murphy explained that such a view persisted notwithstanding "the fact that the history of the physical sciences bristles with important ideas which have borne only a remote relation to the existing laboratory situation, and that the history of biology has been much more a history of learning to observe, either in the woods or under the microscope, than of controlling every relevant variable in the manner of pure experimentation." Murphy further offered the accomplishments of astronomy and geology in support of a broader image of science than was sanctioned by that advocated by strict experimentalists.[14] Allport also relied on this analogy, as in his observation that "psychology is not exclusively experimental in its method, but neither for that matter is the eminently respectable science of astronomy."[15] Lois Barclay Murphy's sympathy research exemplified work in the natural history mode, relying as it did on the patient and skilled observation of individuals, over time, in their natural environments. As Murphy tartly commented in *Social Behavior and Child Personality,* statistically obsessed research had led to a picture of personality "about like a child's kindergarten Christmas tree – an outline with dozens of decorations of different sizes, shapes, and colors pasted on." This is why Murphy proposed, in contrast to this immature rendering, to "study the soil" in which "the varieties and the nature of sympathetic behavior . . . grows."[16] Where Allport and the Murphys differed from many of their colleagues was that they did not *start* from the ideal of pure experimentation and then design simplified research questions that could be adapted to such an ideal; instead they started with questions about complex constructs such as individuality and contextuality and then cast about for methods that would make such phenomena apparent.

Allport's and the Murphys' resistant attitude toward reductionist dictates was related to a contrast that James had earlier posited between "the classic-academic and the romantic type of imagination." The classic-academic perspective, James explained, exhibits "a fondness for clean pure lines and noble simplicity in its constructions. It explains things by as few principles as possible and is intolerant of either nondescript facts or clumsy formulas." For the mind imbued with a view of nature derived from classical physics, "the facts must lie in a neat assemblage, and the psychologist must be enabled to cover them and 'tuck them in' as safely under his system as a mother tucks her babe in under the down coverlet on a winter night."[17] And yet, James observed, "behind the minute anatomists and the physiologists, with their metallic instruments, there have always stood the out-door naturalists with their eyes and love of concrete nature." Although James allowed that the former thought the latter superficial, he found this evaluation to be unpersuasive, asserting instead that "there is something wrong about your laboratory-biologist who has no sympathy with living animals," and with psychologists who dissect away the "varieties of mind in living action."[18]

James judged such abstractionist practices to be philosophically suspect, and he found the oft-repeated claim that "science only stands for a method and for no fixed belief" to be disingenuous. The case was rather, James argued, that "as habitually taken, both by its votaries and outsiders," the word "science" signifies "a certain fixed belief – the belief that the hidden order of nature is mechanical exclusively, and that nonmechanical categories are irrational ways of conceiving and explaining even such things as human life."[19] In this dissatisfaction with mechanism as an overarching explanatory principle, James's radical empiricism coincided with aspects of romanticist thought. James, in fact, defended romanticism in several places in his works, relating it to his radically empiricist view of reality. He offered the comment, for example, that "the personal and romantic view of life has other roots besides wanton exuberance of imagination and perversity of heart. It is perennially fed by *facts of experience,* whatever the ulterior interpretation of those facts may prove to be."[20]

As James indicated, to approach the study of nature as a radical empiricist was, in some senses, to engage with romanticist critiques that conceived of science as being closer in spirit to natural history than to mechanical empiricism. Given the musty aura that "natural history" still retains in academic discourse as a "discipline emeritus," it is important to recognize that the prism of natural history was being deployed by James toward modernist ends in a quest to clarify the relationship between subjective modes of knowing and the nature of experience.[21] This incorporation of the idioms of American natural history within the discourse of radical empiricism is one that survived James's own era, as can be seen in the rhetoric of such later thinkers as Allport and the Murphys.

As part of the modernist assault on the belief in fixed and universal truth, the radically empiricist emphasis on experiential reality challenged conventional strictures on how and where to find authoritative knowledge, as in James's acceptance of "epistemological uncertainty into the core of his scientific and philosophic thinking." Eugene Taylor has further pointed out that James's radical empiricism, beyond providing a theoretical perspective on "the general relation of experience to scientific knowledge" also presents an epistemological critique of "the shortcomings of the experimentalist attitude."[22] References to natural history within radically empiricist discourse both moved such modernist goals forward and modified them, most particularly by bringing romanticist insights regarding emotion into the discussion, advancing an image of the scientist as a "man of feeling" as well as a "man of disciplined mind."[23] This configuration of conceptual stances is one that I characterize below as "experiential modernism," in an attempt to differentiate the focus and goals of radical empiricists such as Allport and the Murphys from those of social scientists who worked within modernist frameworks that devalued emotion and feeling.[24] Before taking up this point, it will be useful to first sketch the sorts of romanticist traditions that Allport and the Murphys found of value in this regard. The forms of romantic thought that Allport and the Murphys encountered as young adults were somewhat different, with British and American sources being of primary significance for the Murphys, and German sources for Allport.[25]

Romantic Traditions and Modernist Values: Radical Empiricism as "Experiential Modernism"

When Lois Murphy spoke of the "deep mutuality" that began between her and Gardner Murphy when they met in the 1920s, she pointed to an intellectual reciprocity "fed by our joint love of the poet Blake, Beethoven's Seventh Symphony, the philosopher Bergson, and their ilk."[26] Elsewhere she underscored the symbolic significance that these Romantic touchstones held, noting that "Shelley, Keats and Wordsworth were so precious to [Gardner] that his discovery of my love for those poets created an instant bond."[27] Gardner Murphy, in an essay on the mutuality of his and his wife's thinking, entitled "Story of Dyadic Work and Thought," singled out a related aspect of Lois's intellectual and emotional development that held great interest for him: the "intense" episodes "of an ecstatic or mystical sort which we may roughly call Wordsworth-type experiences" that she underwent during her late childhood and early adolescence.[28]

Of profound impact as well for Gardner Murphy was the transcendentalist legacy of Concord, the home of his mother's family. Murphy's grandfather, George A. King, was a member of Concord's select "Social Circle" and served on the Concord Library Committee with Ralph Waldo Emerson, taking over as chairman upon Emerson's death.[29] "Concord was always the home base and rallying point," Murphy later recalled; "it was as if my life were a pure culture of the Emerson, Thoreau, Alcott world in which we were steeped."[30] Lois Barclay Murphy observed that "Walden was not just Thoreau's world and creed of living with nature, it was Gardner's retreat and playground as well."[31]

Murphy's choice of Concord as his spiritual "home base" was a deliberate act, not merely a matter of inherited tradition – Gardner's brother, DuBose, felt more of an affinity for the South, a sentiment reflective of the youngsters' childhood years in San Antonio, Texas (their father's birthplace) and Montgomery, Alabama. Although the Murphy boys later spent their summers in Concord, they actually lived there for only two years, when Gardner was seven and eight years old.[32] But Lois Murphy accurately reflected her husband's passionate affiliation with the Concord milieu by titling a chapter of her biography of Murphy "Growing Up in Emerson's Concord."[33]

Gardner Murphy would strike a decidedly Emersonian chord, for example, in his description of the manner in which his moment of mature scientific dawn occurred. The incident centered on a trip that the Murphys made together to visit the Grand Canyon in the summer of 1931. The trip was both a reunion – Gardner Murphy had been teaching at the University of California at Los Angeles while Lois tended their newborn son, Al, in New York – as well as a spiritual journey. Lois Murphy later described the experience by saying the "Grand Canyon was glorious, majestic beyond human imagination. . . . I found myself melting into it all with an intense sense of oneness with the universe and with Gardner."[34] Gardner Murphy elaborated on the event's significance in this way: "In 1931, at the age of 35, I had begun a career in psychology, but I had not found an area of specific psychological research to which I was willing to make a commitment. During that summer when Lois and I were in the Grand Canyon of Arizona, it came to me that I wanted to spend my life studying human personality." Inner and outer nature coalesced as Murphy achieved a vision of his calling, stating that "the black cliffs and deep blue shadows of the Canyon were symbolic to me of the intense sensory richness of the inner world, a world which I believed could be more effectively observed, investigated, described, and understood." Murphy related that he had been considering this conception of his psychological vocation for several years, but that "it was with Lois at my side and the Canyon above, as we followed its upward graded trails, that gave this vision of personality its commanding appeal. I remember as I attended a music service at St. Paul's Chapel, Columbia University that fall, saying to myself: 'So it's to be personality, isn't it; that's what I shall do, isn't it?'"[35]

If the Murphys' sensitivity to the scientific sublime was grounded in Wordsworthian epiphanies and Concord transcendentalism, the roots of Allport's engagement with Romantic thought are to be found instead in German *Naturphilosophie.* Tokens of this encounter can be found, for example, in Allport's choice of a quotation on individuality from Goethe to open his work on *Personality.* Elsewhere, in a critical review of a work on personality in 1932, Allport called upon the authority of the German poet and philosopher to drive his point home, observing that "when Goethe gave it as his opinion that personality is the supreme joy of the children of the earth, he could not have foreseen the joyless dissection of his romantic ideal one hundred years hence. There is certainly nothing poetical, metaphysical, or even humanistic in the modern psychological approaches to personality which are so conveniently surveyed in this volume."[36] When Allport first discovered an affinity with Goethe is hard to say, although it is certainly true that his passion for the German poet was sealed by the end of the first of his two post-doctoral years abroad; Allport spent the first of these in Germany, principally at Hamburg and Berlin, and the second at Cambridge in Great Britain. The impact of his German *wanderjahre* exerted itself into Allport's second year as well: Allport confessed that, while in residence at Cambridge, he "chiefly ruminated on my German year and enjoyed myself by studying Faust with Professor Breuel."[37]

Allport cast his exposure to German psychological research as "an intellectual dawn," where words such as "*Ganzheit* and *Gestalt, Struktur* and *Lebensformen,* and *die unteilbare Person* were new music to my ears." For Allport, such exotic philosophical fare represented "the kind of psychology I had been longing for but did not know existed." As someone who "had been brought up in the Humean tradition," Allport was fully aware that the enticements offered by this latter-day *Naturphilosophie* were not to be indulged in without due care: "Of course I realized that romanticism in psychology could poison its scientific soil. . . . At the same time it seemed to me that the high quality of experimental studies by the Gestalt school, the original empirical investigations at [William] Stern's Institute, and the brilliance of the [Kurt] Lewinian approach (which I came to know at second hand) gave safe anchorage to the kinds of concepts that I found congenial."[38] In brief, Allport came away from his travels believing that "Germany gave me support for the structural view of personality that I had pieced together for myself."[39]

Perhaps nowhere was that sense of support more concrete than in the work of Stern, whose theory of "personalistic" psychology Allport came to admire greatly. Stern's own early training, at the University of Berlin in the 1880s, came during a time when, in Stern's words, "the collapse of speculative philosophy after the death of Hegel had a paralyzing effect, [and] the triumphal procession of natural science a downright hypnotic one. The word 'metaphysics' was in disgrace."[40] Metaphysics was in disrepute, Stern recalled, for "the mechanistic categories of the scientific philosophy were regarded as self-evident and unshakable." The only approved option for scholars in regard to these mechanistic categories was "to interpret them and supplement them from the standpoint of the rational sciences, as well as to deduce from them some general consequences for cosmology – never to take an independent attitude toward them and affect them from some higher peak."[41] Stern saw his own intellectual odyssey as one taken against the "passive resistance of the rest of the scientific world."[42] Stern's sense of fighting for the space to build a picture of nature upon non-mechanist premises is one from which Allport took heart.

When Allport took up his first departmental appointment in psychology at Dartmouth in 1926, his intellectual encounter with German philosophical and psychological thought was still pronounced, as can be seen in a pair of lectures he crafted, "Intuition as a Method in Psychology" and "Intuition and the Aesthetic Attitude."[43] In these two pieces, the young Allport used the German perspectives he had acquired as a mental prism, taking his own particular questions and viewing them first through one facet of the German *Romantik* tradition and then another. Allport noted, for example, that during the last quarter-century – and especially since World War I – there had been "a great surge of interest in scientific method, particularly in Germany, and the contemporary trend in German psychology is strongly antagonistic to the exacting and limiting methods of biological and physiological sciences." In particular, Allport maintained, there existed "an urgent demand for what has been called a *psychological psychology,* that is, a psychology which will attack directly and fearlessly those problems which are uniquely psychological, and which might be summarized under a single heading as problems of *Meaning.*"[44]

Given sanction by the German example to work on problems of "meaning," Allport in turn championed the method of "intuition." In speaking of intuition, Allport was particularly concerned with presenting certain precepts of the *Verstehen* school (*Verstehen* being a term referring to "the mental activity that 'grasps events as fraught with meaning in relation to a totality'").[45] Allport characterized intuition as the apprehension of wholes – that is, "particularities, not parts," "essences," and "gestalts" – in opposition to "analysis," or the dissection of wholes into parts.[46] Allport also equated "intuition" with the "aesthetic attitude": "When for the moment we say, 'Hold, this form is good for its own sake' we are using that particular type of intuition which I have called the 'aesthetic attitude.'" Allport maintained that "the method of poetry" – which he held to be one aspect of the aesthetic attitude – "should be not merely an antidote to science, but should, to my mind, be an integral part of the scientific method."[47]

The vigorous intellectual cross-currents that Allport had experienced in Germany that fell under the heading of the doctrine of *Verstehen* were no mere fringe movements from within psychology, but had emerged, as Allport noted in *Personality,* from an engagement with philosopher Wilhelm Dilthey's influential theoretical work. Dilthey, Allport stated, had proposed the pursuit of a new descriptive psychology that would "elevate the individual to the central place of interest," thus posing methodological questions based on finding ways to "understand" the phenomenon of individuality rather than to "analyze" it.[48] Allport encountered an engagement with Dilthey's views in the writings of such German scholars as Karl Jaspers and Eduard Spranger, as well as in Stern's work. One of Stern's basic teachings was that human beings study the world around them "by an active sympathy or empathy" – a process Stern called "sympathetic introception" – where the "self does not seek to eliminate itself, but to identify itself with its object."[49]

In an article written soon after his time as a student under Stern in Germany, Allport stated that "when we fail to empathize with a person we fail to understand him; conversely, the greater our empathic ability the greater is our capacity for understanding."[50] Such views animated Allport's perspective a decade later as he summed up his final points in the closing pages of *Personality.* Allport argued that the practices of contemporary psychology, in which proper procedure was "to observe fragments of behavior, and then to reason by analogy (which really means binding these fragments to the lives of *other* people), will never yield an understanding of individuality." What were needed instead, he insisted, were methods allowing one to grasp "events as fraught with meaning in relation to a totality" – the goal of German *verstehende Psychologie.* It was only, Allport stressed, "when the life and actions of another are intimately and intelligibly bound together that I *understand* him."[51]

Although Allport acknowledged that "hard-headed" proponents of scientific naturalism were unlikely to be convinced by such arguments, he warned his colleagues against asserting "that this controversy is merely an epistemological snowball fight, and that psychologists should forget it and return to their tests and measurements to collect *facts.* Facts do not speak for themselves."[52] The typical American psychologist had sought information about "populations of people" rather than "concrete personalities," Allport observed, and this adherence to "restricted standards of scientific experimentation have blinded him to certain obvious truths to which the doctrines of intuition call attention."[53]

How should scientists reconcile, however, the seemingly incompatible aims of studying the particular and the universal? During the 1920s, Allport showed the fruits of his German tutelage, pointing to what he characterized as the philosophy of the *Sturm und Drang:* "The two souls that dwell, alas, within my breast (or in psychological language, the conflict of my aesthetic and theoretic attitudes) must wrestle until a resolution is discovered. This resolution will be in accordance with my own personality. Indeed it will be my personality." In a very real sense, Allport's belief about how this resolution would play out for individuals in general pertained as well to his own task of forming a scientific personality. It was a struggle without prescription, for "it is the fate of each enlightened individual to work out for himself the synthesis. It may lead him to disaster, to a self divided against a self, or to the strength of genius. It is his own adventure."[54] A decade later, in *Personality,* he diagnosed his peers as suffering from a "dissociation . . . in mental operations," for they relied on the use of intuition in their personal lives, while banning it from their professional practices.[55] In outlawing methods designed to facilitate direct and immediate understanding, Allport asserted, scientists had distorted experiential reality.

A belief that the mechanistic and reductionist practices characteristic of a strict experimentalism interfered in the full apprehension of reality was shared by American romantics in the nineteenth century, as in Thoreau's acerbic remark that "the ancients, one would say, with their gorgons, sphinxes, satyrs, mantichora, etc., could imagine more than existed, while the moderns cannot imagine so much as exists."[56] The remedy for such a circumstance could be found, Thoreau suggested, by understanding that "we do not learn by inference and deduction, and the application of mathematics to philosophy, but by direct intercourse and sympathy."[57] As Allport drew on a German philosophical legacy that supplied him with the concept of intuition, the Murphys similarly drew on an American school of thought that foregrounded the idea of "sympathy" as a means of thinking about how to access experience directly. For figures such as Emerson, Thoreau, and Whitman, as Roy Male notes, "sympathy expresses not just 'feeling' (that is, life) but also 'feeling with' (that is a relation)."[58] Although the Murphys would not have agreed with Thoreau in his negative assessment of mathematics, they did not discount, as did many of their scientific colleagues, the use of "direct intercourse and sympathy" as a legitimate avenue of knowledge-seeking. Lois Murphy's interest in "sympathy," as seen in the topic of her first research study, is one indication of a continued interest in this relational form of knowing.

For those nineteenth-century thinkers championing "sympathy" as an avenue of knowledge, the "sympathetic experience" was held to be "one of identification and incorporation, a relating to and an absorbing of an external object."[59] Sympathy was both a working tool for the student of nature, and "a necessary instrument" in the general "process of growth."[60] Walt Whitman caught this sense of incorporation as an aspect of growth in his poem "There Was a Child Went Forth":

> There was a child went forth every day,
> And the first object he look'd upon, that object he became,
> And that object became part of him for the day or a certain part
> of the day,
> Or for many years or stretching cycles of years.[61]

The concept of sympathy played a pivotal role in the search of American romantics for a new stance toward the investigation of nature. As Male argues, "Emerson called for a return to the original motive of science, which was 'the extension of man, on all sides, into nature, till his hands should touch the stars, his eyes see through the earth, his ears understand the language of beast and bird, and the sense of the wind; and through his sympathy, heaven and earth should talk with him.'"[62]

James also found the restrictive tenets of scientific naturalism to be problematic – especially, he stated, when an individual allowed it to become "one's only way of thinking," arguing that this perspective represented a "violent breach with the ways of thinking that have played the greatest part in human history." Here James had in mind such forms of thought as the religious, the ethical, the poetic, the teleological, the emotional, and the sentimental: "what one might call the personal view of life to distinguish it from the impersonal and the mechanical, and the romantic view of life to distinguish it from the rationalistic view." Such personal, emotional, and romantic forms of thought, James concluded, "have been, and even still are, outside of well-drilled scientific circles, the dominant forms of thought."[63] In fact, James argued that "theoretic knowledge, which is knowledge *about* things, as distinguished from living contemplation or *sympathetic acquaintance* with them, touches only on the outer surface of reality."[64] Such a sensibility was not alien, of course, to scientific circles which embraced natural history, where a scientific orientation could still include a focus "on the behavior of the animal in question, concern for the natural units of behavior, and an undisguised emotional response."[65] If it was difficult to find a place for Emersonian visions and radically empiricist Jamesian arguments within a science based on neobehaviorist psychologies of the 1930s, such was not wholly the case for psychologies that intersected with the sphere of natural history, where metaphysical considerations regarding the relationship between humans and the natural world could still be voiced.

But what made natural history appealing as an alternative model of science also made it vulnerable to dismissal as unscientific: the sense in which its practices incorporated "feeling" as an aspect of "knowing." By the 1930s, much ink had been spilled over the idea that the successful pursuit of science was held to depend on the suppression of emotion and the cultivation of an impersonal and technical attitude toward the objects of study. Barbara Laslett has shown how the eminent sociologist William Ogburn, for example, found emotion to be inimical to the pursuit of science, as in his claim that it was "necessary to crush out emotion and to discipline the mind so strongly that the fanciful pleasures of intellectuality will have to be eschewed in the verification process."[66] And as Lois Barclay Murphy observed of narrowly restrictive frames of thinking, during "an authoritarian period, when a highly formalized structure of thought and behavior is the accepted basis of life, emotions are generally tightly reined within this structure."[67]

For a science to place a positive value on such concepts as feeling, intuition, and sympathy was to lend it a decidedly Victorian and antiquated air. In this sense, the designations "scientific" and "modernist" share much in common. Literary theorist Suzanne Clark points out that "modernism has given us an ideal of an impersonal, serious art, a poetics separated from rhetoric" that assumes an estrangement between author and audience that is marked by a sense of exile, rather than that of familiarity and community.[68] And yet Allport and the Murphys pursued a modernist set of goals bequeathed to them by James within contexts that refused to disparage the place of feeling in scientific practice. Allport, for example, ended a tribute to Stern by envisioning a time "when psychologists will be less afraid than they now are to delve into problems of personal emotion (lest they seem emotional), or into the study of sentiments (lest they appear sentimental), or into the riddles of personhood (lest they become personal)."[69] Gardner Murphy struck a similar theme toward the end of his life, when he suggested that his "love of psychology" derived from his "love of human experience for its own sake." He added that he would not be "in the least injured if these thoughts are regarded as sentimental," and he claimed "the company of [novelist] Laurence Sterne, William James, and Harry Murray in selecting the term."[70]

The analysis that I am suggesting for "experiential modernism" in relation to science takes up a task similar to the one that Clark proposes for literary theory. Clark identifies the modernist moment as one that foreclosed certain avenues of social and political change at the same time that it claimed to rebel from the conventions of bourgeois life; one of the most vigorously policed threats to the purity of avant-garde culture by those constructing the canon was a fear of the "sentimental." Clark contends that literary modernists buried "the richness and complexity of the sentimental" underneath a series of distorted depictions in which the sentimental was seen to undermine the serious.[71]

Although most scientists no doubt would have identified with middle-class respectability rather than with the bohemianism of Greenwich Village, their emphasis on science as knowledge on the cutting edge of intellectual life bears striking resemblances to the claims of literary modernists. Clark's point that modernist intellectuals saw "sentimentality as both a past to be outgrown and a present tendency to be despised" is one that held true for arbiters of correct scientific method as well. Scientific maturity demanded that its practitioners outgrow emotionality.[72] In this view, the seeds of progressive change, whether in the artistic or the scientific realms, were to be found within modernist discourses that escaped the contamination of the sentimental. And yet, as a cohort of feminist literary critics have begun to point out, arguably the most disruptive and powerful discourse in the service of social and political change in nineteenth-century America was precisely the sentimental discourse – symbolized by Harriet Beecher Stowe's novel, *Uncle Tom's Cabin* – so despised by early twentieth-century modernist critics.[73]

The radically empiricist inclinations of Allport and the Murphys mark them, I am suggesting, as "experiential modernists," that is, as individuals who were engaged in a search for a form of scientific knowing that would unsettle conventional intellectual categories without simultaneously divorcing reason from feeling, and thus from the realm of moral sentiments. From this perspective, to strive for intellectual clarity and intuitive discernment demanded that a connection be established between researchers and the subjects of study. The goal was not to achieve a state of disembodied "unfeeling," but rather, as Keller has described in regard to the work of McClintock, the ability to develop "a feeling for the organism."[74] It is not surprising, then, that the experiential modernism of Allport and the Murphys found expression within research formats that relied heavily on a field studies approach, in which they attempted to evaluate psychological phenomena as they occurred "naturally," within complex contexts that shifted, depending on whether that context was viewed from the vantage point of the subject or the investigator, or through more abstract perspectives such as the historical, societal, or cultural. But these psychologists' research projects did not rely solely on a non-laboratory field approach, for they also strove to bring values associated with the naturalist point of view to bear on their experimental work.

Experimenting with Experimentation

Conventional experimental practices posed a problem for Allport and the Murphys, for, as currently defined, such practices appeared to foreclose knowledge of the very phenomena that they found of interest, such as individuality and contextuality. Each of these psychologists, therefore, sought to reshape the experimental scaffolding that they were expected to use into forms that would be more supportive of the radical empiricist reality that they were attempting to make manifest.

Lois Barclay Murphy's work started from an observational standpoint, with the basic tenets of experimental practice reversed. The point of experimentation for most researchers was that the experimenter "actively" manipulated an artificial situation in order to "control" the conditions under investigation. For Murphy, in contrast, the investigator was a "passive" – although alert and interpretive – witness to an "active" subject responding to a naturally occurring situation that ideally was left "uncontrolled." In using miniature life toys as a research tool, for example, Murphy sought to arrange for the spontaneous use of the toys by the children so that the "experimental" situation receded as an event into the contextual "field" of her subjects' lives.

Essentially, from the Murphyian perspective, "life" itself was conceived as a complex tangle of ongoing "experiments" crying out for scientific attention. In the introduction to the first volume of *Experimental Social Psychology* both Murphys argued that much of what should be of interest to social psychologists "would not or could not occur in an artificial situation," and that appropriate investigative practice would therefore often entail watching "Nature as she makes (usually rather clumsy) experiments." The Murphys added that, if observational methods were "well thought out, there need be no inferiority of such results to those obtained by controlling the phenomena we wish to observe."[75] Indeed, a large proportion of the work that the Murphys discuss in *Experimental Social Psychology* is *observational* in nature, a fact that does not conform to a more restrictive definition of what is properly to be counted as *experimental.* Using the example of an astronomer who "gets his telescopes and cameras ready before an eclipse, and uses them in the right way at the right time," the Murphys argued that "no invidious distinction in favor of the 'experimental' method" over methods of observation need be made. "The essential thing in science," the Murphys maintained, "is to know *the conditions under which an event occurs.*"[76]

The crux of the difference between the advocates of "pure" experimentalism and a naturalist-oriented scientist such as Lois Barclay Murphy lay in the different stances that each took toward their research subjects. From the behaviorist perspective, as Lehrman contends, "scientific explanations of, and statements of scientific insights into, behavior, *consist of* statements about how the experimenter gains control over the behavior, or about how the actions of the subject can be predicted from the actions of the experimenter."[77] From the natural history orientation, in contrast, the researcher's primary focus of attention is "the behavior of the animal considered as an aspect of the way the animal is related to its natural environment." The distinction between an investigator adopting a behaviorist orientation and one who adopts a natural history orientation is that the first, as Lehrman remarks, is attempting to discover when an organism such as a kitten "can learn something that the *experimenter* wants it to learn" and the second "is trying to find out when the kitten can learn something that the *kitten* wants to learn!"[78]

The autonomy that the subject retains in an observational study makes the scientific status of observational techniques problematic. Opponents of observational methods have typically objected to the lack of "control" that observational investigators possess in "manipulating" their "subjects." The "subjective" nature of interpreting what is being observed also makes such critics uneasy. Historian of science Simon Schaffer has discussed how, in the nineteenth century, certain sectors of the scientific community decided that, "because of factors such as personality, observers separated in space and time had to be calibrated with complex social and material technology." If feats of observation were to be counted as fully accredited scientific knowledge, then the observers themselves had to be treated as "part of the 'instrument' to be calibrated." Indeed, academic psychologists were directly involved in this quest to mechanize the practice of observation, through their efforts to render the "personal equation" irrelevant. Such acts, Schaffer claims, should be properly understood "not as an obvious and inevitable advance in understanding and precision, but as one among many contested moves to measure, control, and discipline the act of observation."[79]

The demands for precision and rigor issued by those who saw themselves as representing the cutting edge of experimental practice in psychology carried little weight with Lois Murphy, who adamantly rebutted the priority her colleagues placed on such criteria. In a key treatise on method, for example, Murphy asserted that, "in this monograph, rigidity of procedure is definitely secondary to insight, and traditional patterns of control are followed only in so far as they are helpful to the development of insight into child personality." While stating that she and her co-workers aimed "for ever greater orderliness in our quest for insight," she added firmly that "we are *not* interested in orderliness as such, if and as long as orderliness would stand for mere compulsiveness – in basic ignorance of what we are to be orderly about."[80] Lawrence K. Frank supported Murphy's claim, remarking that the ideology of precision had become anachronistic, for "there are numerous precedents in both the physiological and biological sciences for utilizing procedures that depart rather sharply from the older idea of exact quantitative measurement and favor methods that aim rather at sensitivity and qualitative discrimination."[81]

Ultimately, the greatest source of discomfort toward the scientific use of observational techniques was one that was rarely articulated explicitly: the difficulty that this practice presented in distinguishing "amateurs" from "experts."[82] Scientists, could, by elevating laboratory-based experimentalism to nearly mandatory status as a mark of legitimate scientific research, effectively restrict access to the production of scientific knowledge to a professionally credentialed minority. To the extent that "science" was held to be synonymous with classical physics, the ideology of "pure" experimentation was often held by professionals and nonprofessionals alike to need no further justification.[83]

One way to contest such a representation of the sciences as nothing more than successive approximations of classical physics was to call on a countermodel, as did biologist William Ritter, founder of the Scripps Institute of Oceanography, when he provocatively saluted "that greatest of all amateur zoologists, that 'intellectual colossus,' Charles Darwin." Holding that "Mr. Darwin was temperamentally an amateur naturalist all his life," Ritter credited "this predisposition for some of the deepest insights into human life and conduct, and into the life and conduct of all living things, that have ever been obtained."[84] This line of argument is one that had been employed as well by turn-of-the-century American schoolteachers, who demanded that their classroom-based research be accorded scientific recognition. Historian Leila Zenderland discusses, for example, how such teachers insisted "that they were capable of emulating Charles Darwin when he claimed he had 'without any theory collected facts on a wholesale scale' and thus transformed his field." These Darwinian-sanctioned investigators, "by objectively recording their observations about children," Zenderland remarks, "could thus function as psychology's 'naturalists.'"[85]

A few decades later, in *Child Life in School,* Lois Barclay Murphy and her collaborators claimed a similar affiliation, asserting that "our approach . . . has more in common with the problems and methods of ecology, of regional studies, of topological psychology, than it has with the approach which led to development of intelligence scales on the basis of age norms."[86] During this same period, Murphy recommended that the ecologically tinged construct of "life-space" be used as the appropriate unit of investigative analysis for psychologists. This construct, which Murphy derived in part from the work of Gestalt psychologist Kurt Lewin, was intended to emphasize the subjective nature of the "individual-in-social-context" formulation with which she had been working. Murphy defined the life-space of the child as "the world he has unconsciously selected as significant, [the] awareness of which he carries about with him; the objects, space-relations, relations with people, of his objective world as experienced, interpreted, assimilated by him." The concept of the life-space was one in which internal and external relations merged, for it had "its own geometry, created by the child's perceptions of his own size, strength, and what he counts for in his world. It may have sharp boundaries created by consistent taboos, or fuzzy boundaries created by inconsistent taboos, or no perceptible boundaries, as in the case of children who move freely through space in accordance with limits of which they are constantly conscious."[87] The human organism's habitat was bounded as much by cultural considerations as by geographical ones.

Murphy had in mind here taking into account the subjective nature of the spatial and temporal relations of each child's life – the experience of having one's own room, for example, or living with a father who leaves in the morning and is absent in the child's day until after dinner. Murphy believed that it was unfortunate that the clinical and analytic literature had elevated such phenomena as sexual conflicts and sibling competition to the status of the "chief determiners of personality development" and hoped that attention to the life-space concept would stimulate observers to a "broader awareness."[88] The projective methods with which Murphy preferred to work were intended to place the researcher within the subjective stream of experience of the children being studied, recognizing nevertheless that "the world which the child perceives is not the world which the adult perceives as he watches a group of children." Apprehension of, to paraphrase Carson, "the full flavor of child life" in turn required researchers to use their own subjectivity "to project themselves vicariously" into the childrens' subjective experiences. This Murphy sought to do by the construction of "contextual records."

Murphy's version of recording behavior contextually in *Social Behavior and Child Personality* had entailed the use of running protocols that presented the words, actions, and emotional overtones of children's interactions in naturally occurring situations. In *Child Life in School,* this procedure was carried through in greater detail in a study of ten seven-year-old children. Records of the individual children were taken on a rotating basis, in three cycles of two weeks' duration during the months of April and May of 1937. By the project's end, a total of six weeks of study had yielded sixty hours of observation and 200 records. The lengths of the recording intervals ran from ten to sixty minutes, depending on the nature of the class activity under inspection (the reading period, the writing period, the roof play period, etc.).[89]

The authors stated that they knew "of no short-cut techniques or methods" that would give "a realistic account of the quality, tone and content" of the children's actions. The observers produced "minute by minute behavior records" in prose form of the "actions and speech of an individual child, recorded in their context as they occurred throughout the whole of the observation period." Their goal was to depict on paper "as accurate and lifelike a picture as possible of the total behavior of the child during the time covered by the record," which included striving to capture the "feeling of the behavior observed, as well as to note the succession of events."[90] Recorders would sometimes add a summary statement at the end of their narrative if they felt that their running account had somehow failed to convey an important aspect of the contextual background. Since the goal of the descriptive record was to render the totality of behavior in its context, there was no restriction on what should be included. A recorder did not wait for a certain behavior to appear before making a notation but "wrote steadily whatever she saw, described as accurately as she could." If the pace of activity overwhelmed the recorders, they were given leeway to summarize a set of actions and to note the amount of time that had been missed.[91]

Each observer was given basic guidelines "to help the recorder focus her attention and select her content." Instructions for the lunch period, for example, advised that the social context be noted (that is, which children were at the table with the subject); what the quality of the relation of the subject was to the group (was the child a "listener only, participant, leader, withdrawn, ignored by others, response elicited by others"); as well as recording what topics of conversation occurred and what verbatim contributions were made by the subject. The child's physical behavior was to be recorded, in terms of "general body position, skill with tools, neatness, dispatch" and the attitude toward the food indicated.[92] In the analyses that made up the bulk of the text, summaries of the children's behavior as individuals and as members of a group were presented, as well as quoted excerpts from the contextual records. The longest descriptive report given verbatim was a sixteen-page annotated record of a group experience: the day the children went on strike for shorter rest periods.

Granting that the production of such contextual records entailed selectivity, the authors of *Child Life* remarked that they had chosen to focus their observations "in the directions which we considered most valuable," knowing that "some of the conditioning factors in any situation must inevitably be lost since there are so many and their interrelations are so complex." Such a lack of "control" on the part of the investigators – as well as such an acknowledgment of preferential selection – was anathema to the experimental elite. Nevertheless, the Bank Street researchers believed that "a method of recording which is contextual rather than classificatory and observations made in real, natural situations, are most likely to reveal the significance of behavior." In choosing a research method, they stated, the goal is to adopt a "method of observation most adapted to noting the full stream of experience in free, uncontrolled situations."[93]

In this concern with taking an "ecological" approach, Murphy's perspective paralleled that of certain naturalists during this era. Consider the similarity between Murphy's study of the "whole organism" under natural conditions – in which she presented long excerpts in protocol form of the activity of individual children that she had placed in cultural and social context – and the approach of Margaret Morse Nice, a pioneering ornithologist, during this same period. In her monograph, *Studies of the Life History of the Song Sparrow,* Nice explained that her method had been "almost entirely that of observation with a minimum of experimentation and no collecting, the hope being to find out what actually happens in a population of wild birds. The first year was devoted to intensive study of two pairs."[94] Nice characterized her work as "phenomenological" and insisted that "a necessary condition for success is a continuous sympathetic observation of an animal under as natural conditions as possible. To some degree one must transfer oneself into the animal's situation and inwardly take part in its behavior."[95] From Nice's perspective as a naturalist – and as similarly expressed by Carson during this same period – observational knowledge required more than an "objective" eye: it demanded intuitive insight as well.

At the time she published her researches, the French ornithologist Jean Delacour wrote that Nice's work was "perhaps the most important contribution yet published to our knowledge of the life of a species." Konrad Lorenz observed that Nice's research was "the first long-term field investigation of the individual life of any free-living wild animal."[96] Nice herself commended to her colleagues the advice of the British naturalist, Edmund Selous, who contended that a belief in "uniformity of action is in proportion to paucity of observation" and that "the real naturalist should be a Boswell, and every creature should be, for him, a Dr. Johnson."[97] But if Nice was clearly a member of the naturalist community, it should not be overlooked that she had been first a member of the psychological community as well: as an undergraduate at Mount Holyoke, she had majored in psychology and acquired an intellectual debt to psychologist Helen Thompson – the same figure who, under her married name of Woolley, was an inspiration to Lois Barclay Murphy.[98] During the 1920s, as she was publishing the results of her first researches on various bird populations, Nice also published studies in psychological journals on language acquisition in children.[99]

Another name for such "phenomenological" or "ecological" research, is, of course, "field studies." The concept of the "field" as used by social scientists during this era is one that has yet to be fully examined.[100] As discussed most typically in psychology, the "field" analogy is traced back to its roots in physics, as in the 1954 *Handbook of Social Psychology,* where Morton Deutsch's article "Field Theory in Social Psychology" – which was centered on the work of Kurt Lewin – is introduced with a brief survey of the conceptual contributions of James Clerk Maxwell, Christian Oersted, Michael Faraday, and Albert Einstein.[101] Although the field concept undoubtedly owes much to developments in physics and Gestalt theory, historians should not disregard the fact that the concept carries multiple meanings, one of which derives from the naturalist practice of conducting research "in the field." By the 1930s, in Allport's words, there had developed within psychology "a notable schism between the psychology constructed in a laboratory and the psychology constructed on the field of life."[102]

The naturalist concept of "fieldwork" was brought to prominence from an additional angle during the 1920s and 1930s, through the anthropological research of Franz Boas's students, published in such texts as Margaret Mead's 1928 study, *Coming of Age in Samoa: A Psychological Study of Primitive Youth for Western Civilization.*[103] The rise of fieldwork in anthropology during the first decades of the century paralleled another movement in which observers turned their focus inwardly upon America: the documentary photography of individuals such as Lewis Hine, whose pictures of the daily lives of child laborers, published in such periodicals as *Survey Graphic,* had a profound impact in reform circles. As Alan Trachtenberg suggests, work such as Hines's gave special emphasis to firsthand experience, and "upon the artist's role as *witness,*" as exemplified in Walt Whitman's declaration that "the true use for the imaginative faculty of modern times is to give ultimate vivifications to facts, to science, and to common lives."[104]

All these strands of field study informed actions taken during the 1930s at progressive colleges such as Sarah Lawrence to institutionalize fieldwork in the undergraduate curriculum, a movement that counted Lois Barclay Murphy as a vigorous participant.[105] Murphy persuaded the college to appoint Ralph Steiner – who, with Pare Lorentz, created the noted social documentary film, *The Plow That Broke the Plains* – to work with her in designing an observation course for her students.[106] "The idea," Murphy later explained, "was to learn how to use a camera well and then to have assignments to go into New York and photograph what you see."[107] These fieldwork excursions led to reports such as "Making a Living on the Streets," in which a Murphy and Steiner student interviewed and photographed shoe-shine laborers and fruit-sellers.[108]

William Stott argues that the documentary mode served as the "central rhetoric" of 1930s America.[109] The efforts of radically empiricist scientists such as Allport and the Murphys to represent the reality of the "individual-in-social-context" were directly related to the cultural ubiquity of documentary sensibilities. An important documentary imperative is the fostering of "identification," a process that was enhanced by the newly minted media opportunities then available. As Warren Susman notes, "the whole idea of the documentary – not with words alone but with sight and sound – makes it possible to see, know, and feel the details of life, its styles in different places, to feel oneself part of some other's experience."[110] In dealing, as Stott remarks, with "the almost unimaginable, the all but unimaginable – in short, the still unimagined," documentaries offered, in part, an answer to Thoreau's complaint that the "moderns cannot imagine so much as exists."[111] In the preface to his monograph, *The Use of Personal Documents in Psychological Science,* Allport commented upon the popularity of documentary films, sidewalk interviews, and magazines like *Life,* remarking that the social scientist, like "the layman," had become "caught up in the general cultural tide" of documentary expression.[112] The appearance of this cultural tide is one that Allport believed gave support to the radically empiricist views of experience that he had been advocating over the course of the 1930s, for documentary expression, like natural history, favored the concrete over the abstract, and temporal particularities over decontextualized universals.

Like Lois Barclay Murphy, Allport would also have to decide how to respond to the fact that such concerns evoked little interest from those in conventional experimental circles. In general, Allport was suspicious of knowledge claims derived from experimentation and statistical manipulation, owing to the decontextualizing ends to which scientists typically put such practices. In an essay on Dewey's thought, Allport stated approvingly that "when the laboratory wheels turn and the knives cut, and some exuberant investigator holds up an excised segment of behavior for acclaim, Deweyites are not edified. They know that true statements cannot be made about fragments snatched from their natural context."[113] This wrenching of fragments from natural contexts occurred on a number of levels; Allport was particularly alert to the practice prevailing in "general psychology of drawing the blood and peeling the flesh from human personality, leaving only such a skeleton framework of mind as is acceptable to the sparse canons and methods of nomothetic science." In violating the integrity of the whole organism, which happened "by stripping the person of all his troublesome particularities," all that psychologists bent on decontextualization had done, Allport contended, was to destroy "his essential nature."[114]

If experimental practice was inadequate to the task, how, then, were "essential natures" to be known? In an unpublished essay on "intuition as method," Allport argued that it is "direct apprehension which permits us to grasp the essences of things." Defining intuition as "that sympathetic attitude toward the reality without us that makes us seem to enter into it, to be one with it, to live it," Allport suggested that intuition "is the method of naive every-day knowledge. Its extreme difficulty arises only when scientists trained in fixed concepts attempt to employ it."[115] Elsewhere, Allport stated that, "considered as a psychological process, intuition might be defined merely as the tendency of a mind to structure its content into coherent wholes."[116] Scientists resisted adopting practices that would allow them to apprehend wholes, because they "fear it will make their thinking less exact." In Allport's view, if a loss in precision was occasioned by the use of "intuitive" techniques, it would be more than made up for by scientists gaining more scope and clarity in their quest to understand the natural world.[117] Restoring a sympathetic relationality between scientist and subject would thus restore individuality to view.

Whereas Murphy's response to the lack of fit between her research interests and the experimental techniques then in fashion was to work at a tangent to the mainstream – through the use of observational and projective studies – Allport initially chose a different option. Lacking the supportive environment that Murphy had solidified for herself at Sarah Lawrence, backed as it was with foundation monies, Allport and his work faced a greater danger of marginalization if he detoured around the experimental expectations of neobehaviorist members of the profession. Allport's answer was to use the apparatus of experimental methodology, but to modify it in such a way that it would validate values to which strict methodologists were hostile.

A case in point is Allport's 1933 text, *Studies in Expressive Movement,* where he attempted, through the manipulation of various pieces of experimental hardware such as the kymograph (a device which graphically records pressure or motion) to demonstrate the validity of "intuition," or the "tendency of a mind to structure its content into coherent wholes," by showing that the gestalt of individuality could be physically displayed in the laboratory.[118] Allport's intention in this work was to take a phenomenon that he believed was easily discernible in the "field of life" – personality – and to find a way to make it visible within the laboratory. If he could not convince his experimentalist colleagues to venture out into the contextual world of everyday realities, he would strive to bring an aspect of this larger world to them by making it materialize in the laboratory.

Allport saw the phenomenon of personality as a question of "form," a concept that had been developed by the Gestalt psychologists and other German scientists during the preceding decades. Allport suggested that this perspective was exemplified by thinking of any ordinary entity, such as a table. An object such as this, Allport stated, is "something more than so many billions of atoms of similar physical properties in a state of cohesion. The table possesses a definite structure, peculiar to itself – a *Dinghaftigkeit.* Such a structure is in essence something unique and non-analysable, and something which is more than the sum of its constituent parts."[119] In a like manner, a melody retains its essential structure, whether or not its pitch changes, because the relations of the parts to each other stays constant. What was wrong with the conventional practice of reducing wholes to parts – the analytical method – Allport explained, is that the *form* of the whole object is lost, and therefore the object itself. Allport's militancy on the matter of "form-quality" in regard to conventional experimental technique can be seen in the following comment: "If it is objected that one cannot *measure* pattern, form-quality, or style, the proper answer is, so much the worse for measurement."[120]

As early as 1924, Allport had determined that research into the structuring of personality would entail investigating "just what this *form-quality* is and what role it plays in the apprehension of personality."[121] Allport suspected that the place to begin was in the recognition that "the *Weltanschauung* of all men, however humble, is being continually revealed in their style, bearing, manner, gesture, attitude and speech"; this being the case, such things as "handwriting, gesture, postural sets, and manner of speech" should be "active carriers of the form-quality."[122] Whereas American psychologists had little interest in this line of thinking, this was not true for the German context, as Goodwin Watson reported to the profession in 1934. Watson explained that it was customary for German psychologists to assume that gestures, for example, were bound up with personality "in such a way that no kind of person except this one kind could make this particular sort of gesture or normal handwriting." Watson pointed out that, "as a consequence of this point of view, the relation of personality to physique, to motor expression, and to intellectual productions of all sorts has been more investigated by German psychologists than by our own group."[123] Thus, although Allport, in studying expressive movement, was bending American experimental practice in a direction to which it was unaccustomed, he was not working without any visible means of support. In the topic of expressive movement, Allport believed that he had hold of a theoretical lever that, when applied to the ritual customs of the experimental laboratory, would move American experimentalists to acknowledge – as had their German counterparts – the scientific reality of concepts such as individuality.

Allport proposed that personality could be studied in the "most direct fashion" by examining expressive movement, which he defined as "those aspects of movement which are distinctive enough to differentiate one individual from another."[124] Allport took care to point out in his introduction to *Studies in Expressive Movement* that he was not treading down what he characterized as the tired road of the psychology of individual differences, but was instead pursuing the psychology of personality. Allport argued that what had passed for the psychological study of individuals had not "treated *individuals* at all, but merely differences between reaction times, fatiguability, intelligence, or similar functions abstracted from the total mental life of the individual." Questions of structure and organization in regard to individuality – the true focus, Allport held, of the study of personality – had yet to be explored.[125]

Allport administered a battery of physical tests to his subjects in which he compiled such measures as walking and strolling speed; the drawing of shapes estimating various standard objects such as coins and dollar bills;

estimations of distances and angles; speed and pressure of finger tapping, leg tapping, and writing; and muscular tension under various conditions.[126] Although a number of mechanical instruments were used to record these data, many of the experimental situations were extremely simple in design. For example, writing pressure was determined by interleaving ten sheets of plain paper with ten sheets of carbon paper, placing a single piece of opaque paper on top, and then pinning the set to a drawing board in such a way that the underlying papers were hidden. The researchers then determined the pressure of the impressions by counting how many underlying pieces had been penetrated by subjects copying passages of written material, writing their signatures, and so forth. In other experimental measures assaying speed and pace, subjects were asked to read various excerpts aloud in a natural manner, or they were timed as they did such mundane activities as walking away from the laboratory door. In a session that studied size and distance estimation, subjects were asked to close their eyes and to move their hands apart to a distance they believed to be equal to two feet.

Within the plethora of quantitative data they had compiled, Allport, and his student Philip Vernon, were confident that their statistical results revealed a telling degree of correspondence between the measures taken for each individual. They went on to argue that "wherever reliability or internal consistency is found, we have a presumption of some kind of harmony or integration in the expressive behavior of the subjects."[127] Through such means Allport intended to persuade his readers that "the discovery of well-integrated and consistent expression in the motor field would establish a presumption that similar patterning is to be expected in all aspects of personality."[128] In limning a picture of individual mechanical men, Allport, of course, was seeking to show the soul within the machine. In his own mischievous words, what he proposed to paint for his neobehaviorist colleagues was "a psychomotor portrait" of individuality.[129]

The orthodox tone that Allport was hoping to strike can be glimpsed in a review of *Studies of Expressive Movement,* which reported that "the emphasis in the whole volume is upon experimental technique, and it is certainly an interesting approach to a problem which is very largely slighted by American psychologists." The reviewer noted that "the approach is a behavioristic one which ought to appeal to many psychologists who have objected to the methods of impression and introspection which heretofore have been used in the study of personality."[130] On the whole, though, Allport's effort at rehabilitating experimental technique met with profound indifference; for those for whom experimentalism was a rigorous ideal, Allport's modest experimental manipulations were of little interest, especially since they were employed in the service of advancing a radically empiricist point of view uncongenial to the restrictionist frame of reference adopted by these same experimentalists.

Borrowing the cloak of neobehaviorism could not disguise the fact that, in the end, Allport was using the favored tools of his opponents to try and induce them to sanction concepts that they felt were unscientific. As a sympathetic reviewer of *Studies in Expressive Movement* observed, it is "in the interpretation of the results that the authors, presumably Allport in particular, take up the cudgels for a view which is not popular in American laboratories, viz., the doctrine that our expressive behaviour is conditioned by the structure of our personality." The reviewer found Allport's "pointed" criticism of mechanistic and reductionist practices "unanswerable"; proponents of such methods, however, simply ignored the experimental conversation that Allport was trying to initiate with this volume.[131]

Like Lois Barclay Murphy and Allport, Gardner Murphy was impatient with the experimentalist presumption "that the important variables in behavior are to be found there in the laboratory, waiting to be controlled, and that if life reactions belie the laboratory, the trouble is somehow with life reactions."[132] Murphy insisted that experimentation should not be thought of as the "pathfinder but the crowning completion of long, arduous, and penetrating analysis." Such was especially true, Murphy insisted, "when the uncontrollable variables so vastly outweigh the controllable ones, as is the case in astronomy, geology, and our own field of social psychology." In establishing the age of the Grand Canyon, for example, Murphy pointed out that "experiments in uranium decomposition are critical exactly because thousands of hours of systematic observation have paved the way."[133]

During the late thirties, Murphy offered the work of his student, Muzafer Sherif, as an example of enlightened experimentalism. Sherif was interested in probing the processes by which individuals and groups, when faced with situations which are unstructured and ambiguous, produce standardized norms. In his effort to study the emergence of norms under conditions of fluidity and uncertainty, Sherif's intention was to take a naturally occurring phenomenon in the public world and to induce that same phenomenon in the laboratory. Sherif chose to design his experiments around the "autokinetic effect," which refers to the apparent erratic movement of a stationary light source in an unilluminated room. In a variety of experimental situations Sherif asked individuals and groups to come to a decision about how far the light had "moved." In his experiments, Sherif found, for example, that when individuals who had first experienced the stimulus effect alone were transferred to group settings, they abandoned their personal frames of reference for the norm that emerged as the group's frame of reference – and retained this norm as their own when subsequently tested again when alone. Sherif argued that his research showed how "our experience is organized around or modified by frames of reference participating as factors in any given stimulus situation."[134]

Murphy found the design of Sherif's experiment to be crucial to the production of what he considered to be valid and legitimate results. It was not that the experiment was analogous to an event in the larger world: it was that the experiment was, in its essentials, that same event, incorporating its primary characteristics. Murphy argued that "the laboratory must be utilized in such a way that it is recognized that "the important thing is that a problem in the world outside, a cultural situation, be carefully analyzed and an experiment framed which embodies all the essential elements of the cultural situation"; this, Murphy argued, Sherif had achieved. Such a research effort, was, Murphy stated, "exactly what the astronomer does. He brings the solar system, the galaxy, the island universe into his observatory . . . He makes crucial tests, again travels through space with his telescope and spectroscope, to see whether the laboratory findings illuminate what is beyond his experimental control." Similarly, the experimental practice of the geologist was such that, "in a sense, the earth is brought into his laboratory."[135] With efforts like Sherif's pointing the way, Murphy argued that psychologists could now join their counterparts in disciplines based on natural history in doing likewise.

Murphy emphasized that social psychologists needed to pay special heed to the contextuality of human behavior. He argued that "a man brings into the laboratory not only his eyes and his muscles, but his prejudices and his aspirations; his attitude toward the experimenter and toward experimental psychology in general; his fear of not giving a desired result and his fear of disapproval." In short, what any individual brings into the laboratory is "his culture, or as much of it as he can drag through the door." Any "social psychology which would strip him of his culture as he enters," Murphy maintained, "would be no social psychology at all."[136] Experimental practice thus needed to be reconstructed so that the social psychologist would be "capable of learning at the same time both from culture and from the discipline of the laboratory."[137] The lessons offered by the "discipline of the laboratory," if divorced from cultural context, would be for naught. Experimentalists, Murphy contended, should operate under the assumption that they must bring the world *inside* the laboratory, as he believed Sherif had done in studying the process of norm emergence under the influence of frames of reference.

Murphy's attitude toward the "discipline of the laboratory" was a complicated one, both approving and wary at the same time. Sherif's work offered Murphy the best of both worlds: the fact that Sherif's experimentation focused on the "autokinetic effect" made it an authentic example of current experimental practice, while his focus on the "frames of reference" of his subjects demonstrated the inextricability of social fields from perceptions of reality. Such a double-edged argument was precisely what Allport had sought to achieve in regard to individuality in his *Studies in Expressive Movement.*[138]

While pleased at such a favorable occurrence as Sherif's research, Murphy nevertheless was adamant in arguing that his peers insisted on too restrictive an understanding of experimental method. For example, the attempt by the Murphys in *Experimental Social Psychology* to render observational methods and experimental technique equivalent was one that many of their readers disregarded. In their 1937 revised edition, the Murphys noted with regret the fact that "the publication of our book was unfortunately assumed to be further evidence that the experimental method must always come first and that all problems must fall willy-nilly into a form recognizable by the laboratory worker." The Murphys advised their readers that, this time, the "book is not to be 'written around' experimental method as such"; rather, "experimental method is to be used in such a way as to illuminate our most pressing problems." The authors explained that "one of our most serious duties now is to point out that many important contributions to social psychology since the publication of the first edition have been made by men and women who have seen human problems broadly in historical or comparative form – for example, ethnologists and psychiatrists who have seen in the growth of personality vital processes which they feel obliged to describe, whether they can control and measure them or not."[139]

Although Murphy was eager to make use of experimental and quantitative methods, he was always equally insistent that there was no reason to give them investigative priority over qualitative methods, especially if the phenomena under investigation were not such that they could be reproduced by following the tenets of "rigor" as insisted upon by neobehaviorists. Although it was important to "bring the world into the laboratory," it was not always possible to reconstruct the experimental situation in such a way as do so; if the laboratory disrupted the natural context in which phenomena were displayed, then the scientist would have to leave the laboratory and go out into the field. This concern of Murphy's held not only for the contextual and cultural phenomena in which he was interested as a social psychologist, but also for the psychic phenomena he was simultaneously engaged in researching. Murphy was only too keenly aware that psychical researchers were being held to the strictest possible standards of experimental proof by their critics, while trying to investigate activities that seemed to fade away under laboratory conditions.

For example, a critical methodological question with which Murphy was struggling in regard to telepathy was whether this phenomenon should be studied using "objective" symbols – such as cards marked with circles and triangles – or whether "qualitative" material should be used, such as pictures and poetry extracts. The card-style experiments were typical of J. B. Rhine and the Duke University researchers in the 1930s; the qualitative approach was favored by French researcher René Warcollier, with whom Murphy had worked since the 1920s. Using Warcollier's methods, Murphy and he conducted transatlantic experiments in which evocative stimuli were chosen, as in one trial in which the phrase "Oh, that this too, too solid flesh would melt" from *Hamlet* was used.[140] As Seymour Mauskopf and Michael McVaugh point out in their history of the growth of psychic research, although the use of neutral symbols such as geometric designs allowed an investigator to easily judge whether or not the subject had made a correct match, such methods would reveal little about how the mental operations that could be involved might actually work.[141]

In an essay on Rhine's research, Murphy expressed his impatience with the fact that psychologists had declined to challenge "the standard worldview" of contemporary science.[142] The standard world view of physics and biology did not include paranormal experiences – experiences that, as James had observed, were "capricious, discontinuous, and not easily controlled; they require peculiar persons for their production; their significance seems to be wholly for personal life."[143] Or, as Murphy later strikingly put it: "But psychical research – what a difference! It is Banquo's ghost at the feast, the pariah at the garden wall, the real threat to inner peace."[144] But Banquo's ghost could be vanquished in the laboratory, far from "the raw stuff of everyday telepathic impressions, as they occur in dreams or in the waking state." In his research in parapsychology Murphy was caught between the demands of his critics for experimental rigor and his own intuition that studying parapsychology "in the field" would be the more successful route, for there one could start from "one very simple, very fundamental fact: nature throws at us a great deal of material in the form of impressions, usually pictorial, regarding what other people are thinking." A natural history of psychic phenomena, in this sense, would "take nature as she is rather than as we think she ought to be."[145]

"Natural History" and the Hierarchy of the Disciplines

Before historians can begin to fully grasp what matters were at stake when scientists dissented from the reductionist strictures of their colleagues by adopting, in part, values and methods derived from natural history, they will need to gain a clearer idea of the standing of natural history within the larger culture. For those psychologists who repeatedly called upon the example of Galileo and Newton as justificatory shorthand for their endeavors, it is clear that an appeal to the natural history tradition carried little persuasive weight. But it would be a mistake for historians to conclude, therefore, that the minimal authority accorded by scientific purists to natural history was shared by the larger public. Marginality, to a certain extent, lies in the eye of the beholder.

For those who presumed that professional expertise derived from training and actions that occurred apart from the world of the lay public, it is not surprising to find that an activity associated with the public domain would be perceived as possessing minimal interest for those seeking to display scientific expertise. But for radically empiricist scientists such as Allport and the Murphys who aspired to speak as public intellectuals in the Jamesian mode, the significance accorded to natural history as a widely accessible mode of meaning-making within the larger culture suggested that its value be weighted differently. But beyond this general image, there were specific reasons during the New Deal era that natural history was understood as a powerful form of discourse, due to the fact that documentary expression had emerged as the primary vehicle for contesting and representing reality in 1930s America.

Over the course of the twentieth century, the scientific community has rarely accorded natural history the same prestige that it bestows on those sciences considered to be "basic" or "fundamental," such as genetics and nuclear physics. To some extent, this is because natural history has had the misfortune of being characterized by those hostile to it, which has led to a confused picture of its place within scientific cultures during the modern period. As historian Paul Farber argues, the prevailing image that scholars carry with them is one of "natural history as a discipline emeritus and its practitioners as dowager scientists."[146] Although there is evidence that a recent cohort of historians of science are rethinking this evaluation,[147] the image of the naturalist is not one to which most historians of science are alluding when they discuss twentieth-century "science."[148]

Naturalists themselves knew of their standing within the scientific hierarchy, perhaps better than anyone else. Writing on the "naturalist as a social phenomenon" in 1940, Francis B. Sumner remarked that the word "naturalist" aroused "feelings of antipathy or contempt in the mind of the modern biologist," and that it was "difficult to escape the popular caricature: a wizened, bespectacled individual, who spends his time sorting and naming pickled specimens or endless rows of insects – an eccentric whose chief outdoor diversion consists in sweeping the air with a butterfly-net or gathering snakes and spiders in his wife's cooking utensils."[149] Even less appealing than the designation "naturalist," however was the term "natural history," Sumner related, for it had acquired the connotation of "something altogether primitive and out-of-date – a pursuit suited only to children and amateurs."[150] A generation later, not much had changed, as is evident in Theodosius Dobzhansky's assertion that "naturalists are said to be doing classical or old-fashioned biology; the nonnaturalists are modern and up-to-date." In contrast to those who styled molecular biology as fundamental, "alleg[ing] that the molecular level of biological phenomena is the one in terms of which all other levels must be understood in 'new' biology," Dobzhansky instead suggested with some asperity that "with equal methodological justification it could be dubbed 'low level biology.'"[151]

Such dismissive attitudes on the part of scientists and those who study their activities accounts, in part, for the fact that relatively few studies exist of the mutually existing relationships between the social sciences and "natural history." This is not to deny the appeal that mechanistic metaphors, objectivist ideals, and the glamour of the "exact" sciences have held for sizable contingents of social science researchers. It is time, however, to take serious issue with the assumption that intellectual movements running counter to the "physics manqué" sensibility had nothing like the *cultural* authority of Newtonian mechanics upon which to draw. The devaluation of "natural history" that may occur at the hands of those at the laboratory bench or in the academy does not accord with the views of a significant cohort of individuals: the general public.

In fact, it is highly ironic that natural history, for commentators on American science, does not carry with it the cachet that laboratory experimentalism possesses, given that the identity of the United States as a nation derives in striking ways from the encounters of its citizens with the indigenous environment, and the interpretations that they have drawn from these encounters. As the intellectual historian Perry Miller indicates, symbolic representations of the United States as "Nature's Nation" served as powerful conduits of cultural meaning-making for a large part of the nation's history.[152] Expanding on this theme, Barbara Novak has argued that nineteenth-century nature worship, "despite its international complexion," was perhaps "more strongly nationalistic in America than elsewhere." To be "one nation, under God" was to be "one nation, under Nature," for God was seen *as* nature by large numbers of the American people. Novak suggests that Americans found, to their satisfaction, that "nature's text, like the Bible, could be interpreted with Protestant independence."[153] It is not surprising that natural history was the most widely pursued activity in nineteenth-century America, and that there was great public interest in America's natural inheritance. The widespread nature of this interest can be seen in the enthusiastic public response to the natural marvels presented by P. T. Barnum, in his American Museum in New York and in his "Great Museum, Menageries, Circus and Traveling World's Fair." The natural knowledge purveyed by Barnum is, as Neil Harris remarks, one aspect of what the citizens of Jacksonian America were advocating in the placing of "all authority – social, moral, aesthetic, even religious – in the hearts and minds of . . . the much-celebrated common man."[154]

But even if historians concede the significant cultural place that "natural history" can hold for the larger public in exemplifying the honorific term "science," they may still doubt that the term possesses any definitional power for the twentieth century. The conventional wisdom holds that "modern science" derives a large part of its symbolic prestige from the fact that the laity laud the achievements of technology, which are presumed to be the legitimate issue begat by the "hard" or "basic" sciences. The stature of a figure such as Thomas Alva Edison, then, in the eyes of the public, benefits the image of "science" as a whole, even though Edison – in present-day terminology – belongs to the category of "inventors" rather than to that of "scientists." Historians have yet, however, to recognize the elasticity that the classificatory term "science" has possessed among the general public during the twentieth century; because "science" as a "category" has acquired a certain rigidity within academic circles during this same period, we have lost a sense of the ways in which popular views of "science" function and what they signify.[155]

The plant breeder Luther Burbank, for example, today retains the image of an "amateur" – an image that is consistent with his association with the domain of natural history (natural history being held by professionals, as Sumner stated in 1940, as something seemingly "altogether primitive and out-of-date – a pursuit suited only to children and amateurs"). And yet Burbank the "plant wizard" joined Edison the "wizard of Menlo Park" in achieving enormous popular esteem as a *scientist* for much of the first half of this century: indeed, the "iconic" image of Burbank easily encompassed the conflation of Burbank as an "inventor," as a "naturalist," and as a "scientist." Natural history carried an authoritative cultural weight even as did technology; in representations of an individual such as Burbank, the two sets of metaphors could even be used interchangeably.[156] If one makes a serious effort to confront the changing cultural meanings that scientific practice held for the general public in America, one can then grasp some of the world that the image of Burbank represented to the ordinary citizens of whom this scientist is speaking in 1938: "When I explain to people that I am a geneticist, it means nothing to most of them outside of a very small scientific group. If I tell them I am a follower of Mendel, the group is only slightly increased. But if I explain to them that I am crossing plants such as Burbank did, their faces light up, and many of them begin to feel quite at home and in a position to talk intelligently about a world that they know very little concerning."[157]

It would be incorrect to draw a line that puts natural history on the side of egalitarianism and experimental science on the side of elitism: the historical permutations of each are so diverse that such a sweeping evaluation would be foolish. In nineteenth-century Britain, for example, the lessons of natural history could serve as a potent catechism on the righteousness of hierarchy.[158] Even so, it is clear that, in the American context, the methods and orientations of naturalists could serve both as an intellectual resource for working scientists, and as a symbolic representation of publicly accessible science.[159] Indeed, models drawn from natural history – scientific formats in which the laity "feel quite at home" – informed the democratic critique that Allport and the Murphys elaborated during the depression era. With natural history deriving its imprimatur more from the public world than did laboratory- and statistics-driven practices favored by those highly placed in the scientific profession's hierarchy, social scientists with an eye toward restructuring the ethical, social, and political basis of science found natural history orientations to be strategically useful resources in their disciplinary battles to represent the real.

6

Exploratory Relativism and Patterns of Possibility

> The recognition of cultural relativity carries with it its own values, which need not be those of the absolutist philosophies. It challenges customary opinions and causes those who have been bred to them acute discomfort. It rouses pessimism because it throws old formulas into confusion, not because it contains anything intrinsically difficult.
>
> Ruth Benedict (1934)[1]

Debates over the place of science in the American democracy turned in large part on the meanings invested in the terms "science" and "democracy" by the participants. As I have discussed in the preceding chapters, Allport's and the Murphys' definitions of science contained assumptions that placed their views at odds with those of their colleagues. So, too, with their understandings of democracy: if Allport and the Murphys agreed with many of their contemporaries as to what the social, political, and economic facts of democratic life were, their interpretation of what these facts signified could diverge sharply. This was especially true in regard to whether or not the general public could be trusted with the responsibilities required of a democratic citizenry. From within the radically progressive political framework of these scientists' views, the idea of "social control" meant the regulation of economic elites by the public; for many of their peers, the idea of "social control" meant the regulation of an irrational public by administrative elites.

At issue was the question of how far the nation had progressed in its quest to become a truly democratic polity. Activists such as Allport and the Murphys belonged among those who believed that American democracy was to be defined as an aspiration yet to be reached as opposed to a goal that had already been achieved. Union Theological Seminary professor Eugene Lyman expressed this view when he argued that "no one fully believes in democracy who is not expecting, and seeking to provide for, a new democracy."[2] But what would this new democracy look like? Would it depend on radical reconstruction, or only strategic modifications, introduced in a gradual way? What changes to the image of an American democracy born at Plymouth Rock were necessitated by such later American realities as the Mason-Dixon Line, Ellis Island, Ludlow, and Wall Street? Where now should the pursuit of democratic realities lead?

For many scientific intellectuals, the prospect of an indeterminate "new" democracy was unsettling. Even though scientists prided themselves on always planning for a "new" science, it was a new science in which innovations were expected to arise from within only a select group of "expert" thinkers, not from an ever-broadening array of diverse participants. The idea of a constantly regenerating search for a new democracy also played against the scientific ideal of the desirability of grounding knowledge in "universal laws," the intention of which was to subsume instances of multiplicity and unpredictability within the "higher" reality of generalized laws that revealed the uniformity behind *apparent* heterogeneity. In the question of "e pluribus unum," how was the emphasis on the many to be balanced against the emphasis on unity? Allport and the Murphys would downplay the scientific quest for universality by choosing to stress instead the claims of pluralism and of a never-ending obligation to test the validity of one's principles against a continuously changing reality. As a consequence, their theorizing most often displayed relativist rather than foundationalist lines of analysis.

Many social theorists believed the feasibility of a new democracy was tied to questions about the nature of the city within American society. Indeed, as a concept, the "problem of the public" was virtually interchangeable with the "problem of the city." In the twentieth century, as in preceding centuries, most American social theory was overwhelmingly critical of the modern city and urban culture, reflecting both the nation's tenacious agrarian myth and an uneasiness with immigrants.[3] As historian George Lipsitz has observed, "The problem of the city is the problem of difference," for within a city's boundaries, differences in class, race, interests, and taste are forced upon our attention. This is why, "as a stage for such confrontations, the city often takes on a menacing presence in the popular imagination." And yet, as Lipsitz concludes, "the city is also the site of mutuality and reciprocity, the locus of 'politics' in the best sense of the word. It is the place where people see their destinies as interdependent, where they fashion institutions for mutual advancement, and where collective imagination and effort create new possibilities."[4] For those scientists working from the premise that diversity was a natural reality, the multiplicity and unpredictability characteristic of the city was unproblematic: it was, in fact, democracy exemplified. The heterogeneity of the city was literally "natural" and was a matter for exploration, not eradication.

Cities were radically empirical phenomena, incapable of being contained within abstractions that impoverished the experience of vitality, variety, and unpredictability characteristic of large urban environments. Not surprisingly, James's experiential modernism was a framework well suited to this task, given his impatience with conceptual categories that rendered reality in static terms, in defiance of the fact that "perceptual life fairly boils over with activity and change."[5] Indeed, these words are not far removed from a 1907 description James gave of his experience of New York City, in which he wrote of vibrating with its "magnificent rhythms," and of reveling in "the courage, the heaven-scaling audacity of it all . . . the great pulses and bounds of progress so many in directions all simultaneous that the co-ordination is indefinitely future, giv[ing] a kind of *drumming background* of life."[6] Literary theorist Ann Douglas has recently argued that James's depiction "of American culture as a plural and heterogeneous affair of simultaneous affects" was one of the key intellectual sources feeding the transformative culture that emerged in New York City during the Jazz Age. Douglas casts James, who was born in New York City in 1842 and who spent most of his formative years there from 1847 to 1855, as a philosopher of the city, with New York exemplifying "his own vitalist and buoyant discourse, a complete epistemology of curiosity, motion, and experimentation."[7] Indeed, James's vision of a mutable and changing reality was an important aspect of his legacy; as Allport remarked, in James's thought "the universe seemed but loosely joined, 'filled with copulas.' The word 'and' abounds, trailing along, he said, after every sentence that is spoken. In such a universe we must expect mind to be many things, and truth to be many-sided."[8] This viewpoint offered the possibility of a real world that could be truthfully taken in many ways, by minds that James believed to be "at every stage a theatre of simultaneous possibilities."[9] Historian of psychology David Leary notes that James's belief in the "fundamental reality of alternative and supplemental perspectives permeated [his] entire system of thought." Likewise, philosopher Richard Bernstein highlights James's emphasis on the fact that "what falls within our field of vision" is always "more than we can articulate or capture in our conceptual schemes [because] concrete reality and experience are richer, more dynamic, and thicker than can possibly be expressed by our concepts." It is not, Bernstein points out, that our "concepts *distort* reality; distortion enters only when we slip into thinking that reality itself is exhausted by our descriptions and conceptual analyses."

For most intellectuals during the interwar years, however, it was not the city's possibilities but its perceived debilities that struck them as significant. "The public" appeared to these intellectuals as an irrational and dangerous entity in need of control; democracy needed to be protected *from* the people, and its operations could only be entrusted to experts. Such worried theorists proceeded to implement "a new structure or system of authority," forms of control that historian William Graebner portrays as "democratic social engineering."[10] A "participant democracy" was to be feared unless the "participants" were guided by an elite cadre of experts.

Indeed, the behaviorist strain in psychology can be seen as emerging from a sense of anxiety about the city and a quest for control over the strangers who inhabit it. Consider the following statement by a social scientist in 1928, who is positing that he is "a stranger scientist just down from some distant planet. I know nothing of human beings as they exist on this earth. Suppose, further, that I am in a balloon situated above the center of New York so that I can watch the city. . . . I see millions of people hurrying. . . . [T]he movements are rapid, confusing. There seems to be no more system in these movements of ants than when their nest has been disturbed."[11] The author of this thought experiment was John B. Watson, who was not, as David Bakan notes, a stranger from Mars, but rather a "country boy from rural South Carolina, looking on at the activities of people in the big city who are culturally alien to him." For those scientists who experienced the city as threatening and confusing, one response was to aim at predicting and controlling the conditions of urban life at their initial point of strangeness: individuals different from oneself. As Watson himself declared, "The most fruitful starting point for psychology is not the studying of our own self, but of our *neighbor's* behavior."[12]

Nor were such concerns exclusive to the social sciences. In *The Intelligent Individual and Society,* physicist P. W. Bridgman – the progenitor of "operationism" – confessed that his "desire to lead an intelligently well ordered life" had grown "to an almost physical intensity."[13] What caused Bridgman his greatest anxieties were his "relations with other people. The irrationality of the relations of people to each other obtrudes itself more and more." Because human beings were "irrational," Bridgman held, life needed to be "intelligently ordered," "apprehended with complete vividness," and "purified by scrutiny," in a process marked by "the greatest possible efficiency."[14]

The urge to control one's neighbor's behavior cannot simply be attributed to conservative political commitments. For example, in a review of leftist sociologist Robert Lynd's *Knowledge for What?* Harvard historian Crane Brinton pointedly asked social scientists how they intended to "reconcile their libertarian, equalitarian, and democratic traditions with their growing feelings of contempt for the ordinary American, the man who tunes in on Father Coughlin, reads *The Saturday Evening Post* or *True Story,* throws orange peels and cigarette stubs out of his Ford, and even votes Republican?" If, from a distance, Crane suggested, such social scientists loved "the people," their work nevertheless evinced a disgust with the rank-and-file's "vegetable ways of rooting."[15]

If Brinton overstated his case with Lynd, he did so only slightly. Lynd referred to "individuals struggling . . . in ant-heaps" and of "rootless people wandering from farm to city in quest of gain," adrift in a "chaotic institutional world too big for them." Lynd was greatly worried that the "gap between sophisticated knowledge and folk-thinking is so wide" and that the "curve of intelligence . . . reveals the sharp limitations on the ability of many of our people to learn a large number of complicated things." If this picture wasn't grim enough, Lynd sternly admonished his readers that these rootless, ant-like, bewildered individuals were "increasingly building gross, cliché behavior back into the culture."[16] Lynd maintained that it was "the intractibility of the human factor, and not our technologies, that has spoiled the American dream." This same intractibility was what was also causing the nation to drown "in a sea of disabilities." Lynd therefore exhorted social scientists to acknowledge that only "*active continuous planning and control*" could reverse the fact that "urban living represents one of the backward areas of our culture." Lynd argued that scientists must face the reality that "urban living operates seriously at present to confuse and to devitalize our culture," and that they thus needed "to discover ways to knit these loose population masses into living communities of interest, before this degenerating tendency renders the culture impotent."[17]

Not all interwar intellectuals were convinced that such alarms in regard to the citizenry were well placed, and Allport and the Murphys were among this company. Many of the voices raised in defense of the competency of the public were members of the Social Gospel and progressive education movements. In *The Educational Frontier,* theorist William Kilpatrick noted that "some among us assert that, irrespective of age, most people by nature cannot think reliably, and they further add that the trend of affairs is to leave all important matters more and more to the determination of experts." Such critics, Kilpatrick remarked, "call on us to renounce democracy in all its forms as now outmoded." Taking issue with this view, Kilpatrick stated that the position exemplified by those contributing articles to *The Educational Frontier* was "in conscious and intentional antagonism to that of these critics." Kilpatrick denied that "either the psychology of thinking or the distribution of native intelligence or the history of social and political institutions warrants the conclusion that the many should live in unthinking subservience to any group of experts however selected."[18]

Similarly, former UTS professor George Coe asserted that what lay at the center of discussions of "the public mind" was "the question, Have the masses the capacity for intelligent guidance of our social destiny? Is it fated that the few shall make the decisions, and that the many shall be compelled or induced to comply?"[19] Coe believed that to answer yes to this question spelled doom for democracy. Minister Harris Franklin Rall likewise judged that the question of "controlled social change" came down to a choice between two methods: "one is democratic, working from below upward, from within outward, relying primarily upon spiritual forces. The other is autocratic, working from above downward, operating from without, and relying ultimately upon compulsion."[20]

Such thinkers were indeed interested in "social control," as attested to by Walter Rauschenbusch's closing words in *Christianity and the Social Crisis:* "If the twentieth century could do for us in the control of social forces what the nineteenth did for us in the control of natural forces, our grandchildren would live in a society that would be justified in regarding our present social life as semi-barbarous."[21] But such individuals were concerned not with the control of the general *public,* but with the control of those who possessed the most economic power in society. Bishop Francis McConnell, in *The Christian Ideal and Social Control,* maintained that "a social force as powerful as patriotism cannot be allowed to run loose in the world," and that "the patrioteers, by which term I mean those who profit by patriotism, always know when the Christian ideal threatens their interests."[22] McConnell also commented on the assumption "in orthodox political circles that property ha[s] rights all its own over which society itself ha[s] no control." But McConnell argued that it was now being "recognized that the social force is a creator of the individual's property rights. Some values are created outright by the fact that people live together." As an example, McConnell pointed out that "living together makes necessary the building of roads for the most ordinary intercourse, and the road gives the leader in commerce or industry his extraordinary opportunity."[23] It was from such circumstances, then, that the public derived its right to exert "social control" over its economic elites.

Participation in the polity, for those who advocated the sort of "social control" favored by McConnell, most emphatically did *not* mean acquiescing in an already codified set of policy decisions as laid out by the directives of professional experts. Members of a truly democratic society would instead search *together* for a program that would only emerge in form as part of a process of debate. As progressive educator Boyd Bode asserted, "the primary concern of a democratic educational procedure is to stimulate a reconstruction of our beliefs and habits in the light of their mutual relationships rather than to predetermine the nature of this reconstruction."[24] The benefit to be gained from following such a course, Bode argued, was not that of a simple unification. Indeed, it was likely that such a project "would not lead in every case to the same kind of social out-look," but that "differences in attitude or points of view that exist among our population would tend to become more sharply accentuated and defined." Education for a new democracy would result in carrying "people further apart in some respects," for such discussions would accentuate "certain differences in points of view." However, such methods "would also do much toward cultivating common understandings and purposes."[25] The cultural demands faced by democratic societies, Bode maintained, were inherently unlike those that confronted societies based on classical forms of governance. Inheriting the tradition of democracy was "like inheriting a lawsuit."[26] There would always be new voices to be heard, and new claims to be adjudicated.

During the Great Depression, outspoken commentators in science, politics, literature, and the arts pressed forward a bold claim: that diversity was the bedrock of experience and not merely its surface aspect. Those offering relativist perspectives took variety to be a fundamental "fact" of both the natural and the social worlds, a fact containing moral and political lessons as well as scientific ones. Such a convergence is seen, for example, in the thinking of biologist Alfred Kinsey. Before embarking on the study of human sexual behavior in the late 1930s, Kinsey had first made a name for himself as an entomologist, specializing in the study of wasps.[27] Kinsey believed in the primacy and irreducibility of variation, and to demonstrate this principle he collected more than 300,000 specimens of the gall wasp family in a single one-year research trip.[28] Indeed, as Regina Morantz remarks, Kinsey was "so sensitive to individual uniqueness and the endless possibility of variation as fundamental biological principles that it is a wonder he managed to generalize at all."[29]

In Kinsey's wasp-derived taxonomic philosophy can be found the relativist conviction that holding variety as the fundamental reality of nature demanded new sociopolitical premises. In his study of insects, Kinsey found that some structural characteristics varied as much as 1200 percent, while "in some of the morphologic and physiologic characters which are basic to human behavior which I am studying, the variation is a good twelve thousand percent." Despite this fact, Kinsey noted, "social forms and moral codes are prescribed as though all individuals were identical; and we pass judgments, and heap penalties without regard to the diverse difficulties involved when such different people face uniform demands."[30] Kinsey urged Americans to recognize that the extreme variation that existed in nature falsified dichotomous thinking in terms of normal and abnormal and right and wrong, for there existed rather "a continuous curve on which there are no sharp divisions between normal and abnormal, between right and wrong." It was through differences that "nature achieves progress," Kinsey argued, resulting in "the world's diversity of plants and animals, and the endlessly different kinds of men in it." Any hopes of "a changing society" therefore lay "in the differences between men."[31] The seeds of progress would be nurtured by relativist rather than universalist thinking.

Unlike their more conventional peers, scientists who worked from relativist premises did not strive to "see past" the world's diversity to the presumably uniform entities of which it consisted and the presumably universal foundation on which it rested. In doing so, such social scientists participated in an ethnographic discourse that gained great visibility in 1930s America. American culture during this era, as Guenter Lenz argues, "manifested itself as a *multiculture,* as a culture that was characterized even more by variety, heterogeneity, tensions, and alternative traditions than by the strong drive toward national identity and consensus." Large numbers of cultural critics such as Ruth Benedict and James Agee, Lenz points out, "worked out radical new methods and strategies of cultural *critique* and ethnographic *writing* in the study of American cultures, in the plural."[32] What these new rhetorical forms were being molded to advance was a stance I believe can best be understood as "exploratory relativism."

Exploratory Relativism and Cultural Critique

Varieties of relativist theorizing flourished across the disciplinary spectrum during the 1930s, as in historiography, expressed in the perspectives of Carl Becker and Charles Beard; in "Legal Realism," as represented in the thinking of New Dealers Thurman Arnold and Jerome Frank; in the exegetical work of modernist schools of theology; and, most famously, in Boasian anthropology.[33] By the 1930s, in fact, Boas's students controlled the "emergent institutional power base of anthropology" in the United States. From this platform, relativist anthropologists entered into contemporary political debates and exerted interdisciplinary influence.[34]

Ruth Benedict's theorizing in *Patterns of Culture* served as a focal point for much relativist discussion during the 1930s and 1940s, and I will treat it as exemplary of a viewpoint I characterize here as "exploratory relativism." Proponents of exploratory relativism saw themselves as pioneering new theoretical grounds on which to build a diverse and interconnected world. Furthermore, they believed that they were exploring necessary first steps, not expounding a final point of view. In sum, by surveying other worlds, "exploratory relativists" sought to destabilize familiar and uncontested practices and ideas: their goal was, most often, to critique the *home* culture.[35]

The mission supported by such an outlook was to forge a new society that affirmed the pluralistic nature of experiential reality rather than denying it; the task at hand was to explore how judgments were to be made without recourse to the intellectual shortcut of relying on "universal" criteria that were the expressions of a circumscribed group of people. As Benedict remarked in 1932, "Ethics has always been an apologia for customary behavior, and to derive Absolute Right from our local schemes is to measure the cosmos by a provincial footrule."[36] Boasian anthropologists had long been using this form of argument in their polemics against imperialism and racist ideology. During the 1930s, however, this caution against hierarchical assumptions of nationalist superiority was also used in another sense, as a way to challenge complacent parochialisms and to focus attention on rationalizations of the status quo that served as barriers to meaningful social and political change at home. Such thinkers believed, as Fred Matthews observes, that "by spreading awareness of possibility, one could help to make the possibility real."[37]

The particular relativist moves from the 1930s that I will be discussing here sharply diverge from currently accepted descriptions of relativism that equate the term with the idea of "anything goes." Clifford Geertz, in his essay "Anti Anti-Relativism," argues that most common characterizations of relativism trade on definitions that have been framed by those intellectuals most hostile to it.[38] Whereas Geertz is contending that such loaded characterizations of relativism distort what is being argued in contemporary debates, I believe that such widely current representations also make it difficult to reenter relativist discussions from the past. Benedict's theoretical venture was, in fact, a "relativism" that did not flinch from moral, political, and ethical judgments.

Geertz has been one of the few people to note the "vein of iron" in Benedict's work, a temper bearing little resemblance to a framework attributed to her by "her followers and her critics . . . intuitive, gauzy, sanguine, and romantic." As Geertz understands Benedict's discursive style, it is the "juxtaposition of the all-too-familiar and the wildly exotic in such a way that they change places" – a strategy he terms "self-nativising."[39] By means of "self-nativising," Benedict intended to startle the reader into seeing "our forms of life" as the "strange customs of a strange people."[40] Rather than simply being marked by a superficially nonjudgmental "to-each-their-own" philosophy, Benedict's rhetoric, in the manner of Jonathan Swift, in fact possesses a biting satirical edge. Far from abandoning values, Benedict's work is a call to values, through the process of examining those groups living in worlds different from our own. From Benedict's perspective, to state that a custom had cultural moorings was not to sanction its presence as desirable. One of the cultural customs revealed by the inspection of American life was that of racism. Benedict's stance toward this culturally sanctioned turn of mind – and the rituals that accompanied it – was one of censure, and her activism against it was the most serious item on her relativist agenda.[41]

In her self-nativising rhetoric, Benedict sought to loosen the bonds of conventionality by demonstrating the arbitrary nature of cultural custom, and therefore the possibility of cultural revision. "It is in cultural life as it is in speech," Benedict stated in *Patterns of Culture:* "Selection is the prime necessity." Benedict argued that a culture exhibiting all possible traits would be as unintelligible "as a language that used all the clicks, all the glottal stops, all the labials, dentals, sibilants, and gutturals from voiceless to voiced and from oral to nasal."[42] Reliance on arguments from universal criteria had, to Benedict's mind, served as an intellectual dodge in the matter of social values, sanctioning the view that "culture" was something inherited, rather than something that was actively made.

Since selection of cultural values was unavoidable, Benedict asserted that there was no reason that it should not be done as consciously and forthrightly as possible. What Benedict's exploratory relativism was intended to challenge was the idea of the "impossibility of men's functioning without [their society's] particular traditional forms." Foreign cultures were not the only ones that could be based on interrelationships of traits different from those Americans found customary: so could their own, since cultures were the result of "the historical process of acceptance or rejection of traits."[43] The burden of choice necessitated the exercise of judgment. Writing during the depths of the depression and the rise of fascism, Benedict pointed out that cultures were always open to revision, but usually by the costly process of breakdown and revolution. By means of the "difficult exercise" of becoming increasingly "culture-conscious," Benedict hoped that Americans could train themselves "to pass judgment upon the dominant traits of our own civilization."[44]

In using ethnographic description as a means to self-nativise American citizens, exploratory relativists such as Benedict were advocating that Americans should enter into a culturewide debate that was no longer weighted on the side of a predetermined set of "universal" values that assumed a homogeneous society. Benedict urged Americans to each become anthropologists of their own society, and she challenged them to acquaint themselves with the diversity amidst which they lived and to construct new values informed by this circumstance, rather than positing an ideal reality and then ignoring the lack of fit.[45] By reevaluating and reweaving American patterns of culture, they could create a new democracy.

In the United States, this relativist discourse did not spontaneously arise during the 1930s but had been incubating throughout the Progressive Era. One earlier source for this mode of thinking can be found in the community of scholars who made up the "Chicago School of Sociology," a circle of intellectual thought that has been well scrutinized by academics.[46] Little remarked upon, however, have been the contributions of that community of intellectuals who made up what I term the "Vassar School of Social Critique," a circle that included such figures as English professor Laura Wylie; English and drama professor Gertrude Buck; history professor Lucy Salmon; and economics professor Herbert Mills.[47] These intellectual allies trained a number of women who would forge relativist cultural critiques during the 1930s, among them Benedict herself; Constance Rourke, a founder of American studies; Caroline Ware, who took a leading role in promoting the study of American cultural history;[48] and Lois Barclay Murphy.[49]

There is still much to learn about this particular concentration of critical activity. It is clear, however, that this cohort of students absorbed lessons in egalitarian politics along with critical theory. Wylie, for example, asserted in 1918 that the American people, "democratic in little more than name, have at least, till very lately, been in the main satisfied with an ideal of democracy as superficial as shortsighted; have considered its workmen chiefly as 'hands' and have seen in its workmen's children future employees rather than future citizens."[50] Egalitarianism was not only theorized about but also modeled, as when Salmon "set an example of democracy by returning, unopened, letters addressed to her as anything but 'Miss Salmon' and by objecting to the seating of faculty by rank at college dinners." In the student newspapers, Salmon also argued against "the awarding of commencement honors to seniors with high grades and opposed the establishment of a Phi Beta Kappa chapter."[51] Much of this egalitarian ethic had been replicated at Sarah Lawrence, the progressive college where Lois Barclay Murphy taught. At Sarah Lawrence, as President Constance Warren explained in 1940, "the faculty functions on a democratic basis," with each faculty member holding "equal rank"; as a consequence, there were "no professors, associates or assistants, and, conversely, there are no underpaid graduate 'tutors.'" At Sarah Lawrence, Warren reported, "with a few exceptions, the academic faculty has had extensive graduate training and many of them are Ph.D.'s, but increase of salary, which is the only type of advancement recognized, does not depend upon degrees. No one is ever addressed as 'Doctor.'"[52]

Along with lessons in social and political equality at Vassar went an emphasis on teaching forms of social critique with a relativist cast. Salmon argued in *What Is Modern History?* that "it must be self-evident that the facts of the past are as numerous as are the sands of the sea, that in and of themselves they have no stable or intrinsic value." Salmon asserted that facts derived their importance "from their combination and recombination with other facts," for any one fact could "have no value in combination with one group of facts and may have supreme value in combination with a different group of facts." Salmon further contended "that an endless succession of combinations and re-combinations of kaleidoscopic variety is not only possible but inevitable in all descriptions of the present."[53] In English, Buck championed "a vitalized, democratized conception of literature," in which a book was never again to be viewed as "a barren, finished product, a scholastic abstraction, but a living activity of more than writer and reader, a genuine function of the social body."[54] The critic's valuation, Buck argued, could never be regarded as having "more than a present validity and a relative truth." For example, Buck contended that a critic's estimate of a book such as *The Swiss Family Robinson* as a child "probably differs widely from his grown-up verdict upon it. But his second judgment is not necessarily a truer judgment than the first, nor the first than the second. Each opinion, if indeed it is not a mere parroting of other people's ideas, but honestly his own, is as 'true' as the other – and no truer; since each precisely records the value of the book to him at a given stage in his development."[55]

What students such as Murphy, Rourke, Ware, and Benedict brought from their Vassar educations to their work during the depression years was a commitment to searching for what Rourke called the "social forces beneath rhyme and rhythm and metaphor."[56] All would start from Benedict's premise that "life cannot be seen by an unmoved spectator," and, as she expressed it in *Patterns of Culture,* that no individual "ever looks at the world with pristine eyes."[57] In the strands of "exploratory relativism" that could be found in certain aspects of these women's work during the 1930s, they were asking their colleagues to acknowledge the culture-laden nature of their endeavors, and to establish new modes of inquiry that took into account the possibility of multiple truths. For such cultural critics during the 1930s, the concept of "the immigrant," for example, was used as a symbolic representation of the relativity of knowledge claims. Looking at America with an "immigrant's eye," as opposed to scrutinizing immigrants with an "American eye," was a reversal that was possible only for those thinkers who were ready to build a new America from the point of view of an urban future as opposed to that of a rural past.

"We Are All Immigrants"

The intellectual and social dynamics of relativist thinking in the United States during the interwar period were embedded within the nation's increasing awareness of the polyglot character of American society. While the percentage of aliens was small, there was a large immigrant second generation. Together, first- and second-generation immigrants totaled forty million, or one-third of the white population by 1930.[58] In addition to the large immigrant population, African-Americans were increasingly pressing their claims for full citizenship in the American polity.[59] The heterogeneous nature of American society was especially obvious in the nation's large cities, where the majority of the American population now resided. Questions of cultural definition and ethnic democracy became prominent, encapsulated in a common query: What did it mean to be an "American" in such a diverse society?

Whereas the 1910s and 1920s had seen much anti-immigrant sentiment, during the 1930s, as Richard Weiss indicates, "many Americans changed their attitudes toward minorities and their relationship to American culture." Although racism survived, its acceptance declined, and "by World War II many had come to view ethnic variety as the essence of American nationality." The federal government's Office of Education even issued a "philoethnic" brochure that declared: "*We are all immigrants.*"[60]

Louis Adamic, a Yugoslavian immigrant, became one of America's most noted social critics during the 1930s, and a symbolic personification of America as a multiethnic nation of nations. Social reformer Carey McWilliams, in a pamphlet published in 1935, characterized Adamic as a man who owed his keen interpretations of America to his immigrant status. "The immigrant," wrote McWilliams, "not only comes to us with an idea of what America is like but he quickly comes to form a working notion of America based upon his experience." McWilliams argued that immigrants were thus forced to engage in an intense appraisal of their new world, and in the process must of necessity try "to formulate a concept – America."[61] This process thus allowed immigrants to obtain a clearer and more accurate "picture" of America than was available to "native stock" Americans who did not consciously reflect on the nature of the society in which they lived. It was to Adamic that a frustrated Franklin Roosevelt issued a dinner invitation, in the hopes that he could explain to Winston Churchill that the United States was not an Anglo-Saxon nation.[62] Those citizens who before had been encouraged to disappear into the "melting pot" were now being accorded a special warrant for speaking to the question of American identity.

In the 1930s, the idea that knowledge of such matters as the pilgrims and western expansion was not enough to understand what McWilliams called the "character" of the country became widespread in academia. In *The Cultural Approach to History,* Ware credited Adamic with promoting the realization that "it is to Ellis Island rather than to Plymouth Rock that a great part of the American people trace their history in America. More people have died in industrial accidents than in subduing the wilderness and fighting the Revolution. It is these people rather than frontiersmen who constitute the real historical background and the heroic tradition of the mass of urban Americans." For Ware it was in the still unexplored history of nondominant cultural groups that the "dynamic cultural frontier of modern America" would be found.[63] As Weiss observes, by the 1930s, the melting pot had become "a more apt symbol of exploitation than assimilation. . . . [I]t was necessary to see the immigrant not as an object to be transformed, but as a contributor to a culture still in the process of definition."[64]

As this intellectual reevaluation of the meaning of cultural heterogeneity for American identity was played out during the depression era, the documentary movement – especially as nurtured by the various New Deal arts programs – served as a central channel for communicating and explicating these new ideas. In many instances, those who participated in the various arts projects were engaged in domestic ethnography, asking, "What is an American?" One commentator remarked that the WPA's documentation of America resulted in "a sort of road map for the cultural rediscovery of America from within."[65] What was discovered, in William Stott's words, was that a belief in a "standardized America" could no longer be maintained, for the constellation of documentary projects revealed that there was "no entity to call America." It would be more accurate to say that there were regional Americas, except, as Stott remarks, "if one looked hard enough, the regions gave way and one had communities – which themselves became, on further scrutiny, classes, factions, groups." This documentation of the nation had "turned up such an abundance of what one educator called 'localized information' that no generalization with teeth or vigor held. Each town became so unique that the main thing that joined it with the next was the road."[66]

The arts projects struggled with aspects of American culture that had previously been given scant attention, such as the experiences of African-Americans. A text such as *The Negro of Virginia,* part of a series of field studies initiated by Howard University English professor Sterling A. Brown, who had been appointed "national editor of Negro affairs" for the Federal Writers' Project, was unprecedented. Another arm of the Writers' Project collected more than 200,000 oral histories from former slaves, excerpts of which were published as *Lay My Burden Down: A Folk History of Slavery.* In New York City, twenty-seven writers formed the "Living-Lore Unit," their task being to interview taxi drivers, longshoremen, and garment workers, paying particular attention to what such informants could tell them about their ethnic culture.[67]

The philoethnic stance seen in the WPA projects had been nurtured in part during the previous decades by the thinking of a number of young intellectuals, who, as Matthews notes, perceived "in the immigrant subculture a richer, more human alternative to that Anglo-Saxon Americanism from which they felt themselves increasingly alienated." Rejecting the idea that "ethnic communities should integrate in[t]o a mouldering log cabin," Matthews argues that such figures as Horace Kallen and Randolph Bourne saw "in the transplanted immigrant cultures the hope for a more colorful, varied and creative national life in the future."[68] Cultural rejuvenation would thus spread outward from the nation's metropolitan immigrant enclaves, rather than from the small towns of the inner "heartland."

Lois Barclay Murphy displayed this same philoethnicism in her own work. In reflecting on her life, Murphy recalled how, "as a child in Chicago, I used to leave early to go to school so I could stop in on children of immigrant families who had fascinating things for breakfast like strudel or sausages and piccalilli instead of the boring cream of wheat and muffins that my mother thought were good." For the young, white, Anglo-Saxon, Protestant girl, the ethnic children seemed to have not only a more exciting home life but also a more intimate and loving one. Remembering her envy, Murphy remarked that "their mothers and fathers held them on their laps even when they were six or seven, and I was supposed to be a big girl. Even if they didn't have a

Steinway piano or thousands of books as my parents did, I didn't think they were so deprived."[69]

Murphy echoed such early sentiments in her later work as a psychologist. In a 1936 article on emotional development, for example, Murphy questioned her colleagues as to why there existed so much discussion of "traumas," and so little discussion of "the emotionally shallow person" – a personality type that many philoethnic intellectuals attributed to middle-class America – whose "tragedy may be even greater than that of the scarred victim of emotional tragedy."[70] To combat this plague of emotional shallowness, Murphy advocated that those who nurtured young people should be "stimulating and encouraging spontaneous responses" in their charges. Similarly, Murphy counseled that youngsters should not be pressured "to conform to a pattern," or overprotected from emotion, "and this means avoiding rigid routine." Perhaps most important, Murphy believed, was the necessity of "allowing the child to have warm personal relations with vividly responsive personalities."[71]

In Murphy's evaluation, the fact that American homes and nursery schools were currently lacking in such values was partly due to society's fear of the immigrant. Indulging in racial stereotyping to make her point, Murphy argued that "perhaps the greatest charm of the United States is that which has come from the variety of mood and feeling it has drawn from different nations; it includes the hot-headed Italian, the shrewd Scot, the gay Irish, the warm German, the dignified Briton, the vigorously intellectual Jew, and the quickly rhythmic Negro." Stating that "culture which has been fed from these streams cannot grow by an arbitrary channeling or stereotyping to emotional responses," Murphy lamented that what one heard most about in nursery school circles were concerns about "adjustment, freedom from emotional reactions, [and] well-ordered routine." To Murphy's mind, middle-class nursery schools and homes were inculcating in children an "absence of intensity in all its forms."[72] Those who would be counted among the healthiest of America's future citizens would be those who had absorbed the values characteristic of immigrant life.

Murphy's own passionate idealization of ethnic family life was one she returned to again and again. Later in the decade she sent blank survey forms to teachers from a diverse array of communities who had agreed to cooperate in sharing information about their pupils. At the 1939 national convention of nursery school educators Murphy reported on the "exciting material which has come in from the Tennessee hill country, from large cities and smaller factory and mill towns in Ohio, from migratory camps and a Chinese nursery school in California, from a nursery school in Boston and other sections of the country." Murphy contended that, "if we sift out the constructive and destructive influences of which this material gives us a picture, we find, over and against the strain, frustration, isolation, constraint of the fledglings from industrialized insecure or suburban well-to-do families, certain recurrent echoes of healthy family life." For positive results, Murphy pointed to the findings contained "in the records of the children from the Tennessee hill country, from Chinese-American families, [and] from Bulgarian-Italian families," where the alert psychologist could find examples of "healthy family life in those groups whose own tradition of warm family living has not yet been poisoned by the drug of 'status' and of economic worry, or shattered by patterns enforced by adjustment to factory and mill or office life."[73] Murphy's moral to her audience was that "zeal for American patterns of behavior and manners" must not be allowed to destroy the "springs of vitality and sources of warmth in family experience" that existed in the lives of those who did not conform to the values demanded by the consumer economy, bureaucratic rationality, and industrial Taylorism.[74]

Murphy shared this attitude with Ralph Steiner, the documentary photographer with whom she taught a course in observation at Sarah Lawrence. In recalling the thirties, Steiner related how his attitudes about immigrants caused him to come into conflict with socialist orthodoxy, leading to his being fired from a Communist-dominated photo league in which he was teaching. For their first assignment, Steiner asked his students to record with paper and pencil "what they saw in the seventies and eighties on fashionable Park Avenue. Then they were to do the same on Park Avenue near Harlem underneath the New York Central Railroad elevated tracks. There, for block after block, pushcarts with food and clothing were lined up. There the poor did their buying." At the next session, to Steiner's dismay, his students one "after another read almost identical notes: 'Along expensive Park Avenue live the capitalist rich in unearned luxury and elegance. Under the railroad the downtrodden victims of the rich live in misery.'"[75] Steiner was just as provoked by his students who saw immigrants as abstract representations of class exploitation as was Murphy by proselytizers of middle-class Americanism who saw immigrants as crude and unhealthy. Steiner's response was to ask his class: "Yes, but what did you SEE?" The photographer told his students that what he had seen were "monotonous rows of expensive apartment houses [that] seemed a city of the dead. . . . The sidewalks were almost deserted . . . But uptown, under the railroad, life bubbled. Children chased each other under pushcarts. Mothers screamed at them and went on bargaining for merchandise. Two boys with sticks and garbage-can covers dueled like knights of old." Steiner concluded his parable with the remark that he was fired as a teacher because "I could not see the two Park Avenues through a thick volume of *Das Kapital*."[76]

Like Murphy, Steiner saw the two neighborhoods filtered through a lens compounded of anticapitalist sentiment fused with xenophilia. Images of immigrant life were used by such critics as a means to place "mainstream" values under scrutiny: to see Anglo-Saxon Americanism as the strange customs of a strange people. The intention was to literally bring forth a vibrant world of imagination and activity from what now existed in the image of a "city of the dead." But the image of "the immigrant" served as more than a rhetorical device that could be used to point out alternative values and to self-nativise the American public. As Ronald Goodenow observes, for many progressive educators and social science activists, "although they were not always explicit on this point, racism and the ways in which immigrants were Americanized were seen as problems in the sociology of knowledge."[77]

If, for example, one looked through an "immigrant" perspective at American history, then what constituted historical "knowledge" of the past could look quite different, as Ware pointed out, from what one gained from an "Anglo-American" perspective. By emphasizing the consequences of switching perspectives, as with the immigrant case, relativist-oriented intellectuals asserted that inquiry into what constituted knowledge was always accompanied by the questions of "Whose knowledge?" and "Knowledge for what?" As psychologists, Allport and the Murphys participated in this exploration of the issues raised by the sociology of knowledge in their discussions of the concept of "frames of reference."

Frames of Reference

The relativistic stances that Allport and the Murphys displayed in their theorizing and in their research stemmed from their taking seriously the idea of "frame of reference." All three of these psychologists shared a belief that knowledge is perspectival; in following through on the ramifications of this shared viewpoint, however, Allport and the Murphys would move in opposite directions. Allport pushed the idea of relativity to an extreme on the individual level, leading his opponents to charge him with creating a science without universals (and, therefore, no science at all). The Murphys, on the other hand, focused on relativity at the group level, stressing the cultural and subcultural contextuality of human activity, leading their detractors to censure them for insufficient commitment to the search for universal laws. While Allport and the Murphys placed their efforts at different focal points, their dissent, nevertheless, shared a belief in the experiential reality of a radical diversity that challenged orthodox understandings of the "scientific."

For Lois Barclay Murphy, the logic of cultural relativity extended to the preschool populations whom she was studying: she attempted to describe children's lives as a subculture, and an ethnographic sensibility permeates *Social Behavior and Child Personality.* Murphy's approach in studying the nature of sympathy in nursery school students was in fact more akin to Margaret Mead's anthropological research on adolescent female sexual behavior in Samoa than it was to the neobehaviorist research practices in her own discipline of psychology. Indeed, Murphy was careful to note that "the world which the child perceives is not the world which the adult perceives as he watches a group of children."[78] Murphy therefore placed her introductory comments on "the cultural setting" in *Social Behavior and Child Personality* in two separate chapters: "The World of Adults" and "The World of Children in the Nursery School." As Murphy explained in the introduction to the latter chapter:

> As soon as a child becomes a member of a nursery-school group, he becomes a member of a world quite different from that of the adults in average homes in the community. He becomes a member of a small world of children, all near his own age, although differences in age may vary with the arrangement in different groups. This world of children furnishes, from this time on, a major part of his social diet; he is constantly assimilating this, by direct imitation of other children, by spontaneous reaction to things which they do or to situations which they create, by resistance to this pressure, by repeated experience of tensions aroused by their mere presence.[79]

In the Bank Street text, *Child Life in School,* which was about a seven-year-old group, the authors developed this theme further, stating that they had "taken a small sector of child life and attempted to describe it in terms of a psychological community. . . . It has been helpful to us by analogy to think of this study as a sort of psychological 'Middletown.'"[80] The authors considered their analysis of these seven-year-olds to be that of "a culture pattern, a community of people living their group life under a given set of regulations with a given latitude of freedom within a conditioning ideology." This "culture pattern" was therefore "necessarily different from that of any other group of children in some respects, and likewise different from the pattern these same children would have evolved several years later."[81]

Recognition of the existence of "culture patterns" was of significance not simply for the study of *others'* behavior: the Murphys also insisted that their scientific colleagues recognize the cultural nature of the frameworks through which they themselves approached their work. The Murphys asserted that "one observes 'situations' and 'social behavior' through the colored glasses furnished by one's own civilization. After another twenty or thirty years of fieldwork by anthropologists – and by the psychologists of the U.S.S.R. – we may see more completely just how these principles will apply in a radically different kind of social order." Extending this theme directly to themselves, the Murphys mused that "perhaps even our fundamental principles about organisms in their social environment are merely useful generalizations which seem axiomatic to us because of the twentieth-century prevalence of Darwinism, Weismannism, Gestalt theory, mental hygiene, economic individualism, and faith in science."[82]

Allport also contended that human beings know things always through what he termed "frames of reference," and he underscored the necessary relation, therefore, of interpretation to understanding. The significance that Allport attached to the concepts of "interpretation" and "frame of reference" is shown by the fact that the name of Allport's *magnum opus* – *Personality: A Psychological Interpretation* – carried the first, while the title of his presidential address, "The Psychologist's Frame of Reference," highlighted the second. The "advent of dynamic psychology" had made it "common knowledge," Allport maintained, that "what is accepted as fact depends very largely upon the individual's sense of the importance of fact, each individual carrying with him convictions concerning what is important for him."[83] In stressing that factual knowledge was generated through particular frames of reference, Allport placed the *experience* of individual knowers at the center of his epistemology, emphasizing that the acquisition of knowledge was not simply an act of cognition, but of meaning-making. Allport held that what "are sometimes thought to be neutral-perceptual judgments, even the judgments of scientists themselves, are not wholly determined by an objectively established frame, but are entangled deep in the web of personality."[84] In Allport's view, objects of analysis came to "roost" on a researcher's "mental perches," and "if he is Marxist he sees significance in the class membership of the subject; if he is Kretschmerian, he bears heavily upon the implications of constitutional types; if he is Freudian, it is the toilet training or the Oedipus situation that captures his attention. More subtle, but just as effective, are other favorite theories: conditioning, frustration-aggression, cultural-determinism, compensation, functional autonomy."[85]

In his concerted emphasis on individuality, Allport matched Kinsey in philosophical commitment to diversity as a basic fact of nature. Allport's conception of "traits," for example, exemplified a Kinseyesque understanding of nature as constituted by variation. Allport summarized his determinedly heterogeneous view on traits in his 1936 monograph, *Trait-Names: A Psycho-lexical Study.* The bulk of the monograph consisted of a listing of 17,953 terms that, by linguistic convention, had been used to refer to traits, both historically and in the present.[86] As a feat of inductive empiricism, Allport's effort could hardly be characterized as controversial. What he had to say in his introductory essay, however, was.

Allport opened this piece on a relativistic note, remarking on "the tendency of each social epoch to characterize human qualities in the light of standards and interests peculiar to the times." As an example, Allport stated that the contemporary meaning of such terms as "devotion," "pity," and "patience" were not established "until the Church made of them recognized and articulated Christian virtues."[87] Allport concluded that "it is therefore certain that trait-names are not univocal symbols corresponding throughout the ages to fixed varieties of human dispositions." Instead, Allport argued, "they are invented in accordance with cultural demands, their meaning often varies, and some fall rapidly into disuse."[88]

If each "trait" was a cultural phenomenon, for Allport it was also a phenomenon that was *personal* and *unique.* Allport hypothesized that "no two men possess precisely the same trait. Every life has a unique history, and in the course of its developmental struggles it attains correspondingly unique patterns of mental organization."[89] Allport allowed that a small number of traits – such as "ascendance," "submission," and "perseverance" – "seem common enough to be regarded as comparable from one individual to another." Such "common" traits, Allport maintained, "represent modes of adjustment which most individuals are *forced* to develop (albeit in differing degrees) in our standard civilized environment." Even in commonality, however, Allport detected diversity, for "in reality, of course, each life lends to each trait – even to these 'common' traits – an altogether peculiar coloring."[90]

Allport's radically diverse stance was carried through in his theory of the "functional autonomy of motives," which he used to argue for the "unique patterns of mental organization" peculiar to each individual. In opposition to doctrines that "refer every mature motive of personality to underlying original instincts, wishes, or needs, shared *by all men,*" Allport proposed a theory of motivation that sought to tell how uniqueness comes about.[91] Allport contested the position of contemporary psychoanalytic theorists, for example, who, he contended, proceeded "from the assumption that the structure of all personalities is essentially the same," necessarily deriving as each one presumably did from the Super-ego, the Ego, and the Id, and the uniform mechanisms accounting for their functional interrelationships. In such a system, Allport stated, "it is believed that individuality emerges only from altering the stress and proportions of these basic structures and mechanisms. The *constitution* of personality is uniform."[92] Allport argued that "what motivates each person is not some element common to all individuals, but his *own particular pattern* of tensions."[93] Allport submitted that adult motives should be regarded as "infinitely varied, and as self-sustaining, *contemporary* systems, growing out of antecedent systems, but functionally independent of them."[94] In adopting such a principle, Allport believed that psychologists could abandon their search for the uniform "elements" of personality, and instead focus on studying "the capacity of human beings to replenish their energy through a plurality of constantly changing systems of dynamic dispositions."[95]

Such relativistic forays were too much for some of Allport's colleagues. In reviewing *Trait-Names,* J. P. Guilford protested that "a trait that is possessed by one individual and by him alone is of little scientific interest. Science looks for common elements or similarities and for common variables or differences. As science it is not interested in the unique event; the unique belongs to history, not to science." Guilford advised Allport that it was the proper job of "the psychologist as scientist to put order and significance into this seeming chaos; to reduce to the simple that which is complex; and to describe in economical terms the multifarious aspects of man."[96] Upon the publication of *Personality,* Guilford felt it incumbent upon him to restate his warnings, noting that in Allport's latest text his tone was that of "a revolt against science," and that "again and again he encourages a literal anarchy in the realm of description of personality."[97] Guilford declared his distress with Allport's "emphatic and repeated rejection of all old points of view for their efforts to find general laws that apply to all individuals or to find common properties or traits."[98] "If the history of science has meant anything," Guilford lectured, "it has shown a trend toward *common* observations among men and *common* terminology with which to describe and a terminology reduced to its simplest and most economical limit."[99]

While Allport was assaulting universality from one side, the Murphys were delivering their own blows from the other. In 1929 Gardner Murphy had speculated that "vast possibilities" not only remained in regard to "the discovery of more adequate intellectual instruments than those known to contemporary 'science,' but at least a reasonable possibility that our whole conception of quantities may, as we have long been told so earnestly, turn out to be just as relative as 'knowledge' itself."[100] Two years later, in stressing the inseparability of objects from the fields in which they were embedded, both Murphys accepted cultural relativity as a fact, going so far as to argue that the so-called "universal laws of social psychology" were always necessarily accompanied by an "all things being equal" rider. What are claimed to be "universally valid laws," insisted the two, "have the ring of generalizations about certain *special* conditions characteristic of *occidental* civilization in the nineteenth and twentieth centuries."[101] Elsewhere, Gardner Murphy noted that "the relativity of all scientific schemes arising out of the nineteenth-century middle class family has been more and more apparent."[102]

Gardner Murphy held that current research in social psychology was being hindered by "serious errors or limitations due to failure to see the individual clearly in his full cultural context."[103] As a consequence, Murphy argued, "social psychologists need to know, and to use, anthropology at least as well as geologists know and use chemistry."[104] In his editor's foreword to the new interdisciplinary journal *Sociometry,* Murphy asserted that knowing and using anthropology meant recognizing that the time had passed "when a sociologist, historian, or cultural anthropologist could describe the stream of culture as something disembodied, a stable or a shifting pattern of customs bathing the individual organism in its waters but moving on its course indifferent to the bather." "Culture," Murphy declared, was "a name for a special and complex type of biological reality, a name for very complex integrations of habit, attitude and value, never really stable, never really impersonal, sensitive always to individual caprice, accidental exaggeration or deformation, and the production of new emergents arising unexpectedly under certain circumstances of personal contact."[105] In such descriptions of the conditions of the cultural domain, Murphy had moved as far away as was perhaps possible from the stable and antiseptic laboratory world in which the majority of his colleagues wished to conduct research.

The Murphys even went so far as to render such relativistic views in what they could expect would be their most scrutinized text: their revised *Experimental Social Psychology.* Giving off echoes of Heisenberg's uncertainty principle, they maintained that "the wishes of the social scientist . . . *do* alter the way in which he behaves and therefore the subject matter which he and other social scientists study. . . . [T]o some slight degree [therefore] social scientists indirectly *alter the social order* they attempt to describe."[106] They suggested, however, that of even greater consequence was the fact that "to an infinitely larger degree, the wishes of a Lenin, a Mussolini, a Gandhi *change the 'basic laws'* operating in society." The consequences flowing from this phenomenon were dizzying: "If it is true that individual men can not only discover but literally *make* the laws of cultural development, and if the laws which they exploit in such creation of cultures depend on a complex interaction between their own personalities and social situations emerging from time to time, it seems very improbable that abstract laws . . . will have any usefulness and relevance."[107]

In regard to personality itself, Gardner Murphy similarly argued that it was "not a stable entity capable of being pinned to a table and analyzed," for personality "interacts constantly with situations in such a way as to make it difficult to talk about personality traits as inherent only in the organism. What is really inherent in the organism is rather a wide range of *potentialities* more or less unique for each person." Such potentialities were known, Murphy commented, "only when we test them. But here a paradox arises, for the test situation is itself a social situation, responded to in very different ways by different persons." Once again, Murphy called on the quantum theory analogy: "Just as the physicists' instruments for studying atomic structure alter the structure, so the psychologists' method of studying the potentialities of a personality changes the personality."[108] Thus, Murphy pointed out, "we never deal directly with an organism but only with an organism-in-an-environment." Citing his wife's work on social behavior in children, Murphy maintained that "it would, therefore, be a mistake to say that 'the personalities interact'; it would be nearer the truth to say that personality is being redefined from one situation to another."[109]

Would-be universalists, according to this view, must contend not only with the diversity of cultural circumstance, but with the constant *mutability* of culture, and of the supposedly stable entities within it. Allport also incorporated commitments to diversity and mutability that are similar to those of the Murphys in his work, as in his emphases on the unique structuring of each individual, and on the ever-changing nature of that structure: Allport's enlargement of the internal ecology of individuals essentially made them microscopic "cultures" within a macroscopic culture. Such arguments regarding the overwhelming presence of variety and change rested uneasily within a disciplinary discourse that sought to produce invariant laws of human behavior.

This is not to say that either Allport or the Murphys dismissed the possibility of universal laws in a dogmatic manner. At the very least, such a move would have been highly impolitic, given that much of science's power was held to derive from the revelation of nature's laws. Furthermore, Gardner Murphy knew well that the opportunity to explain psychic phenomena in a compelling way could indeed hinge on ascertaining the existence of some relationship between phenomena such as extrasensory perception and accepted physical laws. Allport and the Murphys saw no reason to prejudge the matter of universals. But neither did they see any reason to abandon their insistence on the reality of atypicality and inconstancy because others insisted on defining science as the quest for universals. To restrict theoretical and investigative activity within such narrow channels, they believed, would be to renounce the exploration of new realities.

In *Personality,* for example, where Allport discussed his theory of the functional autonomy of motives at length, he preceded his comments with a rhetorical turn, in which he imagined the response of his critics to the fact that "the stress in this volume is constantly on the ultimate and irreducible uniqueness of personality."[110] Allport envisioned the words of assorted upholders of convention as going something like this: "'But how,' cry all the traditional scientists, including the older dynamic psychologists, 'how are we ever to have a *science* of unique events? Science must generalize.'"[111] Allport's paradoxical answer both supported and undermined their view: "Perhaps it must, but what the objectors forget is that *a general law may be a law that tells how uniqueness comes about.*"[112]

The Murphys took a similar rhetorical tack in their revised *Experimental Psychology,* although they were more conciliatory toward their critics than Allport was to his. Well into their introduction they claimed to "hear the startled reader protest: 'What, must we worship forever at the shrine of particulars? Have we no place for universal laws, clear exceptionless principles?'" In response, the Murphys maintained that "we are indeed eager to find such laws, but we doubt whether our generation will live to see them established."[113] But given that a number of their other comments – as reviewed above – insisted on the priority of relativistic considerations, it would not be surprising if readers discounted this statement, suspecting that the idea of universality might be rendered irrelevant if scientists proceeded upon relativistic premises for any length of time. Such was the judgment, at least, of sharp-eyed upholders of scientific propriety. Yale psychologist Leonard Doob, in his review, expressed dismay with the Murphys' statements on cultural relativity, urging them to "abandon the deplorable, defeatist attitude they now display toward laws and principles," and stating in rebuke that "dodging the responsibility and discouraging others in this fashion is really dangerous and almost anti-scientific."[114]

Could laws, as Allport asserted, legitimately be about uniqueness? Could research be pursued without an obsession with universality, putting its imperatives aside to first take into account diversity and mutability, as the Murphys advised? Or was Guilford right that Allport's work was a "revolt against science," and Doob that the Murphys' theoretical stance was "dangerous and almost anti-scientific"? The answers to such questions depended, of course, on who was defining "science," and for what ends. For Allport and the Murphys, the cast of their science was constituted in large part by their understanding of the nature of democracy. Scientists such as Allport and the Murphys dissented from conventional scientific premises in part because such premises offered little support for recognizing variety, complexity, spontaneity, and change – a constellation of values that they believed defined democracy. Relativist-oriented thinking, on the other hand, presented a promising frame of reference from which to affirm the pluralistic aspects of experiential reality, and of the existence of multiple forms of truth.

For exploratory relativists, the goal of social science was not transcendence, but engagement. Such a view was an updating of the arguments of those such as Jane Addams, who had asserted in *Democracy and Social Ethics* that the individual taking "the betterment of humanity for his aim and end must also take the daily experiences of humanity for the constant correction of his process." Addams insisted that an individual embarked on such a mission would succeed only insofar as "he has incorporated that experience as his own," for it was necessary "to know of the lives of our contemporaries, not only in order to believe in their integrity, which is after all but the first beginnings of social morality, but in order to attain any mental or moral integrity for ourselves or any such hope for society."[115] Similarly, for exponents of a radicalized Social Gospel, social salvation could only be achieved by a national commitment to the expansion of democratic ethics to all spheres of the social world. What stood "against all exploitation by class or vested privilege, against those ancient social evils of slavery and the assumed inferiority of women and the defenselessness of childhood, against persistent feelings of race superiority and contempt for others," Rall declared, was the "conviction of democracy . . . that it is man as man that counts, not man as male or white or nordic or rich or wise, and that nothing less than the well being of all men can be the goal of social change."[116] Preparation for a new democracy meant placing presumed universals at risk by enlarging the terms of the debate. The radically empiricist premises of Allport and the Murphys' research brought these politically charged views to bear on psychological science.

Although it is true that neobehaviorist perspectives were pervasive in various sectors of the social sciences, it is equally the case that relativist-soaked social science possessed footholds of its own. The arguments that Allport and the Murphys put forward on behalf of such phenomena as "individuality" and "contextuality" must therefore be understood not only as expressions of an oppositional and minority sensibility but also as contributions to a discourse of considerable moment. Allport and the Murphys, in the course of their quest to capture the realities of "individuals-in-social-contexts," were traveling well within the contours of an intellectual migration that was leaving behind the search for eternal reality, and instead moving toward the explication of contingent realities.

To trust the democratic process to each other entailed risk and allowed for an uncertain future. In a letter to Adamic, Allport expressed his feelings on the haphazard nature of democratic change in the following way:

> Of course the rebirth of militant democracy is the only possibility of steering the social re[v]olution that is under way and of avoiding the horrors of all the alternatives. But one cannot produce a revolutionary movement synthetically, and until reading your book [*Two-Way Passage*] I had no real hope of a spontaneous rising of the people. Your brilliant work as spokesman of the non-Anglo Saxon American, seems to me *half way* toward the spontaneous uprising. It is still however, a bit synthetic and on the pamphleteering side, but I hope it will turn out to be a *prophecy* of a spontaneous rebirth of the democratic revolution on a world scale. . . . I am a bit fearful of the "synthetic solutions" in advance. The final form of the movement may surprise and even disappoint you and me. But no revolution writes its blueprint in advance. Only the Idea can be emphatically expressed, and it must be expressed again and again.[117]

Conclusions

> The world is full of partial stories that run parallel to one another, beginning and ending at odd times. They mutually interlace and interfere at points, but we can not unify them completely in our minds.
>
> William James (1907)[1]

How to confront the complexity of the natural and the social worlds was one of the most highly charged questions that the generation of Allport and the Murphys faced during the 1930s. Many of their colleagues sought to manage this complexity by reducing the subject matter they studied to the simplest terms possible, whether by searching for clues to the nature of human life in the behavior of rats or pigeons or microscopic worms, or by restricting the form of scientific questions to those that could be answered in the language of statistical regularity. But others, Allport and the Murphys among them, argued for the intellectual, moral, and political benefits of working one's way through complexity, rather than seeking to make it disappear.

As their efforts took shape over the course of the 1930s, these three psychologists produced a diverse body of work from within social and personality psychology that challenged dimensions of scientific life that the discipline's arbiters were striving to present as matters of settled fact. When Lois Barclay Murphy labeled her own work an "exploratory study," or Allport declared that an aspect of his research represented "a declaration of independence," or Gardner Murphy spoke of the need to "shut our eyes and jump as far as the structure of the human mind permits," they were each making plain that they considered the rules of the game to be open to challenge. In an era when the nation's leaders had pledged that "the pattern of an outworn tradition" would be cast off in order to initiate a "New Deal" for the American people, these three psychologists critiqued science and society in a similar spirit, in the expectation that they could persuade their peers to do what Franklin Roosevelt had asked the nation to do: join together in a program of "bold, persistent experimentation" in the pursuit of new realities.[2] Their belief that self and society were mutually constituted and continuously being remade led them to place themselves at the crossroads of the social and the personal during the 1930s, as they searched for new ways to understand the nature of such phenomena as individuality and contextuality.

As the nation was plunged into the destabilizing world of economic crisis and political uncertainty during the depression era, American science was experiencing a crucial transformation. What had been a fairly intimate set of professional networks had begun to expand into a more truly transcontinental web of participants as successively larger cohorts of recruits to the academy's ranks underwent training. Disciplinary life was marked by heated intramural debates regarding the grounds of knowledge as old alliances disintegrated and new ones were formed. Many scientists sought to mitigate or neutralize disruptions to the status quo by arguing for the need for stability, rigor, and hierarchy. Others such as Allport and the Murphys instead approached the unsettled character of scientific life during this period as an opportunity to contest previously agreed-upon assumptions about the nature of reality, the proper attributes of legitimate research practices, and the relation of science to society. As intellectual heirs of a previous generation of thinkers who had elaborated intellectual and social critiques, Allport and the Murphys introduced aspects of what I have characterized as experiential modernism and exploratory relativism into their scientific work during this period in ways that they hoped would contribute to new forms of democratic inquiry. As the decade drew to a close, their emergent critique had gained visibility, momentum, and support.

The 1930s represent a pivotal moment in the development of American science, with institutions, ideas, and social identities being juxtaposed in ways that unsettled custom and generated tensions that were productive as well as inhibitory. The era's social scientific knowledge was constructed in significant measure from the struggles over definition occasioned by these juxtapositions, and the academic cohort who came to scientific maturity as participants in these debates set in motion countercurrents to scientistic trends that would reverberate and ramify throughout the 1940s, 1950s, and 1960s. The audacity and creative intensity characteristic of American science during the 1930s has been overshadowed, however, by presumptions that World War II had such a profound influence on science that what came before is of little consequence for events that followed the war's end. Although it is clearly the case that World War II had a powerful impact on the career of the sciences, and that the atomic age contained distinctive cultural imperatives of its own, it would be historically shortsighted to consign the intellectual activity of 1930s science to footnotes.

Within the discipline of psychology, concerns articulated during the 1930s by psychologists such as Allport and the Murphys certainly continued to play out during subsequent decades.[3] Volumes by Gardner Murphy in the post–World War II period such as *Human Nature and Enduring Peace* and *In the Minds of Men: The Study of Human Behavior and Social Tensions in India* were continuations of internationalist and political concerns to which he devoted increasing attention during the 1930s.[4] The India study was funded as part of a UNESCO project, "Tensions Affecting International Understanding," headed by social psychologist Otto Klineberg – a close personal friend and colleague of Gardner and Lois Murphy from the 1930s whose research had attacked the idea of racial differences in intellectual ability.[5] African-American social psychologists Kenneth and Mamie Clark, whose work would achieve public prominence during the civil rights era, had studied with both Klineberg and Murphy during the 1930s. If psychology came to the civil rights movement in the post–World War II period, the movement came to psychology as well, as when Kenneth Clark chaired a distinguished lecture by Martin Luther King, Jr., at the 1967 American Psychological Association convention. King chose to speak on the "role of the behavioral scientist in the civil rights movement."[6] Allport contributed to this discourse as well, notably in his book *The Nature of Prejudice*, published in 1954, the year of the Supreme Court's *Brown vs. Board of Education* decision. It was a work that Clark introduced as "a classic" upon its reissuance at its twenty-fifth anniversary.[7]

Other themes developed by psychologists such as Allport and the Murphys during the 1930s also found expression in the forties, fifties, and sixties, as with psychiatrist Robert Coles's exploration of the subjective experiences of children, recorded in his *Children of Crisis* series. Like Lois Barclay Murphy, Coles presented children's views of what constituted reality as both a set of research results and as an acute critique of American social values. Similarly, the natural history model favored by Allport and the Murphys would be developed in idiosyncratic ways by Roger Barker and Herbert Wright, as they set out to simultaneously study the roots of democracy and to contribute to the literature on child development as they established the field of "ecological psychology" in the forties, fifties, and sixties.[8] Exploration of the relationships that exist between post–World War II developments such as these and their roots in the 1930s would help to restore to view the continuity of the period that extends from the New Deal era through the Great Society.

Nor did the debate against restrictionist neobehaviorist norms in which Allport and the Murphys had participated during the 1930s disappear at the decade's end, although the field of emphasis shifted in ways that have yet to be fully examined. If key neobehaviorist tenets were contested from within the fields of social and personality psychology during the 1930s, this critique was carried forward in later decades in such areas as cognitive psychology, humanistic psychology, and clinical psychology.[9] How this remapping of the intellectual landscape affected this debate, both as a discourse within psychology and outside of it, is a matter that historians are just beginning to consider. Here, it may be necessary to note aspects of continuity in order to appreciate differences in intent: where arguments from within social and personality psychology during the 1930s were aimed at restructuring the disciplinary commons of academic psychology as a whole, later developments may have sought instead to open up intellectual suburbs that would reduce the possibility of continued debate on divisive questions regarding theory, methodology, epistemology, and ethics. Cognitive theorist Jerome Bruner, for example, an Allport student who became one of the pioneers of the "new look" studies in perception during the 1940s and 1950s, wrestled uneasily with the legacy of his former teacher's views even as he set out to leave them behind. As an undergraduate during the 1930s, Bruner later recalled, Allport's *Personality* had shaken his "faith in the finality of research on animals." In the 1940s, however, he would find Allport's strictures on "wholeness" unsatisfying, desiring instead "some penetrating principle that would simplify, would render the surface complexity into something like a crystal"; but as he pursued this goal, Bruner would sometimes imagine that the "shade of Gordon Allport" sat across from him at the table, representing "a voice of social significance who intoned (in my mind at least, if not in fact) uncomfortable questions about the bearing of what I was doing on the human condition."[10]

But the issue of how the concerns of Allport's and the Murphys' generation would play out as their cohort continued to advance their perspectives within changing circumstances is more than a question in disciplinary history. In 1929 Allport had remarked that the conventions of the guild were such that psychologists would be justified in shaking their heads and declaring: "What price science!"[11] This was a question that persisted beyond the 1930s, but not always in ways that were transferred within a genealogy marked by an orderly disciplinary line of succession. It will be difficult, however, to identify and examine the subsequent cultural permutations of the radically empiricist critique mounted by figures such as Allport and the Murphys as long as scholars continue to address social science only insofar as it conforms to expectations that it represents a lesser body of knowledge orbiting the massive sphere inhabited by the natural sciences.

It is not surprising, perhaps, that scholars rarely look to the social sciences for instruction on metaphysics and epistemology, given the presumption that social science possesses a derivative, or "soft," status vis-à-vis the "hard" sciences. In fact, both within the sciences and outside of them, there has been a pervasive tendency to view the social sciences as *inferior* to the physical sciences. For example, the logical empiricist movement – synonymous with philosophy of science for so many years in the United States – advocated that the social sciences could achieve legitimacy to the extent that they borrowed methods from the physical sciences.[12] As Peter Novick has noted in his discussion of the history profession's debates over objectivity, social science epistemology was "almost isomorphic with that of the natural sciences" up through the 1960s.[13] These trends were reinforced by Thomas Kuhn's influential treatise on the structure of scientific revolutions, which held the physical sciences as paradigmatic examples of "mature" science. Indeed, in the preface to this work, Kuhn remarked that his inquiry was inspired, in part, by his puzzlement over why "the practice of astronomy, physics, chemistry, or biology normally fails to evoke the controversies over fundamentals that today often seem endemic among, say, psychologists or sociologists."[14] And, in an ironic twist on the granting of "legitimacy," many of those historians and sociologists of science who sought to demonstrate the reality of social relations in the construction of scientific knowledge rushed to find the "hardest" cases – by which they usually meant classical physics. The social sciences, regarded as fatally tainted by their immersion in the human world, were believed to offer little of import in these debates.[15]

And yet these debates over fundamentals from within the social sciences offer more than a window into the struggle to adapt methods of studying the physical world to the human world. If it has sometimes seemed as if the only story there was to tell of social science is that of its subordination to the physical sciences, my intention in describing the emergence of the radically empiricist science of Allport and the Murphys is to argue the benefits of observing that tides ebb as well as flow – and that tenacious worlds of activity exist among the tidepools if we choose to look.[16] Unlike their orthodox associates, radically empiricist scientists proposed that the assumptions and constraints with which the social sciences must grapple could provide valuable insights in regard to conceptualizing reality. For such thinkers, the social sciences could in fact be imagined as an *advantageous* place from which to commence the search for natural knowledge, precisely because the subject matter forced one to deal more openly with such concepts as the multiple nature of causality, the complexity of open-ended systems, the phenomena of plurality and particularity, the role of subjectivity in research, the arbitrary nature of boundaries between expert and lay knowledge, and the centrality of acts of interpretation to the investigative process.[17]

In contemporary social science, analysts such as anthropologist Clifford Geertz have related this frame of mind to the cross-disciplinary flux that is a consequence of numerous practitioners turning "away from a laws and instances ideal of explanation toward a cases and interpretation one, looking less for the sort of thing that connects planets and pendulums and more for the sort that connects chrysanthemums and swords."[18] Geertz contends that this refiguration of social thought "is a phenomenon general enough and distinctive enough to suggest that . . . something is happening to the way we think about the way we think."[19] That this phenomenon has roots that extend back to the thirties is a question of more than antiquarian interest. Scholars and scientists will have a chance to apprehend this cognitive refiguration most fully and to assess it most usefully when they have a broader and deeper understanding of the dynamics that have shaped it in the past and that continue to do so. The emergence of Allport's and the Murphys' contributions thus represent a few of the partial stories that make up this transformation, running parallel to one another, beginning and ending at odd times, mutually interlacing and interfering. When scholars move too quickly to produce a unified story by relying on highly schematized snapshots or in positing paradigms that presume the existence of a too-sweeping consensus, then partial stories such as these have been lost to sight, along with the unresolved challenges they posed to the scientific status quo.

Ultimately, James's radically empiricist legacy offered social scientists such as Allport and the Murphys a view of scientific life that lightened the burden of scientific tradition in the service of a more expansive view of method. James, for example, acknowledged that science had made "glorious leaps in the last three hundred years, and [has] extended our knowledge of nature so enormously both in general and detail."[20] But he also pointed to "the brevity of science's career. It began with Galileo, not three hundred years ago. Four thinkers since Galileo, each informing his successor of what discoveries his own lifetime had seen achieved, might have passed the torch of science into our hands as we sit here in this room." James doubted whether it could be credibly argued that "such a mushroom knowledge, such a growth overnight as this, *can* represent more than the minutest glimpse of what the universe will really prove to be when adequately understood." James and such prodigal progeny as Allport and the Murphys argued the contrary, claiming that "our science is a drop, our ignorance a sea. Whatever else be certain, this at least is certain – that the world of our present natural knowledge *is* enveloped in a larger world of *some* sort of whose residual properties we at present can frame no positive idea."[21]

Notes

Introduction

1. Lois Barclay Murphy, *Social Behavior and Child Personality: An Exploratory Study of Some Roots of Sympathy* (New York: Columbia University Press, 1937), p. 324; Gardner Murphy, "The Research Task of Social Psychology," *Journal of Social Psychology,* 1939, *10*:107–20, pp. 118–19; Gordon W. Allport, "Conceptions of Human Motivation: Liberalism and the Motives of Men," *Frontiers of Democracy,* 1940, *6*:136–7, p. 137.

2. On this point, see especially Peter Kuznick, *Beyond the Laboratory: Scientists as Political Activists in 1930s America* (Chicago: University of Chicago Press, 1987).

3. Especially important in discussing such thinkers is James T. Kloppenberg, *Uncertain Victory: Social Democracy and Progressivism in European and American Thought, 1870–1920* (New York: Oxford University Press, 1986). Also useful is David Ray Griffin, ed., *Founders of Constructive Postmodern Philosophy: Peirce, James, Bergson, Whitehead, and Hartshorne* (Albany: State University of New York Press, 1993). On the question of modernism, see Dorothy Ross, ed., *Modernist Impulses in the Human Sciences, 1870–1930* (Baltimore: Johns Hopkins University Press, 1994).

4. Alfred North Whitehead, *Science and the Modern World* (New York: Macmillan, 1925), p. 118.

5. Although I will be focusing on these psychologists' stances during the 1930s, I have neither ignored relevant work from the 1920s nor arbitrarily excluded work that was not published until the first few years of the 1940s.

6. The phrase "rebels within the ranks" is one I have appropriated from Gordon Allport. In his text, *Personality: A Psychological Interpretation* (New York: Henry Holt, 1937), Allport used it to characterize Gestalt psychologists as being among "the many rebels within the ranks of general psychology attacking its long-entrenched assumption that elements abstracted from individual experience are the proper data of the science" (pp. 15–16).

7. For the council and presidency information, see Seymour H. Mauskopf and Michael R. McVaugh, *The Elusive Science: Origins of Experimental Psychical Research* (Baltimore: Johns Hopkins University Press, 1980), p. 142. Autobiographical material on these three psychologists can be found in Gordon W. Allport, "An Autobiography," in his *The Person in Psychology: Selected Essays* (Boston: Beacon Press, 1968); Lois Barclay Murphy, *Gardner Murphy: Integrating, Expanding and Humanizing Psychology* (Jefferson, N.C.: McFarland, 1990); Gardner Murphy, "Gardner Murphy," in E. G. Boring and Gardner Lindzey, eds., *A History of Psychology in Autobiography,* vol. 5 (New York: Appleton-Century-Crofts, 1967); Gardner Murphy, "There Is More Beyond," in T. Krawiec, *The Psychologists: Autobiographies of Distinguished Living Psychologists,* vol. 2 (London: Oxford University Press, 1973); Lois Barclay Murphy, "Lois Barclay Murphy," in A. O'Connell and N. Russo, eds., *Models of Achievement: Reflections of Eminent Women in Psychology* (New York: Columbia University Press, 1983); and Lois Barclay Murphy, "Roots of an Approach to Studying Child Development," in T. Krawiec, ed., *The Psychologists: Autobiographies of Distinguished Living Psychologists,* vol. 3 (Brandon, Vt.: Clinical Psychology, 1979).

8. Frank was well known to those in the child study movement because of his supervision of funds from the Laura Spelman Rockefeller Memorial (LSRM) during the 1920s. In part, the LSRM money went toward initiating or assisting child-welfare research at six university institutes across the country. See Elizabeth Lomax, "The Laura Spelman Rockefeller Memorial: Some of Its Contributions to Early Research in Child Development," *Journal of the History of the Behavioral Sciences,* 1977, *13*:283–93, p. 283. Also important are Steven Schlossman, "Philanthropy and the Gospel of Child Development," *History of Education Quarterly,* 1981, *21*:275–99; and Stephen J. Cross, "Designs for Living: Lawrence K. Frank and the Progressive Legacy in American Social Science," unpublished dissertation, Johns Hopkins University, 1994. Frank continued to supervise Rockefeller money for child development research when the LSRM funds were transferred to the jurisdiction of the Rockefeller General Education Board (GEB) in the 1930s. Frank retired from the GEB in 1936 and then joined the Macy Foundation. See Lomax, "Laura Spelman," p. 290.

9. On the Sarah Lawrence research team, see Eugene Lerner and Lois Barclay Murphy, "Editorial Forward," in Eugene Lerner and Lois Barclay Murphy, eds., *Methods for the Study of Young Children, Monographs of the Society for Research in Child Development,* vol. 6, 1941, p. iv.

10. For an overview of the issues in play during the depression years, see William E. Leuchtenberg, *Franklin D. Roosevelt and the New Deal: 1932–1940* (New York: Harper & Row, 1963); Richard Pells, *Radical Visions and American Dreams: Culture and Social Thought in the Depression Years* (Middletown, Conn.: Wesleyan University Press, 1973); Edward A. Purcell, Jr., *The Crisis of Democratic Theory: Scientific Naturalism and the Problem of Value* (Lexington: University Press of Kentucky, 1973); Alan Brinkley, *Voices of Protest: Huey Long, Father Coughlin, and the Great Depression* (New York: Alfred Knopf, 1982); and Warren Susman, "The Culture of the Thirties," in his *Culture as History: The Transformation of American Society in the Twentieth Century* (New York: Pantheon, 1984).

11. Letter from Gordon W. Allport to Dr. [Richard Clarke] Cabot dated September 29, 1937 (Gordon W. Allport Papers, Harvard University Archives, Cambridge, Massachusetts; hereafter Allport Papers).

12. Gardner Murphy, "Dr. Rhine and the Mind's Eye," *American Scholar,* 1938, 7:189–200, pp. 200, 189.

13. Murphy, *Social Behavior and Child Personality,* p. 20.

14. Lerner and Murphy, "Editorial Forward," in Lerner and Murphy, *Methods for the Study of Young Children,* p. x.

15. William James, *A Pluralistic Universe* (Cambridge: Harvard University Press, 1977/1909), p. 125.

16. Considering as mediations the scientific use of such concepts is to suggest, in L. J. Jordanova's words, "that they speak to and contain implications about matters beyond their explicit content." The idea of mediation thus "enables us to lay out the links between different levels, areas, processes and languages." See L. J. Jordanova, *Sexual Visions: Images of Gender in Science and Medicine between the Eighteenth and Twentieth Centuries* (New York: Harvester Wheatsheaf, 1989), pp. 2, 159.

17. John Dewey and John L. Childs, "The Underlying Philosophy of Education," in William H. Kilpatrick, ed., *The Educational Frontier* (New York: Century, 1933), p. 291.

18. Murphy, *Social Behavior and Child Personality,* pp. 17 and 4; Allport, *Personality,* pp. 371, viii. On the rise of the "personality and culture" discourse see, for example, Richard Handler, "Personality and Culture," in George Stocking, ed., *Malinowski, Rivers, Benedict, and Others: Essays on Culture and Personality* (Madison: University of Wisconsin Press, 1986); and Cross, "Designs for Living."

19. For analyses of these issues, see, for example, Franz Samelson, "Organizing for the Kingdom of Behavior: Academic Battles and Organizational Policies in the Twenties," *Journal of the History of the Behavioral Sciences,* 1985, *21*:33–47; Jill G. Morawski, ed., *The Rise of Experimentation in American Psychology* (New Haven, Conn.: Yale University Press, 1988); and Kurt Danziger, *Constructing the Subject: Historical Origins of Psychological Research* (Cambridge: Cambridge University Press, 1990).

20. Ruth Benedict, *Patterns of Culture* (Boston: Houghton Mifflin, 1934), p. 233. Benedict's final chapter is entitled "The Individual and the Patterns of Culture."

21. Caroline F. Ware, "Introductory Note" to "Part One: Techniques of Cultural Analysis," in Caroline F. Ware, ed., *The Cultural Approach to History* (New York: Columbia University Press, 1940), p. 19. This section includes "Society as Viewed by the Anthropologist" by Geoffrey Gorer; "Clio and Psyche: Some Interrelations of Psychology and History" by social psychologist Goodwin Watson; and "Psychology and the Interpretation of Historical Events" by psychoanalyst Franz Alexander.

22. Murphy, "Research Task of Social Psychology," p. 113. In this piece Murphy also emphasized the need for psychologists to study biology.

23. Murphy, *Gardner Murphy,* pp. 117, 173.

24. On Allport's reading list, see the typescript of readings for his 1937 Social Psychology seminar, "Tentative Outline for Psychology 29 Seminary in Social Psychology: Culture and the Individual" and "Psychology 29 Seminary in Social Psychology 1937–1938 Topics for Social Psychology to Consider in Treating the Influence of Culture upon the Individual" (Allport Papers). Regarding sociology students taking Allport's course, see Pitirim Sorokin, *A Long Journey* (New Haven, Conn.: College and University Press, 1963), p. 243.

25. On Lynd, see Murphy, "Gardner Murphy," p. 162; on meeting Allport, see Lois Murphy, *Gardner Murphy,* p. 174.

26. Gordon Allport, Gardner Murphy, and Mark May, *Memorandum on Cooperation and Competition* (New York: Social Science Research Council, 1937). This project also resulted in a companion report, *Cooperation and Competition among Primitive Peoples,* by Mead.

27. Allport, et al., *Memorandum on Cooperation and Competition,* p. 1.

28. Gardner Murphy and Lois Barclay Murphy, *Experimental Social Psychology* (New York: Harper and Brothers, 1931); Gardner Murphy, Lois Barclay Murphy, and Theodore Newcomb, *Experimental Social Psychology: An Interpretation of Research upon the Socialization of the Individual,* rev. ed. (New York: Harper and Brothers, 1937); Allport, *Personality.*

29. Allport, *Personality,* p. vii; Murphy and Murphy, *Experimental Social Psychology,* p. 2.

30. Gordon Allport, "Attitudes," in C. C. Murchison, ed., *A Handbook of Social Psychology* (Worcester, Mass.: Clark University Press, 1935); Gardner Murphy and Friedrich Jensen, *Approaches to Personality: Some Contemporary Conceptions Used in Psychology and Psychiatry* (New York: Coward-McCann, 1932).

31. Gordon W. Allport, "Dewey's Individual and Social Psychology," in P. A. Schilpp, ed., *The Philosophy of John Dewey* (Evanston, Ill.: Northwestern University Press, 1939); Gardner Murphy, "Personality and Social Adjustment," *Social Forces,* 1937, *15*:472–6; Murphy, et al., *Experimental Social Psychology,* pp. 18–24.

32. Allport wrote this work with a former student, Hadley Cantril; *Psychology of Radio* (New York: Harper, 1935).

33. Gardner Murphy and Rensis Likert, *Public Opinion and the Individual: A Psychological Study of Student Attitudes on Public Questions, with a Retest Five Years Later* (New York: Harper and Brothers, 1938); Paul Grabbe, *We Call It Human Nature* (New York: Harper and Brothers, 1939). On Murphy's involvement see his preface.

34. Gordon Allport, "Some Guiding Principles in Understanding Personality," *The Family* (June 1930), pp. 124–8 and "The Nature of Motivation," *Understanding the Child* (January 1935), pp. 3–6. Gardner Murphy, "The Geometry of Mind: An Interpretation of Gestalt Psychology," *Harper's Magazine* (October 1931), pp. 584–93; "Things I Can't Explain," *American Magazine* (November 1936), pp. 40–1, 130–2. See also "The Mind Is a Stage: Adjusting Mental Problems in a 'Spontaneity Theater,'" *The Forum and Century* (May 1937), pp. 277–80. Lois Barclay Murphy, "What a Little Child Needs," *Parents' Magazine* (October 1939), pp. 20–1, 67–9; "How to Help the Shy Child," *Parents' Magazine* (April 1942), pp. 32, 81–6; and "Grown-up Music That Children Love," *Good Housekeeping* (May 1938), pp. 80, 214–16.

35. Stephen Toulmin and David Leary, "The Cult of Empiricism in Psychology, and Beyond," in Sigmund Koch and David E. Leary, eds., *A Century of Psychology as Science* (New York: McGraw-Hill, 1985), p. 601. On Watson and early behaviorism, see Franz Samelson, "Struggle for Scientific Authority: The Reception of Watson's Behaviorism, 1913–1920," *Journal of the History of the Behavioral Sciences,* 1981, *17*:399–425; John M. O'Donnell, *The Origins of Behaviorism: American Psychology, 1870–1920* (New York: New York University Press, 1985); and Kerry Buckley, *Mechanical Man: John Broadus Watson and the Beginnings of Behaviorism* (New York: Guilford, 1989).

36. Toulmin and Leary, "Cult of Empiricism," p. 602. For a sense of the divergences within behaviorism, see Danziger, *Constructing the Subject;* Morawski, *Rise of Experimentation;* Samelson's articles; Laurence D. Smith, "Metaphors of Knowledge and Behavior in the Behaviorist Tradition," in David E. Leary, ed., *Metaphors in the History of Psychology* (Cambridge: Cambridge University Press, 1990); and Laurence D. Smith, *Behaviorism and Logical Positivism: A Reassessment of the Alliance* (Stanford,

Calif.: Stanford University Press, 1986). For an introduction to B. F. Skinner's radical behaviorism, see Daniel W. Bjork, *B. F. Skinner: A Life* (New York: Basic Books, 1993). For the establishment of behaviorism as a scientific language, see Charles Bazerman, "Codifying the Social Scientific Style: The APA *Publication Manual* as a Behaviorist Rhetoric," in John S. Nelson, Allan Megill, and Donald N. McCloskey, eds., *The Rhetoric of the Human Sciences: Language and Argument in Scholarship and Public Affairs* (Madison: University of Wisconsin Press, 1987).

37. John Dashiell, *Fundamentals of Objective Psychology* (Boston: Houghton Mifflin, 1928), p. 8.

38. Clark L. Hull, "Mind, Mechanism, and Adaptive Behavior," *Psychological Review,* 1937, *44*:1–32, pp. 15, 32.

39. J. G. Morawski, "Organizing Knowledge and Behavior at Yale's Institute of Human Relations," *Isis,* 1986, *77*:219–42, pp. 235, 239.

40. For a contemporaneous discussion of these trends, see Jerome S. Bruner and Gordon W. Allport, "Fifty Years of Change in American Psychology," *Psychological Bulletin,* 1940, *37*:757–76.

41. Toulmin and Leary, "Cult of Empiricism."

42. There exists a considerable historiography on psychological testing, befitting its central disciplinary role in the interwar years. See, for example, Franz Samelson, "World War I Intelligence Testing and the Development of Psychology," *Journal of the History of the Behavioral Sciences,* 1977, *13*:274–82 and "Putting Psychology on the Map: Ideology and Intelligence Testing," in Allan R. Buss, ed., *Psychology in Social Context* (New York: Irvington, 1979); Stephen Jay Gould, *The Mismeasure of Man* (New York: W. W. Norton, 1981); Michael M. Sokal, ed., *Psychological Testing and American Society 1890–1930* (New Brunswick, N.J.: Rutgers University Press, 1987); Richard von Mayrhauser, "The Practical Language of the American Intellect," *History of the Human Sciences,* 1991, *4*:371–93; JoAnne Brown, *The Definition of a Profession: The Authority of Metaphor in the History of Intelligence Testing, 1890–1930* (Princeton, N.J.: Princeton University Press, 1992); and John Carson, "Army Alpha, Army Brass, and the Search for Army Intelligence," *Isis,* 1993, *84*:278–309. On issues relating to experimentation and statistics, see Morawski, *The Rise of Experimentation in American Psychology* and Danziger, *Constructing the Subject.*

43. For an excellent introduction to the variations of thought within neobehaviorism, see Laurence D. Smith's chapters on Tolman and Skinner in his *Behaviorism and Logical Positivism.*

44. An important forum for the promotion of a shared set of values regarding a scientific identity based on neobehaviorism appears to be the "psychological round table" (PRT), founded by S. S. Stevens and E. B. Newman in 1936 as a venue for the presentation of cutting-edge results by young experimentalists. Further study of this private annual event would help widen historians' understanding of the development of methodological discourse in psychology during the 1930s and 1940s. For an intriguing description of the PRT, see Ludy T. Benjamin, "The Psychological Round Table: Revolution of 1936," *American Psychologist,* 1977, *32*:542–9.

45. Robert S. Woodworth, *Contemporary Schools of Psychology* (New York: Ronald Press, 1931), p. 88.

46. The "austere" characterization is from Gordon W. Allport, "The Psychologist's Frame of Reference," *Psychological Bulletin,* 1940, *37*:1–28, p. 12; the second quotation is from Gordon W. Allport, *The Use of Personal Documents in Psychological Science* (New York: Social Science Research Council, 1942), p. 37.

47. Letter from Gordon [Allport] to Dear Garry [E. G. Boring] dated October 6, 1939 (E. G. Boring Papers, Harvard University Archives, Cambridge, Massachusetts).

48. Gardner Murphy, *An Historical Introduction to Modern Psychology* (New York: Harcourt, Brace, 1929), p. 278. As illustrative of this point of view, Murphy footnoted Karl Lashley's statement – from his 1923 article, "The Behavioristic Interpretation of Consciousness" – that adequate descriptions of human behavior would ultimately be found at the mechanical and chemical level.

49. Ibid., p. 409.

50. Ibid., p. 415.

51. Allport, "The Psychologist's Frame of Reference," pp. 12, 15–16. The data discussed in this address are detailed in Bruner and Allport, "Fifty Years of Change in Psychology."

52. Gordon W. Allport, "Psychology in the Near Future," undated unpublished manuscript (Allport Papers). The address was delivered in Durham, New Hampshire, to undergraduates; from internal references it appears to be from 1940 or 1941.

53. I think Edward Tolman may be the "close colleague" to whom Allport is referring here.

54. Allport, "Frame of Reference," p. 14.

55. Ibid., p. 26.

56. Kurt Danziger, "The Project of an Experimental Social Psychology: Historical Perspectives," *Science in Context,* 1992, *5*:309–28, p. 310.

57. Mitchell Ash, "Historicizing Mind Science: Discourse, Practice, Subjectivity," *Science in Context,* 1992, *5*:193–207, p. 194.

58. Fred Matthews, "Present Pasts: From 'Progressive' to 'Liberal' in Intellectuals' Understanding of American History, 1925–1950," *Canadian Journal of History,* 1991, *26*:429–51, p. 430.

59. Allport, "The Genius of Kurt Lewin," *Person in Psychology* (the essay was originally published in 1947), p. 363.

60. Murphy, "Gardner Murphy," p. 274; Murphy, "Roots of an Approach," p. 171.

61. Murphy, *Gardner Murphy,* p. 174; Gordon W. Allport, interview in Richard I. Evans, *Gordon Allport: The Man and His Ideas* (New York: E. P. Dutton, 1970), p. 18 (Allport also cites Kurt Lewin and Henry Murray in this respect).

62. Allport made this suggestion in a letter to Roger Barker dated January 13, 1966 (Midwest Field Station Archives, University of Kansas, Lawrence). Allport cited Lois Murphy's "constant productivity in the field of child development" and stated that Gardner Murphy "has contributed to the advancement of more branches of psychology than any other living member of A.P.A." Allport then proposed "a radical and unusual suggestion: HOW ABOUT GIVING TWO AWARDS TO THE MURPHY *EHEPAAR* [marriage partnership]. This would, I think, be a dramatic and highly popular decision."

63. On actions against the war in Spain, see Kuznick, *Beyond the Laboratory,* p. 181; within psychology, see Lorenz J. Finison, "Psychologists and Spain: A Historical

Note," *American Psychologist,* 1977, *32*:1080–4. Copies of the FBI reports on Gardner Murphy are contained in the Gardner Murphy and Lois Barclay Murphy Papers, Archives of the History of American Psychology, University of Akron (hereafter Murphy papers). The papers also contain FBI material related to Wade Crawford Barclay, Lois Barclay Murphy's father.

64. SPSSI was founded in 1936. For an overview of SPSSI's initial phase, see Lorenz J. Finison, "The Psychological Insurgency – 1936–1945," *Journal of Social Issues,* 1986, *42*:21–34. The whole issue is devoted to SPSSI's history.

65. Arthur Jenness, "Gordon W. Allport," in David L. Sills, ed., *International Encyclopedia of the Social Sciences, Biographical Supplement,* vol. 18 (New York: The Free Press, 1979), p. 15.

66. Murphy, *Gardner Murphy,* p. 116.

67. See the FBI records on Lois Barclay Murphy in the Murphy Papers.

68. Murphy, *Gardner Murphy,* p. 137.

69. At times, each of these individuals alluded explicitly to their implicit personal styles. Allport, for example, remarked that "much of my writing is polemic in tone"; Allport, "An Autobiography," p. 405. In "Roots of an Approach to Studying Child Development," Lois Barclay Murphy informs the reader, "I will discuss some of [my approach] in terms of the pioneer image" (p. 167). Gardner Murphy once described his judicious temperament as reflected in the Henry Fonda character in the movie *Twelve Angry Men;* Murphy, *Gardner Murphy,* pp. 344–5.

70. In this work I do not address Gordon Allport's relationship with his brother Floyd, who was eight years older, had a Harvard Ph.D. in psychology, and worked at Syracuse University. Floyd Allport made a name for himself as a social psychologist in the 1920s by conducting experiments on how individual behavior is affected by the presence of other people. For a brief introduction to this work, see David L. Post, "Floyd H. Allport and the Launching of Modern Social Psychology," *Journal of the History of the Behavioral Sciences,* 1980, *16*:369–76. Floyd does not appear to have served as a close mentor of Gordon's: by the latter 1920s, Gordon was striking off on different paths and quickly surpassed his brother in professional prominence. There is little in Gordon Allport's papers that sheds light on the temporal dynamics of their relationship, and it remains an issue that requires further research. For brief comments by the brothers on each other see Gordon Allport, "An Autobiography"; and Floyd H. Allport, "Floyd H. Allport," in Gardner Lindzey, ed., *A History of Psychology in Autobiography,* vol. 6 (Englewood Cliffs, N.J.: Prentice-Hall, 1974).

71. Gordon Allport, *The Nature of Prejudice* (Reading, Mass.: Addison-Wesley, 1954); Gardner Murphy, *Personality: A Biosocial Approach to Origins and Structure* (New York: Harper and Row, 1947); Lois Barclay Murphy and Alice Moriarty, *Vulnerability, Coping, & Growth: From Infancy to Adolescence* (New Haven, Conn.: Yale University Press, 1976).

72. Murphy's projective testing research (initially written with Ruth Horowitz) was the foundation from which psychologists Kenneth and Mamie Clark developed the "doll studies" cited in the Supreme Court's *Brown v. Board of Education* decision. On the "doll studies," see Ben Keppel, *The Work of Democracy: Ralph Bunche, Kenneth B. Clark, Lorraine Hansberry, and the Cultural Politics of Race* (Cambridge: Harvard University Press, 1995); on the *Brown* case in general, see Richard Kluger, *Simple Justice: The History of Brown v. Board of Education and Black America's Struggle for*

Equality (New York: Knopf, 1975); on SPSSI's role, see Otto Klineberg, "SPSSI and Race Relations, in the 1950s and After," *Journal of Social Issues,* 1986, *42*:53–9. Murphy refers to her Head Start and Parent Center work in Murphy, *Gardner Murphy,* p. 144. For Murphy on projective testing, see Ruth Horowitz and Lois Barclay Murphy, "Projective Methods in the Psychological Study of Children," *Journal of Experimental Education,* 1938–9, 7:133–40 and Lerner and Murphy, *Methods for the Study of Young Children.*

73. Regarding Allport, see Roy Schafer, Irwin Berg, and Boyd McCandless, *Report on Survey of Current Psychological Testing Practices* (Washington, D.C.: American Psychological Association, 1951); regarding Murphy, see Kenneth E. Clark, *American Psychologists: A Survey of a Growing Profession* (Washington, D.C.: American Psychological Association, 1957), p. 116.

74. Allport received his Gold Medal Award in 1963 and Murphy in 1972. The Gold Medal is awarded to senior American psychologists with a "distinguished and long-continued record of scientific and scholarly accomplishment" (see the January 1973 *American Psychologist,* p. 75). Other awardees of the G. Stanley Hall Award have included such figures as Jean Piaget, Jerome Bruner, and Erik Erikson.

75. On psychology in particular, see Mitchell G. Ash, "The Self-Presentation of a Discipline: History of Psychology in the United States between Pedagogy and Scholarship," in Loren Graham, Wolf Lepenies, and Peter Weingart, eds., *Functions and Uses of Disciplinary History,* vol. 7, *Sociology of the Sciences* (Dordrecht: Reidel, 1983).

Chapter 1

1. Horace Kallen, "Introduction: The Meaning of William James for 'Us Moderns,'" in William James, *The Philosophy of William James: Selected from His Chief Works* (New York: Modern Library, 1925), p. 44; Walter Rauschenbusch, *Christianizing the Social Order* (New York: Macmillan, 1912), p. 353.

2. The contours of Progressivism have been delineated in such classic texts as Richard Hofstadter, *The Age of Reform: From Bryan to FDR* (New York: Vintage, 1955); Robert Wiebe, *The Search for Order, 1877–1920* (New York: Hill and Wang, 1967); Robert M. Crunden, *Ministers of Reform: The Progressives' Achievement in American Civilization, 1889–1920* (New York: Basic Books, 1982); and Alan Trachtenberg, *The Incorporation of America: Culture and Society in the Gilded Age* (New York: Hill and Wang, 1982). For a trenchant critique of "progressivism" as a category of historical analysis, see Daniel Rodgers, "In Search of Progressivism," *Reviews in American History,* 1982, *10*:113–32.

3. See, for example, Thomas Haskell, *The Emergence of Professional Social Science: The American Social Science Association and the Nineteenth-Century Crisis of Authority* (Urbana: University of Illinois Press, 1977); Alexandra Oleson and John Voss, eds., *The Organization of Knowledge in Modern America, 1860–1920* (Baltimore: Johns Hopkins University Press, 1979); Mary O. Furner, *Advocacy and Objectivity: A Crisis in the Professionalization of American Social Science, 1865–1905* (Lexington: University of Kentucky Press, 1975); and Dorothy Ross, *The Origins of American Social Science* (Cambridge: Cambridge University Press, 1991).

4. Ross, *The Origins of American Social Science,* p. 404. Ross defines scientism as "an effort to make good on the positivist claim that only natural science provided certain knowledge and conferred the power of prediction and control" (ibid., p. 390). Also of relevance on this theme are Edward J. Purcell, Jr., *The Crisis of Democratic Theory* (Lexington: University of Kentucky Press, 1973); Robert C. Bannister, *Sociology and Scientism: The American Quest for Objectivity, 1880–1940* (Chapel Hill: University of North Carolina Press, 1987); William Graebner, *The Engineering of Consent: Democracy and Authority in Twentieth-Century America* (Madison: University of Wisconsin Press, 1987); Peter Novick, *That Noble Dream: The "Objectivity Question" and the American Historical Profession* (New York: Cambridge University Press, 1988); John M. Jordan, *Machine-Age Ideology: Social Engineering and American Liberalism, 1911–1939* (Chapel Hill: University of North Carolina, 1994); and Mark C. Smith, *Social Science in the Crucible: The American Debate over Objectivity and Purpose, 1918–1941* (Durham, N.C.: Duke University Press, 1994).

5. Dorothy Ross, "The Development of the Social Sciences," in Oleson and Voss, *Organization of Knowledge*, p. 129.

6. On the term "radical progressivism," see Robert B. Westbrook, *John Dewey and American Democracy* (Ithaca, N.Y.: Cornell University Press, 1991), p. 189.

7. James T. Kloppenberg, *Uncertain Victory: Social Democracy and Progressivism in European and American Thought, 1870–1920* (New York: Oxford University Press, 1986). Reaching back somewhat further is Paul Jerome Croce, *Science and Religion in the Era of William James: The Eclipse of Certainty, 1820–1880*, vol. 1 (Chapel Hill: University of North Carolina Press, 1995).

8. Kloppenberg, *Uncertain Victory*, p. 3.

9. Ibid., p. 9.

10. Ibid., p. 4.

11. Ibid., p. 174. Kloppenberg refers specifically to William James and John Dewey in this regard.

12. Ibid., pp. 3, 11. See also Kloppenberg's further reflections on pragmatism in his recent article, "Pragmatism: An Old Name for Some New Ways of Thinking?" *Journal of American History*, 1996, *83*:100–138.

13. See, for example, David DeLeon, *The American as Anarchist: Reflections on Indigenous Radicalism* (Baltimore: Johns Hopkins University Press, 1978). DeLeon argues that "it is time to recognize, after two hundred years of political independence, that our native radicalism is fundamentally different from that of Europe, Russia, China, or the Third World" (p. 3). For further discussions that differentiate the American tradition of radicalism from European forms, see George Bernard Cotkin, "Working-Class Intellectuals and Evolutionary Thought in America, 1870–1915," unpublished dissertation, Ohio State University, 1978; John L. Thomas, *Alternative America: Henry George, Edward Bellamy, Henry Demarest Lloyd, and the Adversary Tradition* (Cambridge, Mass.: Belknap Press, 1983); Rick Tilman, *C. Wright Mills: A Native Radical and His American Intellectual Roots* (University Park: Pennsylvania State University Press, 1984); and Deborah J. Coon, "Courtship with Anarchy: The Socio-Political Foundations of William James's Pragmatism," unpublished dissertation, Harvard University, 1988, and "'One Moment in the World's Salvation': Anarchism and the Radicalism of William James," *Journal of American History*, 1996, *83*:70–99. For contemporary examples from the depression period, see Lillian Symes and Travers Clement, *Rebel America: The Story of Social Revolt in the United States* (New York: Harper & Brothers, 1934); and Oscar Ameringer, *If You Don't Weaken: The*

Autobiography of Oscar Ameringer (Norman: University of Oklahoma Press, 1983/1940). James R. Green notes that Ameringer, a German immigrant who published the radical newspaper the *American Guardian* during the 1930s, "did not bring his socialism with him from the old country," but came to it by way of "industrial unionism, Jeffersonian democracy, and Mark Twain's humor before he studied Marxism"; James R. Green, "Introduction," pp. xx, xxiii.

14. Ralph Barton Perry, *The Thought and Character of William James: Philosophy and Psychology,* vol. 2 (Boston: Little, Brown, 1936), p. 668.

15. The phrase is Allport's; see Gordon Allport, "The Productive Paradoxes of William James," in Gordon W. Allport, *The Person in Psychology: Selected Essays* (Boston: Beacon Press, 1968), p. 298 (this piece was originally published in 1943).

16. David E. Leary, "William James on the Self and Personality: Clearing the Ground for Subsequent Theorists, Researchers, and Practitioners," in Michael G. Johnson and Tracy B. Henley, eds., *Reflections on The Principles of Psychology: William James after a Century* (Hillsdale, N.J.: Lawrence Erlbaum Associates, 1990), p. 120; Eugene Taylor, "The Case for a Uniquely American Jamesian Tradition in Psychology," in Margaret Donnelly, ed., *Reinterpreting the Legacy of William James* (Washington, D.C.: American Psychological Association, 1992), pp. 10, 11. See also Richard P. High and William R. Woodward, "William James and Gordon Allport: Parallels in Their Maturing Conceptions of Self and Personality," in R. W. Rieber and Kurt Salzinger, *Psychology: Theoretical-Historical Perspectives* (New York: Academic Press, 1980), and William R. Woodward, "William James's Psychology of the Will: Its Revolutionary Impact on American Psychology," in Josef Brožek, ed., *Explorations in the History of Psychology in the United States* (Lewisburg, Pa.: Bucknell University Press, 1984), especially pp. 176–7.

17. Lois Barclay Murphy, *Gardner Murphy: Integrating, Expanding and Humanizing Psychology* (Jefferson, N.C.: McFarland, 1990), p. 174.

18. Lois Barclay Murphy, "Roots of an Approach to Studying Child Development," in T. Krawiec, ed., *The Psychologists: Autobiographies of Distinguished Living Psychologists,* vol. 3 (Brandon, Vt.: Clinical Psychology, 1979), p. 172. Remarking on those whose work she felt closest to, Murphy lists "Erik Erikson, Harry Murray, Robert Coles as well as Gardner and William James" (ibid., p. 173).

19. William James, *Pragmatism* (Cambridge, Mass.: Harvard University Press, 1975/1907), p. 97.

20. This distinction was first brought to my attention by Eugene Taylor at the 1989 meeting of Cheiron at Slippery Rock University.

21. John J. McDermott, "Preface," in Charlene Haddock Seigfried, *Chaos and Context: A Study in William James* (Athens: Ohio University Press, 1978), p. ix. See, particularly, the essays "Does 'Consciousness' Exist?" and "A World of Pure Experience" in William James, *Essays in Radical Empiricism* (Cambridge, Mass.: Harvard University Press, 1976/1912). For a historical discussion that places James's attempt to grasp the nature of pure experience in a central position, see Daniel W. Bjork, *William James: The Center of his Vision* (New York: Columbia University Press, 1988), especially pp. 215–27, which deal directly with radical empiricism.

22. See Alfred North Whitehead, *Science and the Modern World* (New York: Macmillan, 1925), p. 199.

23. Richard Bernstein, "Introduction," in William James, *A Pluralistic Universe* (Cambridge, Mass.: Harvard University Press, 1977/1909), p. xxiv.

24. McDermott, "Preface," p. xi.

25. David Hollinger, "William James and the Culture of Inquiry," in his *In the American Province: Studies in the History and Historiography of Ideas* (Baltimore: Johns Hopkins University Press, 1985), p. 4. See also Hollinger's "The Problem of Pragmatism in American History" in the same volume.

26. James Livingston, *Pragmatism and the Political Economy of Cultural Revolution, 1850–1940* (Chapel Hill: University of North Carolina Press, 1994), pp. 227, 228. On this point, see also Croce, *Science and Religion in the Era of William James,* p. ix; and James Edie, *William James and Phenomenology* (Bloomington and Indianapolis: Indiana University Press, 1987), p. viii. For an extended discussion of recent appropriations of the pragmatist label, see Kloppenberg, "Pragmatism."

27. William James, "Preface," in his *The Will to Believe and Other Essays in Popular Philosophy* (Cambridge, Mass.: Harvard University Press, 1979/1897), p. 7.

28. Allport, "Productive Paradoxes of William James," pp. 304–5.

29. Ibid., pp. 305, 306.

30. Gardner Murphy, "Introduction," in Gardner Murphy and Robert O. Ballou, eds., *William James on Psychical Research* (New York: Viking Press, 1960), p. 13.

31. James, *A Pluralistic Universe,* pp. 142, 96.

32. Ibid., p. 10.

33. Ibid., p. 26.

34. William James, "Frederic Myers's Service to Psychology," *Proceedings of the Society for Psychical Research,* 1901, *17*:13–23, p. 14.

35. James, *Will to Believe,* pp. 5, 5–6.

36. Ibid., p. 6.

37. The first quoted phrase is from H. S. Thayer, "Introduction," in James, *Pragmatism,* p. xxxviii; the second is from James, *Pluralistic Universe,* p. 26.

38. James, "Frederic Myers's Service to Psychology," pp. 21–2.

39. Arnold Metzger, "William James and the Crisis of Philosophy," in Herbert W. Schneider, ed., *In Commemoration of William James* (New York: Columbia University Press, 1942), p. 215.

40. Kallen, "Introduction," p. 5.

41. Ibid., p. 53.

42. Ibid., pp. 54, 44.

43. Edna Heidbreder, "The Psychology of William James," in her *Seven Psychologies* (New York: Appleton-Century-Crofts, 1933), p. 158. Walter Lippman, "An Open Mind: William James," reprinted in Linda Simon, *William James Remembered* (Lincoln: University of Nebraska, 1996), p. 256. Lippman's piece originally appeared in the December 1910 issue of *Everybody's Magazine.* For an insightful discussion of James's politicization and its relationship to his philosophy see Coon, "Courtship with Anarchy" and "'One Moment in the World's Salvation.'"

44. George Cotkin, *William James, Public Philosopher* (Urbana: University of Illinois Press, 1989), p. 4. For a critical analysis of Cotkin's use of the concept of "public philosopher," see Ross Posnock, "The Politics of Pragmatism and the Fortunes of the Public Intellectual," *American Literary History,* 1991, *3*:566–87. For other views of James, see Bjork, *William James;* Gerald E. Myers, *William James: His Life and Thought* (New Haven, Conn.: Yale University Press, 1986); and R. W. B. Lewis, *The Jameses: A Family Narrative* (New York: Farrar, Straus, and Giroux, 1991).

45. Cotkin, *William James,* p. 153.

46. Ibid., p. 17. See also Coon, "Courtship with Anarchy" and "'One Moment in the World's Salvation.'"

47. Letter from William James to W. Cameron Forbes dated June 11, 1907, in Henry James, ed., *The Letters of William James,* vol. 2 (Boston: Atlantic Monthly Press, 1920), p. 289.

48. Seigfried, "William James's Concrete Analysis of Experience," p. 538. Seigfried has extended her analysis of pragmatism to feminist theory; see her *Pragmatism and Feminism: Reweaving the Social Fabric* (Chicago: University of Chicago Press, 1996). See also the special issue on feminism and pragmatism in *Hypatia,* 1993, vol. 8/2.

49. Eugene Taylor, "Radical Empiricism and the New Science of Consciousness," *History of the Human Sciences,* 1995, *8*:47–60, p. 52. Other discussions of radical empiricism in relation to James's psychology include the following essays from Donnelly, *Reinterpreting the Legacy of William James:* Donald A. Crosby and Wayne Viney, "Toward a Psychology That Is Radically Empirical: Recapturing the Vision of William James"; Charlene Haddock Seigfried, "The World We Practically Live In"; and Wayne Viney, "A Study of Emotion in the Context of Radical Empiricism."

50. James, *Pluralistic Universe,* p. 149.

51. Suggestive in this regard is Jeanne Watson Eisenstadt, "Remembering Goodwin Watson," *Journal of Social Issues,* 1986, *42*:49–52, especially on the importance of Union Theological Seminary and psychology at Columbia University; and Ernest R. Hilgard, "From the Social Gospel to the Psychology of Social Issues: A Reminiscence," *Journal of Social Issues,* 1986, *42*:107–11. Hilgard indicates the importance for psychology of those who received Kent Fellowships granted by the National Council on Religion in Higher Education. This group, Hilgard recalled, reflected Social Gospel concerns and was an organization "in which liberal theologians such as Harry Emerson Fosdick were active" (p. 109).

52. John Hedley Brooke, for example, in *Science and Religion: Some Historical Perspectives* (Cambridge: Cambridge University Press, 1991), p. 322, discusses the intellectual presumption that religion and science coexist in different spheres. Detailed discussions of the issue of religion and science can be found in David C. Lindberg and Ronald L. Numbers, eds., *God and Nature: Historical Essays on the Encounter between Christianity and Science* (Berkeley: University of California Press, 1986). For an example of the merging of Christian social ethics, democratic philosophy, and disciplinary imperatives in biology, see Gregg Mitman, "Evolution as Gospel: William Patten, the Language of Democracy, and the Great War," *Isis,* 1990, *81*:446–63.

53. Martin Rudwick, "Senses of the Natural World and Senses of God: Another Look at the Historical Relation of Science and Religion," in A. R. Peacocke, ed., *The Sciences and Theology in the Twentieth Century* (Notre Dame: University of Notre Dame Press, 1981), p. 245.

54. Henry F. May, "Intellectual History and Religious History," in John Higham and

Paul K. Conklin, eds., *New Directions in American Intellectual History* (Baltimore: Johns Hopkins University Press, 1979), p. 114.

55. William R. Hutchison, *The Modernist Impulse in American Protestantism* (Cambridge, Mass.: Harvard University Press, 1976), p. 165.

56. William McGuire King, "An Enthusiasm for Humanity: The Social Emphasis in Religion and Its Accommodation in Protestant Theology," in Michael J. Lacey, ed., *Religion and Twentieth-Century American Intellectual Life* (Washington, D.C.: Woodrow Wilson International Center for Scholars, 1989), pp. 50, 49.

57. Henry F. May, *Protestant Churches and Industrial America* (New York: Harper and Brothers, 1949), p. 235. For a concise overview, see Sydney E. Ahlstrom, "The Social Gospel," in his *A Religious History of the American People* (New Haven, Conn.: Yale University Press, 1972). The following are also of interest: Donald Meyer, *The Protestant Search for Political Realism, 1919–1941* (Berkeley and Los Angeles: University of California Press, 1960); James Dombrowski, *The Early Days of Christian Socialism* (New York: Octagon Books, 1966/1936); Robert T. Handy, *The Social Gospel in America: Gladden, Ely, Rauschenbusch* (New York: Oxford University Press, 1966); and Peter J. Frederick, *Knights of the Golden Rule: The Intellectual as Christian Social Reformer in the 1890s* (Lexington: University Press of Kentucky, 1976). The classic foundational texts are Walter Rauschenbusch, *Christianity and the Social Crisis* (New York: Macmillan, 1907) and *Christianizing the Social Order.*

58. Rauschenbusch, *Christianizing the Social Order,* p. 322.

59. See Arthur Jenness, "Gordon W. Allport," in David L. Sills, ed., *International Encyclopedia of the Social Sciences, Biographical Supplement,* vol. 18 (New York: The Free Press, 1979), p. 13; and the autobiographical statement of his brother, Floyd H. Allport, "Floyd H. Allport," in Gardner Lindzey, ed., *A History of Psychology in Autobiography,* vol. 6 (Englewood Cliffs, N.J.: Prentice-Hall, 1974), p. 3. After his mother's death, Gordon Allport produced a privately printed volume, *The Quest of Nellie Wise Allport,* which discusses her religious views. The text was written in 1942 and revised in 1944. In this tribute, Gordon Allport notes that, during his youth, his mother "held to Sabbath observance, to strict church attendance, to camp meetings and conversion" (p. 7); Gordon W. Allport Papers, Harvard University Archives, Cambridge, Massachusetts (hereafter Allport Papers).

60. See Hoyt Landon Warner, *Progressivism in Ohio 1897–1917* (Columbus: Ohio State University Press, 1964).

61. Warner, *Progressivism,* p. 23; on George as a Christian socialist, see Dombrowski, *The Early Days of Christian Socialism in America;* and Thomas, *Alternative America.*

62. Gordon Allport, "An Autobiography," in Allport, *Person in Psychology,* p. 379. In a letter dated November 6, 1935, to [Ross] Stagner – who had just taken a position at the University of Akron – Allport remarked: "I know Akron very well. My father was the founder of the People's Hospital and the Akron Pharmacy, both of which institutions I hope you will never need to patronize" (Allport Papers).

63. For information on Allport's career, see "An Autobiography," p. 119. On the Social Ethics Department, see David B. Potts, "Social Ethics at Harvard, 1881–1931: A Study in Academic Activism," in Paul Buck, ed., *Social Sciences at Harvard, 1860–1920: From Inculcation to the Open Mind* (Cambridge, Mass.: Harvard University Press, 1965).

64. For the first set of activities, see Jenness, "Gordon Allport," p. 13; for the summer job, see Allport, "An Autobiography," p. 381.

65. Allport, Harvard Scrapbook (Allport Papers).

66. Ibid.

67. The reference to the Harvard Mission is contained in ibid. On the conflict, see Jenness, "Gordon W. Allport," p. 13.

68. Gordon Allport, "The Spirit of Richard Clarke Cabot," in *Person in Psychology,* p. 372.

69. See Laurie O'Brien, "'A Bold Plunge into the Sea of Values': The Career of Dr. Richard Cabot," *New England Quarterly,* 1985, *58*:533–53.

70. Richard C. Cabot, *Adventures on the Borderlands of Ethics* (New York: Harper and Brothers, 1926), p. 59.

71. Anonymous, "A Physician of the New School," *Christian Century,* 1939, *56*:660, p. 660.

72. Hugh C. Bailey, *Edgar Gardner Murphy: Gentle Progressive* (Coral Gables, Fla.: University of Miami Press, 1968), pp. 82–3.

73. Ralph Luker, *A Southern Tradition in Theology and Social Criticism, 1830–1930: The Religious Liberalism and Social Conservatism of James Warley Miles, William Porcher Dubose, and Edgar Gardner Murphy* (New York: Edwin Mellen Press, 1984), p. 337.

74. Ibid., p. 298; on segregation, see pp. 11, 322.

75. Ibid., p. 333.

76. Gardner Murphy, "Notes for a Parapsychological Autobiography," *Journal of Parapsychology,* 1957, *21*:165–78, p. 166.

77. Letter from Edgar Gardner Murphy to Gardner Murphy dated January 15, 1913 (Murphy Papers).

78. Murphy, *Gardner Murphy,* pp. 81–9.

79. Letter of August 27, 1916, from Florence Lorimer to Gardner Murphy (Murphy Papers); Letter to Dubose Murphy [Spring 1917], Murphy Family Papers, Concord Free Library.

80. On this, see p. 161 of his "Study of Myself," written in 1917 for Robert Yerkes's course on Ontogenetic Psychology (Murphy Papers).

81. Murphy, "There Is More Beyond," p. 328.

82. Lois Barclay Murphy, "The Evolution of Gardner's Thinking in Psychology and in Psychical Research"; Murphy Papers.

83. Murphy, "There Is More Beyond," p. 329. These lectures were based on Fosdick's book, *The Modern Use of the Bible* (New York: Macmillan, 1924). On Fosdick and his audience during the interwar period, see Richard H. Potter, "Popular Religion of the 1930's as Reflected in the Best Sellers of Harry Emerson Fosdick," *Journal of Popular Culture,* 1969–70, *3*:712–28 and Robert Moats Miller, *Harry Emerson Fosdick: Preacher, Pastor, Prophet* (New York: Oxford University Press, 1985).

84. Lois Barclay Murphy, "Lois Barclay Murphy," in A. O'Connell and N. Russo, eds., *Models of Achievement: Reflections of Eminent Women in Psychology* (New York: Columbia University Press, 1983), p. 87; and Murphy, *Gardner Murphy,* p. 41. See also Murphy, "Roots of an Approach." As a young woman, Murphy in fact obtained a license to preach in the Methodist denomination, as indicated in a hometown article on her 1926 wedding (Murphy Papers). Such a license meant that its holder, although ordained, was still a member of the laity; see Gerald O. McCulloh, "The Theology and Practices of Methodism, 1876–1919," in Emory Stevens Burke, ed., *The History of American Methodism,* vol. 2 (New York: Abingdon Press, 1964), pp. 658–9.

85. Robert H. Craig, "The Underside of History: American Methodism, Capitalism, and Popular Struggle," *Methodist History,* 1989, *24*:73–88, p. 86. Robert Moats Miller notes that, "of the individual denominations, Methodism planted its flag ahead – or to the Left – of the general line held by American Protestantism"; see his *American Protestantism and Social Issues, 1919–1939* (Chapel Hill: University of North Carolina Press, 1958), p. 65, and the extended discussion on pp. 65–71.

86. Martin E. Marty, *Modern American Religion: The Noise of Conflict, 1919–1941,* vol. 2 (Chicago: University of Chicago Press, 1991), p. 293.

87. Elizabeth Dilling (Mrs. Albert W. Dilling), *The Red Network: A "Who's Who" and Handbook of Radicalism for Patriots* (Kenilworth, Ill.: Elizabeth Dilling, 1934), p. 263. Dilling discusses the Methodist Federation for Social Service on pp. 190–2, "Christian Socialism" on pp. 28–33, and "Methodists Turn Socialistic" on pp. 33–8. She mentions the epithet "Red Seminary" on p. 232. UTS's most "notorious" professors were Harry Ward and Reinhold Niebuhr. For an overview of the seminary's history, see Robert T. Handy, *A History of Union Theological Seminary, 1836–1986* (New York: Columbia University Press, 1987).

88. Herbert Elmer Mills, *College Women and the Social Sciences: Essays by Herbert Elmer Mills and His Former Students* (Freeport, N.Y.: Books for Libraries Press, 1971/1934), p. xiii.

89. Ibid., p. 293.

90. Murphy, *Gardner Murphy,* p. 103.

91. Ely would marry UTS professor Eugene Lyman and later joined the UTS faculty herself. Lyman was an ardent Social Gospeler. See, for example, his "Religious Education for a New Democracy," *Journal of Religion,* 1923, *3*:449–57, where he declares that "the Bible throbs with democracy, and every true democrat's heart beats faster as he reads its prophecies, its gospels, its epistles" (p. 456).

92. References to her UTS teachers can be found in Murphy, "Lois Barclay Murphy," p. 96; Murphy, *Gardner Murphy,* p. 136; and Murphy, "Roots to an Approach," p. 173. In a letter dated June 22, 1992, from Lois Barclay Murphy to the author, Murphy stated: "When my family moved to Evanston, about 1924, my father was a friend of H. F. Rall and I took a course with him at Garrett in the summer of 1926 I believe. My father, Harry Ward, and Bishop McConnell were close friends all intensely dedicated to the Methodist Federation for Social Service." For a further sense of the MFSS during these years, see George D. McClain, "No Illusions about Capitalism: The Federation on the Eve of the Great Depression," *Radical Religion,* 1980, *5*:38–40.

93. Lois Barclay Murphy, manuscript titled "Tentative Outline," circa 1928 (Murphy Papers).

94. Lois Barclay Murphy's remarks are cited from "Lois Barclay Murphy," p. 96.

95. The phrase is Eisenstadt's, from "Remembering Goodwin Watson," p. 49.

96. Newcomb first met Lois while both were students at UTS. For information on

Newcomb, see his autobiographical essay in Gardner Lindzey, ed., *A History of Psychology in Autobiography,* vol. 6 (Englewood Cliffs, N.J.: Prentice-Hall, 1974). As a Congregationalist youth, Newcomb had planned to be an overseas missionary. Newcomb was particularly impressed with radical Methodist Harry Ward's teaching: "We heard, and most of us came to believe, that Christianity was empty without ethical behavior, which in turn was meaningless without social action" (p. 370). Newcomb briefly became a member of the Socialist party in the 1930s (p. 376).

97. Miller, *American Protestantism and Social Issues,* pp. 99, 100, 100–101.

98. Wade Crawford Barclay, Litany for "Fellowship with Those Who Hunger," in Wade Crawford Barclay, ed., *Challenge and Power: Meditations and Prayers in Personal and Social Religion for Individual and Group Use* (New York: Abingdon Press, 1936), p. 90. This work is discussed in George D. McClain, "Prayer and Class Struggle: Devotional Life within the Federation," *Radical Religion,* 1980, *5*:47–9. McClain's piece is from an entire issue devoted to Harry Ward and the Methodist Federation for Social Service.

99. Barclay, "Prayer for Justice to Labor," in *Challenge and Power,* p. 52.

100. Lois Barclay Murphy and Henry Ladd, *Emotional Factors in Learning* (Westport, Conn.: Greenwood Press, 1944), p. 5. Because Ladd died in 1941, the book was wholly authored by Murphy. She and Ladd were the initial directors of a five-year project funded by the General Education Board of the Rockefeller Foundation.

101. Gordon W. Allport, "The Education of a Teacher," *Harvard Progressive,* 1939, *4*:7–9, p. 8. The *Harvard Progressive* was published by the Harvard Student Union, a group with five hundred members who were committed to "a vaguely, rather diffuse Marxist Left ideology." See Leo Marx, "The Harvard Retrospect and the Arrested Development of American Radicalism," in John Lydenberg, ed., *Political Activism and the Academic Conscience: The Harvard Experience 1936–1941,* a symposium at Hobart and William Smith Colleges December 5 and 6, 1975 (Geneva, N.Y.: Hobart and William Smith Colleges, 1977), p. 32.

102. That the democratic ideals of Allport and the Murphys were essentially underwritten by a profound faith means that they did not feel the same need as some of their peers to search for a rational and universal set of grounds for guaranteeing the theoretical validity of democracy. For an acute discussion of the debates among those who did, see Purcell, *Crisis of Democratic Theory.*

103. On this point, see King, "An Enthusiasm for Humanity"; on the emphasis on "person" in the Social Gospel, see especially p. 53.

104. Edgar Sheffield Brightman, *Is God a Person?* (New York: Young Men's Christian Association Press, 1932), p. 5. For Brightman's Social Gospel views, see Edgar Sheffield Brightman, "From Rationalism to Empiricism," *Christian Century,* March 1, 1939, p. 277. Brightman commented: "It now seems to me that social disorder and injustice and international warfare will almost certainly continue to prevail until the problems of the social and economic order are approached in the spirit of Christian Socialism" (ibid.). Brightman's piece is one of a number on the topic of "How My Mind Has Changed in This Decade."

105. Brightman was an exponent of the philosophical doctrine of "Personalism," as derived from the thinking of Borden Parker Bowne (Brightman was in fact the Borden Parker Bowne Professor of Philosophy at Boston University). American variants of personalist philosophy ground ontological reality in personhood and hold personality to be the ultimate spiritual value. For an introduction to Bowne's thought, see Warren E. Steinkraus, ed., *Representative Essays of Borden Parker Bowne* (Utica, N.Y.: Meridian, 1981). For examples of Brightman's thinking during the 1930s, see Edgar S. Brightman, *Personality and Religion* (New York: Abingdon Press, 1934); and *The Future of Christianity* (New York: Abingdon Press, 1937).

106. Rall, "Social Change," p. 217.

107. Allport, "The Education of a Teacher," pp. 7–8.

108. John C. Burnham, "The Encounter of Christian Theology with Deterministic Psychology and Psychoanalysis," *Bulletin of the Menninger Clinic,* 1985, *49*:321–52, p. 327. Burnham notes that important aspects of Allport's scientific work – such as his insistence on the significance of subjective experience – meshed with features of modernist Protestantism (p. 328). An excellent discussion of the question of "personality" in Social Gospel discourse is Richard Wightman Fox, "The Culture of Liberal Protestant Progressivism, 1875–1925," *Journal of Interdisciplinary History,* 1993, *23*:639–60.

109. Fosdick, *Modern Use of the Bible,* pp. 16–17. Fosdick explained that, as revealed in the Christian Gospels, God "cares for men individually, loving us every one as though there were but one of us to love. And prayer is not now sending a delegate up Sinai to brave his thunders, but going into the inner chamber, shutting the door, and speaking to the Father who seeth in secret" (p. 19).

110. Attributed to "Group Action," adaptation in Barclay, *Challenge and Power,* p. 157.

111. Comments by Edgar S. Brightman in the *College Intelligencer* (a publicity circular from Henry Holt), dated February 10, 1938 vol. 7, no. 18 (Allport Papers).

112. Letter to Edgar S. Brightman from Gordon W. Allport dated January 23, 1939 (Allport Papers).

113. Letter to Gordon Allport from Wallace F. Abadie, dated April 4, 1941 (Allport Papers).

Chapter 2

1. Arthur G. Bills, "Changing Views of Psychology as Science," *Psychological Review,* 1938, *45*:377–94, pp. 377, 393. This was Bills's address as president of the Midwestern Psychological Association. On p. 377, Bills stated that his "philosophical self-complacency about the immutable criteria of a scientific psychology" had been shaken by the "caustic criticisms from the pages of Gordon Allport's latest book [*Personality*]." Bills's citation for the Newton quotation is an 1846 edition of the *Principia* published in New York by Ivison and Phinney.

2. Bills, "Changing Views of Psychology as Science," pp. 391, 377. Bills, a student of neuropsychologist Karl Lashley, received his Ph.D. from the University of Chicago in 1926. Volume 6 (1938) of *American Men of Science* indicates that Bills was an assistant professor at Chicago from 1927 to 1937, when he left to become chair of the University of Cincinnati's psychology department. Bills's area of interest was work and fatigue. See Arthur G. Bills, *General Experimental Psychology* (New York: Longmans, Green, 1934), and *The Psychology of Efficiency* (New York: Harper and Brothers, 1943).

3. Carroll C. Pratt, *The Logic of Modern Psychology* (New York: Macmillan, 1939), p. 57.

4. Michael M. Sokal, "James McKeen Cattell and American Psychology in the 1920s," in Josef Brožek, ed., *Explorations in the History of Psychology* (Lewisburg, Pa.: Bucknell University Press, 1984), pp. 275, 294.

5. Albert R. Gilgen, *American Psychology since World War II: A Profile of the Discipline* (Westport, Conn.: Greenwood Press, 1982), p. 21; for the 1929 statistic, see Ernest R. Hilgard, *Psychology in America: A Historical Survey* (San Diego: Harcourt Brace Jovanovich, 1987), p. 743. Hilgard notes that, initially, APA membership rested largely on the applicant having an appropriate university position; by 1921 the admission standards were changed so that the requirements for APA membership were a Ph.D. and the production of published scientific research. In 1925 two classes of membership were adopted – members and associates – with associates being allowed all the privileges of members save voting and officeholding (pp. 743–4). For the 1940 number, see James H. Capshew and Ernest R. Hilgard, "The Power of Service: World War II and Professional Reform in the American Psychological Association," in Rand B. Evans, Virginia Staudt Sexton, and Thomas C. Cadwallader, eds., *100 Years: The American Psychological Association: A Historical Perspective* (Washington, D.C.: American Psychological Association, 1992), p. 149. On the number of Ph.D.'s, see David Napoli, *Architects of Adjustment: The History of the Psychological Profession in the United States* (Port Washington, N.Y.: Kennikat Press, 1981), p. 47. On the early years, see also Thomas Camfield, "The Professionalization of American Psychology, 1870–1917," *Journal of the History of the Behavioral Sciences,* 1973, *9*:66–75. For details on the APA during the 1920s and 1930s see Franz Samelson, "The APA between the World Wars: 1918 to 1941," in Evans et al., *100 Years.*

6. Laurel Furumoto, "On the Margins: Women and the Professionalization of Psychology in the United States, 1890–1940," in Mitchell G. Ash and William R. Woodward, eds., *Psychology in Twentieth-Century Thought and Society* (Cambridge: Cambridge University Press, 1987), p. 102.

7. Furumoto, "On the Margins," pp. 102, 106. For a general overview of the place of women in American science during this period, see Margaret W. Rossiter, *Women Scientists in America: Struggles and Strategies to 1940* (Baltimore: Johns Hopkins University Press, 1982).

8. Gilgen, *American Psychology,* p. 26. On the role of foundations in the 1920s, see Franz Samelson, "Organizing for the Kingdom of Behavior: Academic Battles and Organizational Policies in the Twenties," *Journal of the History of the Behavioral Sciences,* 1985, *21*:33–47; Robert E. Kohler, *Partners in Science: Foundations and Natural Scientists, 1900–1945* (Chicago: University of Chicago Press, 1991); and Stephen J. Cross, "Designs for Living: Lawrence K. Frank and the Progressive Legacy in American Social Science," unpublished dissertation, Johns Hopkins University, 1994.

9. In this book, Gestalt psychology is presented as it was seen through the eyes of American commentators, who should not be taken as always faithfully rendering the Gestaltists' own interpretations of their research. For a detailed examination of those ideas in their intellectual and cultural context, see Mitchell Ash, *Gestalt Psychology in German Culture, 1890–1967: Holism and the Quest for Objectivity* (Cambridge: Cambridge University Press, 1995).

10. This cultural exchange was a two-way process, as Mitchell G. Ash has argued. See Mitchell G. Ash, "Gestalt Psychology: Origins in Germany and Reception in the United States," in Claude Buxton, ed., *Points of View in the Modern History of Psychology* (Orlando, Fla.: Academic Press, 1985); Mitchell G. Ash, "Cultural Contexts and Scientific Change in Psychology," *American Psychologist,* 1992, *47*:198–207; Mitchell G. Ash, "Emigré Psychologists after 1933: The Cultural Coding of Scientific and Professional Practices," in Mitchell G. Ash and Alfons Söllner, eds., *Forced Migration and Scientific Change: Emigré German-Speaking Scientists and Scholars after 1933* (Washington, D.C.: Cambridge University Press, 1996); and chapter 19 in Ash, *Gestalt Psychology in German Culture.* See also Michael M. Sokal, "The Gestalt Psychologists in Behaviorist America," *American Historical Review,* 1984, *89*:1240–63; and Jean Matter Mandler and George Mandler, "The Diaspora of Experimental Psychology: The Gestaltists and Others," in Donald Fleming and Bernard Bailyn, eds., *The Intellectual Migration: Europe and America, 1930–1960* (Cambridge, Mass.: Belknap Press, 1969).

11. Gordon W. Allport, *Personality: A Psychological Interpretation* (New York: Henry Holt, 1937), pp. 15–16.

12. Gardner Murphy and Friedrich Jensen, *Approaches to Personality: Some Contemporary Conceptions Used in Psychology and Psychiatry* (New York: Coward-McCann, 1932), p. 5. This section is taken from part I, authored by Murphy.

13. Robert S. Woodworth, *Contemporary Schools of Psychology* (New York: Ronald Press, 1931), p. 93. Edna Heidbreder, *Seven Psychologies* (New York: Appleton-Century-Crofts, 1933), pp. 330, 332. For a broad overview of the general contrasts between psychology in Germany and psychology in the United States during this time, see Goodwin Watson, "Psychology in Germany and Austria," *Psychological Bulletin,* 1934, *10*:755–76.

14. Ralph Barton Perry, review of *The Growth of Mind* by Kurt Koffka and *The Mentality of Apes* by Wolfgang Köhler in *The Saturday Review of Literature,* May 23, 1925, p. 773. For a hostile evaluation of Gestalt psychology, see Edward S. Robinson's article, "A Little German Band: The Solemnities of Gestalt Psychology," in the November 27, 1929 issue of the *New Republic.*

15. Gardner Murphy, *Historical Introduction to Modern Psychology* (London: Kegan Paul, 1929); and Murphy and Jensen, *Approaches to Personality.* The discussion of "Gestalt and Type" is also presented in Gardner Murphy, "The Geometry of Mind: An Interpretation of Gestalt Psychology," *Harper's Magazine,* 1931, *163*:584–93.

16. Gordon W. Allport, "The Standpoint of *Gestalt* Psychology," *Psyche,* 1923–1924, *4*:354–61, and "The Study of the Undivided Personality," *Journal of Abnormal and Social Psychology,* 1924, *19*:132–41.

17. Gordon W. Allport, "The Study of Personality by the Intuitive Method: An Experiment in Teaching from *The Locomotive God,*" *Journal of Abnormal and Social Psychology,* 1929, *24*:14–27, p. 14.

18. Gordon W. Allport, "The Leipzig Congress of Psychology," *American Journal of Psychology,* 1923, *34*:612–15, p. 612.

19. Ash, *Gestalt Psychology in German Culture,* p. 1.

20. Ibid., pp. 2–3.

21. Anne Harrington, "Interwar 'German' Psychobiology: Between Nationalism and the Irrational," *Science in Context,* 1991, *4*:429–47, p. 431. See also Anne Harrington, *Reenchanted Science: Holism in German Culture from Wilhelm II to Hitler* (Princeton, N.J.: Princeton University Press, 1996).

22. Harrington, "'German' Psychobiology," p. 433. For a related discussion, see also

Fritz K. Ringer, *The Decline of the German Mandarins: The German Academic Community, 1890–1933* (Cambridge, Mass.: Harvard University Press, 1969).

23. Gardner Murphy, "The Geometry of Mind," p. 586 (emphasis added). Mitchell Ash has pointed out to me that Murphy's claim that the Gestalt theorists had not put their case strongly enough was a misperception that was widely shared by many Americans at this time. Max Wertheimer's assertion that Gestalt theory directly concerned the nature of reality can be seen in his 1925 lecture, "On Gestalt Theory," which is available in translation in the 1944 volume of *Social Research.* This piece is quoted extensively in Mitchell G. Ash, "Gestalt Psychology in Weimar Culture," *History of the Human Sciences,* 1991, *4*:395–415. See also his *Gestalt Psychology in German Culture,* chap. 17.

24. Murphy, "Geometry of Mind," p. 584.

25. Ibid., pp. 588–9 (emphasis added).

26. Ibid., pp. 593, 590. During the 1930s, Murphy worked on a manuscript taking off from the Gestalt position entitled "Organism and Quantity: An Essay on Wholes, Parts, and Relations"; see the unpublished manuscript (with the date 1933 penciled in) of that name in the Gardner Murphy and Lois Barclay Murphy Papers, Archives of the History of American Psychology, University of Akron (hereafter Murphy Papers). This piece was eventually published in a somewhat modified form as "Organism and Quantity: A Study of Organic Structure as a Quantitative Problem," in Seymour Wapner and Bernard Kaplan, eds., *Perspectives in Psychological Theory: Essays in Honor of Heinz Werner* (New York: International Universities Press, 1960).

27. Letter from Gordon W. Allport to Kurt Goldstein, dated August 1, 1939, Gordon Allport Papers, Harvard University Archives (hereafter Allport Papers).

28. Gordon W. Allport, "The Functional Autonomy of Motives," *American Journal of Psychology,* 1937, *50*:141–56, p. 141; Gordon W. Allport, "The Psychologist's Frame of Reference," *Psychological Bulletin,* 1940, *37*:1–28, p. 1.

29. On his stance, see "The Middle of the Road," in Woodworth, *Contemporary Schools of Psychology.* On Woodworth and experimentation, see Andrew S. Winston, "Robert Sessions Woodworth and the 'Columbia Bible': How the Psychological Experiment was Redefined," *American Journal of Psychology,* 1990, *103*:391–401; see also Winston's "*Cause* and *Experiment* in Introductory Psychology: An Analysis of R. S. Woodworth's Textbooks," *Teaching of Psychology,* 1988, *15*:79–83.

30. For a brief discussion of Dorothy Thomas's behaviorism – in relation to her contributions to her husband's work – see Stephen O. Murray, "W. I. Thomas, Behaviorist Ethnologist," *Journal of the History of the Behavioral Sciences,* 1988, *24*:381–91.

31. For an overview of psychology's inferior position to philosophy at Harvard in the 1920s, see Bruce Kuklick, *The Rise of American Philosophy: Cambridge, Massachusetts 1860–1930* (New Haven, Conn.: Yale University Press, 1977), pp. 459–63.

32. Quoted in Sokal, "The Gestalt Psychologists in Behaviorist America," p. 1256. Sokal provides a detailed and insightful account of the vicissitudes of Boring's relationship with Köhler.

33. Kuklick, *Rise of American Philosophy,* p. 460.

34. On the circumstances surrounding this appointment, see Rodney G. Triplet, "Henry A. Murray and the Harvard Psychological Clinic, 1926–1938: A Struggle to Expand the Disciplinary Boundaries of Academic Psychology," unpublished dissertation, University of New Hampshire, 1983, pp. 239–40. Triplet's study has also been published as Rodney G. Triplet, "Harvard Psychology, the Psychological Clinic and Henry A. Murray: A Case Study in the Establishment of Disciplinary Boundaries," in Clark A. Elliott and Margaret Rossiter, eds., *Science at Harvard University: Historical Perspectives* (Bethlehem, Pa.: Lehigh University Press, 1992). Lashley's preferred organism for study was the rat. See Frank Beach, ed., *The Neuropsychology of Lashley* (New York: McGraw-Hill, 1960), pp. xiv-xv; and Darryl Bruce, "Lashley's Shift from Bacteriology to Neuropsychology, 1910–1917," *Journal of the History of the Behavioral Sciences,* 1986, *22*:27–44.

35. K. S. Lashley, "The Behavioristic Interpretation of Consciousness I," *Psychological Review,* 1923, *30*:237–72, p. 244. This article was cited as characteristic of Lashley by Murphy in *An Historical Introduction,* in a note on p. 278. See Lashley's remarks as well in K. S. Lashley, "Experimental Analysis of Instinctive Behavior," *Psychological Review,* 1938, *45*:445–71, where he indicates that all of the major problems of dynamic psychology can be found displayed in the behavior of a worm (cf. Allport, "Frame of Reference," p. 15 note).

36. Lashley, "Behavioristic Interpretation of Consciousness I," p. 272.

37. K. S. Lashley, "The Behavioristic Interpretation of Consciousness II," *Psychological Review,* 1923, *30*:329–53, p. 346.

38. Boring had first met Allport, when, as a graduate student, Allport made his professional debut at a meeting of E. B. Titchener's select Society of Experimental Psychologists. Titchener, who was Boring's mentor, was rudely dismissive of Allport's report of his work on personality. The episode is recounted in Allport, "An Autobiography," p. 385; and E. G. Boring, "Titchener's Experimentalists," *Journal of the History of the Behavioral Sciences,* 1967, *3*:315–25, p. 323. Titchener's relentless advocacy of "pure" science is examined further in Laurel Furumoto, "Shared Knowledge: The Experimentalists, 1904–1929," in Jill Morawski, *The Rise of Experimentation in American Psychology* (New Haven, Conn.: Yale University Press, 1988), and in Ruth Leys and Rand B. Evans, eds., *Defining American Psychology: The Correspondence between Adolf Meyer and E. B. Titchener* (Baltimore: Johns Hopkins University Press, 1990).

39. On the personality course's novelty, see Gordon Allport, "An Autobiography," in Gordon W. Allport, *The Person in Psychology: Selected Essays* (Boston: Beacon Press, 1968), p. 385.

40. John M. O'Donnell, "The Crisis of Experimentalism in the 1920s: E. G. Boring and His Uses of History," *American Psychologist,* 1979, *34*:289–95.

41. B. F. Skinner, *The Shaping of a Behaviorist: Part Two of an Autobiography* (New York: Alfred A. Knopf, 1979), p. 75.

42. Triplet's text, "Henry A. Murray," provides an excellent rendering of Murray's career in this period, and of the impact and ramifications of his struggles for legitimacy.

43. Karl Lashley to President [James Bryant] Conant, dated January 6, 1937 (Allport Papers). Triplet notes that Lashley was a personal favorite of and adviser to Conant; "Henry A. Murray," p. 244.

44. Letter to Mr. [James Bryant] Conant from Gordon W. Allport dated January 5, 1937 (Allport Papers).

45. Allport intended, if necessary, to accept an offer extended to him by Clark University, and Lashley indicated that he would return to Chicago; Triplet, "Henry A. Murray," p. 257.

46. Conant assumed that the agreement was for ten years; Triplet, "Henry A. Murray," pp. 257–8. Lashley's threatened resignation was headed off by Conant's offer to Lashley of a research professorship, which freed Lashley from all teaching and administrative obligations in the department of psychology (pp. 258–9).

47. Gordon W. Allport to Dean [George D.] Birkhoff, dated March 30, 1938 (Allport Papers). Regarding Stevens, Boring wrote to Allport: "I may not get my way at all. Or it may be my turn. Certainly Harry Murray was not my idea of a good appointment!"; Garry [E. G. Boring] to Gordon [Allport], dated October 6, 1937 (Allport Papers). Boring allowed that Stevens "was difficult, selfish, egoistic," that he "neglected his teaching for his research," and "that people generally disliked him." Boring added, however, that he hoped that he "might not live to experience the shame of knowing that Harvard took personality characteristics into account in making the appointment of very able men" (ibid.).

48. Letter from EGB [Edwin Garrigues Boring] to [S. S.] Stevens, dated October 12, 1935 (S. S. Stevens Papers, Harvard University Archives).

49. S. S. Stevens, "Psychology and the Science of Science," *Psychological Bulletin,* 1939, *36*:221–63, p. 223.

50. S. S. Stevens, "The Operational Basis of Psychology," *American Journal of Psychology,* 1935, *47*:323–30, p. 323. For an intriguing analysis of Stevens's philosophy of science, see Gary L. Hardcastle, "S. S. Stevens and the Origins of Operationism," *Philosophy of Science,* 1995, *62*:404–24. Operationism at Harvard is also discussed in Maila L. Walter, *Science and Cultural Crisis: An Intellectual Biography of Percy Williams Bridgman, 1882–1961* (Stanford, Calif.: Stanford University Press, 1990). On the nuances of such theoretical positions during this period, see Laurence Smith, *Behaviorism and Logical Positivism: A Reassessment of the Alliance* (Stanford, Calif.: Stanford University Press, 1986).

51. Stevens, "Psychology and the Science of Science," p. 236; Stevens, "Operational Basis," p. 329; Stevens, "Psychology and the Science of Science," pp. 232, 231; Stevens, "Operational Basis," p. 327.

52. Stevens, "Operational Basis," p. 328.

53. Edwin Garrigues Boring, Herbert Sidney Langfeld, and Harry Porter Weld, eds., *Psychology: A Factual Textbook* (New York: John Wiley and Sons, 1935), p. vii. The editors claim that theoretical discussions had "been to a great extent omitted and controversial points avoided," resulting, therefore, in "a factual text as one should expect from a science" (p. vii). Allport wrote *Studies in Expressive Movement* (New York: Macmillan, 1933) with Vernon, along with several key articles.

54. P. E. [Philip] Vernon review of Boring et al., *Psychology: A Factual Textbook,* in *Character and Personality,* 1935–6, *4*:181–2, p. 181.

55. On Lashley as Bills's mentor see Triplet, "Henry A. Murray," p. 265. See also Bills, "Changing Views," p. 383.

56. Bills, "Changing Views," pp. 392–3.

57. Stevens, "Psychology and the Science of Science," pp. 232, 256.

58. Laurie O'Brien, "'A Bold Plunge into the Sea of Values': The Career of Dr. Richard Cabot," *New England Quarterly,* 1985, *58*:533–53, p. 534.

59. Gordon W. Allport, "The Spirit of Richard Clarke Cabot," in Allport, *Person in Psychology,* p. 372.

60. Ibid.

61. See, for example, Richard C. Cabot, "Better Doctoring for Less Money," *American Magazine* (April and May 1916). Cabot was also out of step with many of his peers by not being a "lab man;" on this point, see O'Brien, "Bold Plunge," pp. 538–9. On the changing values in medicine during this era, see Morris J. Vogel and Charles E. Rosenberg, eds., *The Therapeutic Vision: Essays in the Social History of American Medicine* (Philadelphia: University of Pennsylvania Press, 1979); Charles Rosenberg, "Inward Vision and Outward Glance: The Shaping of the American Hospital, 1880–1914," *Bulletin of the History of Medicine,* 1979, *53*:346–91 and Charles Rosenberg, *The Care of Strangers: The Rise of America's Hospital System* (New York: Basic Books, 1987); Paul Starr, *The Social Transformation of American Medicine* (New York: Basic Books, 1982); and John Harley Warner's review article, "Science in Medicine," *Osiris,* 1985, *1*:37–58.

62. David B. Potts, "Social Ethics at Harvard, 1881–1931: A Study in Academic Activism," in Paul Buck, ed., *Social Sciences at Harvard, 1860–1920: From Inculcation to the Open Mind* (Cambridge, Mass.: Harvard University Press, 1965), p. 121; O'Brien, "'A Bold Plunge,'" p. 546. See also Lawrence T. Nichols, "The Establishment of Sociology at Harvard: A Case Study of Organizational Ambivalence and Scientific Vulnerability," in Elliott and Rossiter, *Science at Harvard.*

63. Potts, "Social Ethics at Harvard," pp. 123–4.

64. Letter from Gordon W. Allport to Dr. [Richard Clarke] Cabot, dated September 29, 1937 (Allport Papers).

65. Ralph Dudley White, Ralph Barton Perry, and Gordon Willard Allport, "Minute on the Life and Services of Richard Clarke Cabot," *Harvard University Gazette,* 1939–40, *35*:12–14, p. 13.

66. Allport, "Spirit of Richard Clarke Cabot," p. 374.

67. White, Perry, and Allport, "Minute on the Life," p. 13.

68. Ibid.

69. Letter from Richard Clarke Cabot to [Gordon W.] Allport, dated July 16, 1937 (Allport Papers).

70. Allport, "An Autobiography," p. 389.

71. Ibid., p. 371.

72. Edward A. Tiryakian, "Pitirim A. Sorokin," in David Sills, ed., *International Encyclopedia of the Social Sciences* (New York: Macmillan, 1979), p. 61; Barry V. Johnston, "Pitirim Sorokin and the American Sociological Association: The Politics of a Professional Society," *Journal of the History of the Behavioral Sciences,* 1987, *23*:103–22 p. 103.

73. Tiryakian, "Pitirim A. Sorokin," p. 62.

74. On the censorious reception to *Social and Cultural Dynamics,* see Johnston, "Pitirim Sorokin and the American Sociological Association"; and Barry Johnston, "Sorokin and Parsons at Harvard: Institutional Conflict and the Origin of a Hegemonic Tradition," *Journal of the History of the Behavioral Sciences,* 1986, *22*:107–27.

75. Allport, "An Autobiography," p. 391.

76. Clifford Geertz, *After the Fact: Two Countries, Four Decades, One Anthropologist* (Cambridge, Mass.: Harvard University Press, 1995), p. 100.

77. Kuklick, *Rise of American Philosophy,* p. 462. From 1924 to 1933, for example, the number of psychology doctorates awarded by Columbia was 89, whereas the number awarded by Harvard was 37; for the years 1934–43 the numbers are 133 and 42, respectively. On the differences in political ambience between Columbia and Harvard from a student's perspective, see Carl Schorske, "A New Yorker's Map of Cambridge: Ethnic Marginality and Political Ambivalence," in John Lydenberg, ed., *Political Activism and the Academic Conscience: The Harvard Experience 1936–1941,* a symposium at Hobart and William Smith Colleges December 5 and 6, 1975 (Geneva, N.Y.: Hobart and William Smith Colleges, 1977).

78. For a thumbnail sketch of Woodworth, see Hilgard, *Psychology in America,* pp. 86–7.

79. On Thorndike and Teachers College see Geraldine Joncich, *The Sane Positivist: A Biography of Edward L. Thorndike* (Middletown, Conn.: Wesleyan University Press, 1968). Heidbreder, "Dynamic Psychology and Columbia University," in *Seven Psychologies,* p. 291.

80. Heidbreder, *Seven Psychologies,* pp. 287, 288, 291–2.

81. Frederick C. Thorne, "Reflections on the Golden Age of Columbia Psychology," *Journal of the History of the Behavioral Sciences,* 1976, *12*:159–65, p. 161.

82. Winston, "Robert Sessions Woodworth and the 'Columbia Bible,'" p. 394.

83. Ibid., p. 397.

84. Winston, "*Cause,*" p. 80.

85. Quoted in ibid., p. 81.

86. Gardner Murphy, "Notes for a Parapsychological Autobiography," *Journal of Parapsychology,* 1957, *21*:165–78, pp. 167–8. Murphy's aim was to get "both range and depth in the matter of spontaneous cases, and in the mediumistic and cross-correspondence materials on survival" (p. 167). Murphy personally visited the SPR in 1921 to read various unpublished items and to talk to the officers.

87. The results of these experiments, Murphy noted, were "formally reported by Warcollier himself at the Third International Congress of Psychical Research, held in Paris in 1927"; Gardner Murphy, "Introduction," in René Warcollier, *Mind to Mind* (New York: Creative Age Press, 1948), p. xv.

88. Murphy, "Introduction," in *Mind to Mind,* p. xvi. On the relationship with Warcollier – and on Murphy's career as a parapsychologist in general – see Seymour H. Mauskopf and Michael R. McVaugh's well-detailed study, *The Elusive Science: Origins of Experimental Psychical Research* (Baltimore: Johns Hopkins University Press, 1980). Murphy is also discussed in R. Laurence Moore, *In Search of White Crows: Spiritualism, Parapsychology, and American Culture* (New York: Oxford University Press, 1977).

89. Gardner Murphy, "Notes," p. 169.

90. The quotations and discussion of the 1924 events regarding Columbia are in an interview of Gardner Murphy by Montague Ullman on March 21, 1967 (part II), pp. 9–10 (Murphy Papers).

91. Murphy, "Notes," p. 169. Murphy apparently was not "blind," but his eyesight had failed to the point that "he could not see to write"; interview of Murphy in *Psy-*

chic, 1970, *1*:4–39, p. 7. Murphy's book, *Historical Introduction to Modern Psychology,* which appeared in 1929, was produced by Murphy dictating to student stenographers. His sight was restored "in 1927 in response to the very unorthodox methods of Dr. Frank Marlow," a Syracuse doctor who "demonstrated and corrected a muscular imbalance"; Gardner Murphy, "Gardner Murphy," in Edwin G. Boring and Gardner Lindzey, eds., *A History of Psychology in Autobiography,* vol. 5 (New York: Appleton-Century-Crofts, 1967), p. 258. See also Lois Barclay Murphy, *Gardner Murphy,* p. 83.

92. Murphy, "Notes," p. 170. Lois Barclay Murphy recalled: "At one point when Gardner asked me what I thought about his devoting full time to psychical research I said that he might do more to support psychical research if he established himself as a respected psychologist"; Lois Barclay Murphy, "The Evolution of Gardner's Thinking in Psychology and in Psychical Research," paper delivered to the Cosmos Club, Washington, D.C., dated April 10, 1987 (Murphy Papers).

93. Murphy, "Notes," p. 170.

94. Ibid., p. 171. The illness, which Lois Barclay Murphy termed "encephalitic influenza," apparently left Gardner Murphy in a generally weakened state, in which he was "extremely fatigable"; Murphy, *Gardner Murphy,* p. 108. Murphy described himself as being a "semi-invalid" at this time (ibid., p. 258). Dr. William Hay ran a sanitarium in New York, which featured a month-long program in which the body was "detoxified" and a "special diet" prescribed (ibid., p. 108).

95. Paul D. Allison, "Experimental Parapsychology as a Rejected Science," in Roy Wallis, ed., "On the Margins of Science: The Social Construction of Rejected Knowledge," *Sociological Review Monograph,* 1979, no. 27, p. 280. H. M. Collins and T. J. Pinch, *Frames of Meaning: The Social Construction of Extraordinary Science* (London: Routledge and Kegan Paul, 1982), p. 159. H. M. Collins makes fascinating use of the debates surrounding experimental parapsychology to explore the nature of "replication" in science; see especially chapter 5, "Some Experiments in the Paranormal: The Experimenters' Regress Revisited," in his *Changing Order: Replication and Induction in Scientific Practice* (London: SAGE, 1985). For a sense of the place of research into psychic phenomena in psychology during the 1930s, see "The ESP Symposium at the APA," *Journal of Parapsychology,* 1938, *2*:247–72.

96. H. M. Collins and T. J. Pinch, "The Construction of the Paranormal: Nothing Unscientific Is Happening," in Wallis, "On the Margins," pp. 253, 243.

97. Gardner Murphy, "Dr. Rhine and the Mind's Eye," *American Scholar,* 1938, *7*:189–200, p. 189.

98. Ibid., p. 200.

99. Murphy, "Notes," p. 168.

100. Gardner Murphy, "There Is More Beyond," in T. Krawiec, ed., *The Psychologists: Autobiographies of Distinguished Living Psychologists,* vol. 2 (London: Oxford University Press, 1973), p. 340.

101. Gardner Murphy, review of Theodore Besterman, ed., *Transactions of the Fourth International Congress for Psychical Research* in *Proceedings of the Society for Psychical Research,* 1930–1, *39–40*:99–104, pp. 101, 102. The 1882 event to which Murphy is referring is the founding of the SPR, which included a number of noted intellectuals and physicists. For an attempt to sketch the SPR constitution and context, see Brian Wynne, "Physics and Psychics: Science, Symbolic Action, and Social Control in Late

Victorian England," in Barry Barnes and Steven Shapin, eds., *Natural Order: Historical Studies of Scientific Culture* (Beverly Hills, Calif.: Sage, 1979).

102. Mauskopf and McVaugh, *Elusive Science,* p. 142.

103. Murphy, "Notes for a Parapsychological Autobiography," pp. 171–2. Pratt's work was reported in volume 1 of the *Journal of Parapsychology* and Taves's in volume 3.

104. René Warcollier, *Experiments in Telepathy* (New York: Harper and Brothers, 1938). The text contained abridged portions of Warcollier's *La Télépathie,* articles from the *Revue Métapsychique,* and recent unpublished studies. A slightly longer version was published simultaneously by the Boston Society for Psychic Research in 1938.

105. Mauskopf and McVaugh, *Elusive Science,* p. 284. Riess was a psychologist at Hunter.

106. Murphy later described his influence as stemming from the "midwife role" he had taken in relation to parapsychology and "in some other branches of psychological work" (ibid., p. 174).

107. Murphy, "There Is More Beyond," p. 334.

108. See, for example, Murphy, *Gardner Murphy,* p. 173.

109. Joncich, *Thorndike,* quotes Woodworth on p. 4. By the phrase, Woodworth meant to signify Thorndike's "eclectic tolerance" as a positivist (ibid.).

110. Danziger, *Constructing the Subject,* pp. 147, 235, 146–7; on the appeal of applied educational psychology during the 1920s and 1930s, see Danziger's chapter, "Marketable Methods."

111. Dorothy Swaine Thomas, "Experiences in Interdisciplinary Research," *American Sociological Review,* 1952, *17*:663–9; for biographical details, see pp. 663–4.

112. Dorothy Swaine Thomas, "Statistics in Social Research," *American Journal of Sociology,* 1929, *35*:1–17, p. 14.

113. Thomas, "Experiences in Interdisciplinary Research," p. 665.

114. Ibid. Thomas is quoting herself here from an earlier work. Thomas added: "Whereas I still push the statistical aspect of all studies to the limit, I no longer relegate the subjective and the descriptive to secondary positions" (p. 669).

115. Alice Marie Loomis, *A Technique for Observing the Social Behavior of Nursery School Children* (New York: Bureau of Publications, Teachers College, Columbia University, 1931), p. 2. On the "control of the observer," see Simon Schaffer, "Astronomers Mark Time: Discipline and the Personal Equation," *Science in Context,* 1988, *2*:115–45. For a further discussion of this issue, see chapter 5.

116. Irving Knickerbocker review of D. S. Thomas, A. M. Loomis, and R. E. Arrington, *Observational Studies of Social Behavior,* in *Psychological Bulletin,* 1934, *31*:364–6, p. 365.

117. Ibid.

118. Ruth Arrington review of Lois Barclay Murphy, *Social Behavior and Child Personality,* in *Annals of the American Academy of Political Science,* 1938, *198*:241–2, p. 241. Arrington was looking particularly for "such important details as methods of computing observational and experimental scores" and "information on the reliability of the observational records" (p. 241).

119. Helen Thompson review of Murphy, *Social Behavior,* in the *Psychological Bulletin,* 1938, *35*:109–12, p. 112.

120. Some of Murphy's work during this period includes the following: "When School Records Display Insight," *Progressive Education* (December 1934), pp. 467–73; "Emotional Development and Guidance in Nursery School and Home," *Childhood Education,* 1936, *12*:306–11; "Social and Emotional Development," *Review of Educational Research,* 1941, *9*:479–501; "The Nursery School Contributes to Emotional Development," *Childhood Education* (May 1940), pp. 404–7; "The Contribution of Development to Morale: The Nursery Years," *Progressive Education* (May 1941), pp. 243–6; "The Evidence for Sympathy," *Childhood Education* (October 1942), pp. 58–63; and Harold E. Jones, Herbert S. Conrad, and Lois Barclay Murphy, "Emotional and Social Development and the Educative Process," in Guy Montrose Whipple, ed., *The Thirty-Eighth Yearbook of the National Society for the Study of Education: Part I: Child Development and the Curriculum* (Bloomington, Ill.: Public School, 1939).

121. Furthermore, by 1935 Murphy already had two significant publications in psychology, *Experimental Social Psychology* with Gardner Murphy in 1931, and the theoretical article, "The Influence of Social Situations upon the Behavior of Children," with Murphy again, for the 1935 *Handbook of Social Psychology.* I discuss these texts in chapter 4.

122. Murphy stated the amount of the original Macy grant, which was for 1932–5, in an interview with the author from November 1988; as a point of comparison, Murphy mentions there that Gardner Murphy's full-time salary at that time was $4,000 per year. The second Macy grant was $25,000 to establish the laboratory, with $5,000 for a secretary to make records possible, and a part-time assistant; interview with Lois Barclay Murphy by Ronnie Walker, March 1988 (Sarah Lawrence College Archives). An article by Murphy, "Nursery School as Laboratory," in the March 1939 issue of the *Sarah Lawrence Alumnae Magazine,* mentions that funds were allocated for a parent consultant and physician as well.

123. See Rosalind Rosenberg's thoughtful discussion of the University of Chicago environment and Woolley's relationship with her professors in her book *Beyond Separate Spheres: Intellectual Roots of Modern Feminism* (New Haven, Conn.: Yale University Press, 1982). The regard accorded to Woolley by her professors was recalled years later by fellow graduate student, John B. Watson: "I received my degree Magna Cum Laude and was told almost immediately by Dewey and Angell that my exam was much inferior to that of Miss Helen Thompson, who had graduated two years before with a Summa Cum Laude. I wondered then if anybody could ever equal her record. That jealousy existed for years" (quoted by Rosenberg, p. 81).

124. See Marguerite W. Zapoleon and Lois Meek Stolz, "Helen Bradford Thompson Woolley," in Barbara Sicherman and Carol Hurd Green, with Ilene Kantor and Harriette Walker, eds., *Notable American Women: The Modern Period: A Biographical Dictionary* (Cambridge, Mass.: Belknap Press, 1980), p. 658. The Vocation Bureau owed its existence to the efforts of M. Edith Campbell, a Cincinnati social economist. Rosenberg suggests that Woolley was the most successful woman of her academic cohort in "wedding her interest in science to a commitment to social reform"; see *Beyond Separate Spheres,* p. 83.

125. Lois Barclay Murphy, "Lois Barclay Murphy," in Agnes O'Connell and Nancy Russo, eds., *Models of Achievement: Reflections of Eminent Women in Psychology* (New York: Columbia University Press, 1983), p. 93; Letter from Lois Barclay Murphy to the author, dated September 1, 1989; interview of Lois Barclay Murphy by the author, November 1988.

126. Murphy, "Lois Barclay Murphy," p. 92; Lois Barclay Murphy, "Roots of an Approach to Studying Child Development," in T. Krawiec, ed., *The Psychologists: Autobiographies of Distinguished Living Psychologists,* vol. 3 (Brandon, Vt.: Clinical Psychology, 1979), p. 171; Letter from Murphy to the author, dated September 1, 1989. For detailed references to the careers of Washburn and Woolley in relation to their contemporaries, see Rossiter, *Women Scientists in America: Struggles and Strategies to 1940.* Murphy was equally admiring of Mabel Fernald, who took over for Woolley when the latter became associate director of the Merrill-Palmer School in Detroit in 1922. The clinic was primarily run by women somewhat older than Murphy, and she speaks fondly of the "warmth" and support that the other members of the testing staff extended to her, as well as of their "humorous warm appreciation of the children;" November 1988 interview and September 1, 1989, letter. In sum, Murphy "felt that both HTW and Mabel Fernald respected me, did not treat me like an inexperienced neophyte, although I was. There was no 'pulling rank' – we were all part of the team" (September 1, 1989, letter).

127. White, et al., "Minute on the Life," p. 13.

128. On Woolley, see Zapoleon and Stolz, "Helen Bradford Thompson Woolley," p. 659; Woolley's article appeared in the December 1925 number of *Pedagogical Seminary and Journal of Genetic Psychology;* an example of a Murphy case study is "Joyce from Two to Five," which appeared in the January 1941 issue of *Progressive Education.*

129. Joyce Antler, "Progressive Education and the Scientific Study of the Child: An Analysis of the Bureau of Educational Experiments," *Teachers College Record,* 1982, *83*:559–91, p. 562. On Mitchell, see also Joyce Antler, *Lucy Sprague Mitchell: The Making of a Modern Woman* (New Haven, Conn.: Yale University Press, 1987); and Lucy Sprague Mitchell, *Two Lives: The Story of Wesley Clair Mitchell and Myself* (New York: Simon and Schuster, 1953).

130. Antler, "Progressive Education," p. 584.

131. Antler, *Lucy Sprague Mitchell,* p. 408.

132. Ibid., p. 327.

133. Antler, "Progressive Education," pp. 581–4.

134. Antler, *Lucy Sprague Mitchell,* p. 317.

135. Letter from Barbara Biber to Lois Barclay Murphy dated April 13, 1935 (Murphy Papers).

136. Barbara Biber, Lois Barclay Murphy, Louise P. Woodcock, and Irma S. Black, *Child Life in School: A Study of a Seven-Year-Old Group* (New York: E. P. Dutton, 1942), pp. 22, 23.

137. Biber, et al., *Child Life,* p. 23.

138. As an experimental college, Sarah Lawrence sought to apply the principles of progressive elementary and secondary schools at the college level. The college sought to individualize instruction as much as possible, and to encourage "self-directed learning." The preferred course format was that of an informal seminar, in which students chose questions of interest to pursue on their own and in common under the guidance of a teacher. There were no grades and examinations, and fieldwork and independent study were encouraged. Evaluation consisted of numerous individual conferences and qualitative assessments. For details, see Constance Warren, *A New*

Design for Women's Education (New York: Frederick A. Stokes, 1940); Constance Warren, "Self Education: An Experiment at Sarah Lawrence College," *Progressive Education*, 1934, *11*:267–70; and Louis Tomlinson Benezet, *General Education in the Progressive College* (New York: Teachers College, Columbia University, 1943).

139. Lois Barclay Murphy, Eugene Lerner, Jane Judge, and Madeleine Grant, *Psychology for Individual Education* (New York: Columbia University Press, 1942); and Lois Barclay Murphy and Henry Ladd, *Emotional Factors in Learning* (New York: Columbia University Press, 1944). Other volumes in this series include Ruth L. Munroe, *Teaching the Individual* (New York: Columbia University Press, 1942); Esther Raushenbush, *Literature for Individual Education* (New York: Columbia University Press, 1942); and Helen Merrell Lynd, *Field Work in College Education* (New York: Columbia University Press, 1945).

140. Allport, *Personality*, pp. 3, 23, 19.

141. Eugene Hartley, "Comments on the Occasion of the Formal Presentation of a Doctor Philosophiae Honoris Causa by the University of Hamburg to Gardner Murphy 9 Dec 1976," typewritten speech (Murphy Papers).

142. Gordon Allport, "Chapel. February 25, 1935," typewritten manuscript (Allport Papers). The text is from Ecclesiastes 7:23–5 and 8:16–7. On the chapel tradition itself and Allport's participation in it, see Peter Bertocci, "Introduction," in Gordon W. Allport, *Waiting for the Lord: 33 Meditations on God and Man* (New York: Macmillan, 1978). Bertocci indicates that the daily prayers in Appleton Chapel were part of a voluntary gathering of students, faculty, and community for a fifteen-minute period before the beginning of morning classes; Allport offered meditations about twice a year from the 1930s to 1966, the year before his death (p. xi).

143. Gordon W. Allport, "Chapel," typewritten manuscript dated March 26, 1936 (Allport Papers); Allport is speaking from Isaiah 55:6–11.

144. Murphy and Ladd, *Emotional Factors in Learning*, pp. 5–6. As indicated previously, due to Ladd's death the book was wholly authored by Murphy. For an introduction to Bewer's writing, see Julius A. Bewer, *The Literature of the Old Testament*, rev. ed. (New York: Columbia University Press, 1933).

145. Gordon W. Allport, "The Psychologist's Frame of Reference," *Psychological Bulletin*, 1940, *37*:1–28, pp. 16, 22.

146. Appended to Gordon Allport, "The Fundamental Law of Psychology," undated typescript (Allport Papers). Allport adapted the "creed" for his local Harvard audience, which gives evidence of its 1930s' vintage:

> I believe in De Gustibus, the one Law of Mental Life, the maker of Schools and of Truth.
>
> And in the method of experiment, its last hope and refuge. Conceived by [blank], born of Whilhelm [sic] Wundt, It suffered through psychophysics, became existential, functional, factorial. It Descended into Freud. Some day it may arise again from the Id, and ascend into Operations, sitting on the right hand of Emerson Hall, whence it will presume to judge both the quick and the dense.
>
> I believe in Partial Identity, the occurrence of requiredness, the innerness of necessity, the positiveness of the time error, the localizing of bright pressures, and in MONISM Everlasting.
>
> Amen.

Chapter 3

1. Franklin D. Roosevelt, *The Public Papers and Addresses of Franklin D. Roosevelt,* vol. 1 (New York: Random House, 1938), p. 624; the citation is from a 1932 campaign address.

2. Jorge J. E. Gracia, *Individuality: An Essay on the Foundations of Metaphysics* (Albany: State University of New York Press, 1988), pp. xi, xii. See also P. F. Strawson, *Individuals: An Essay in Descriptive Metaphysics* (London: Methuen, 1959); and Carlo Ginzburg, "Clues: Roots of an Evidential Paradigm," in his *Myths, Emblems, Clues* (London: Hutchinson Radius, 1990). For a historical discussion of individuality from the standpoint of quantum physics, see M. Norton Wise, "How Do Sums Count? On the Cultural Origins of Statistical Causality," in Lorenz Krüger, Lorraine J. Daston, and Michael Heidelberger, eds., *The Probabilistic Revolution: Ideas in History,* vol. 1 (Cambridge, Mass.: MIT Press, 1987). For a historical discussion from the standpoint of biology, see Gregg Mitman, "Defining the Organism in the Welfare State: The Politics of Individuality in American Culture, 1890–1950," in Sabine Maasen, Everett Mendelsohn, and Peter Weingart, eds., *Biology as Society, Society as Biology: Metaphors* (Dordrecht: Kluwer Academic Publishers, 1995).

3. Gardner Murphy, "The Geometry of Mind: An Interpretation of Gestalt Psychology," *Harper's Magazine,* 1931, *163*:584–93, p. 589. A version of this commentary on Gestalt theory can also be found in chapter 1, "Gestalt and Type," of Gardner Murphy and Friedrich Jensen, *Approaches to Personality: Some Contemporary Conceptions Used in Psychology and Psychiatry* (New York: Coward-McCann, 1932).

4. Gordon W. Allport, *Personality: A Psychological Interpretation* (New York: Henry Holt, 1937), p. 3. A few years before this Allport had lamented: "When Goethe gave it as his opinion that personality is the supreme joy of the children of the earth, he could not have foreseen the joyless dissection of his romantic ideal one hundred years hence"; Gordon W. Allport review of P. M. Symonds's *Diagnosing Personality and Conduct, Journal of Social Psychology,* 1932, *3*:391–7, p. 391.

5. Allport, *Personality,* p. 3.

6. Ibid., p. 215. On the ongoing struggle of psychologists (as well as other social scientists) against the public's claims to know their own minds, see Mitchell Ash, "Historicizing Mind Science: Discourse, Practice, Subjectivity," *Science in Context,* 1992, *5*:193–207, pp. 198–9 and Jill G. Morawski and Gail A. Hornstein, "Quandary of the Quacks: The Struggle for Expert Knowledge in American Psychology, 1890–1940," in JoAnne Brown and David van Keuren, eds., *The Estate of Social Knowledge* (Baltimore: Johns Hopkins University Press, 1991).

7. Kurt Danziger, *Constructing the Subject: Historical Origins of Psychological Research* (Cambridge: Cambridge University Press, 1990), pp. 87, 77. For a discussion of a psychiatrist's defense of individuality during this period, however, see Ruth Leys, "Types of One: Adolf Meyer's Life Chart and the Representation of Individuality," *Representations,* 1991, *34*:1–28; and Ruth Leys and Rand B. Evans, eds., *Defining American Psychology: The Correspondence between Adolf Meyer and E. B. Titchener* (Baltimore: Johns Hopkins University Press, 1990). A related issue is the rise of such constructs as "the average man" in public opinion polling; Warren Susman touches on this phenomenon in "Culture of the Thirties" and "The People's Fair," in Warren Susman, *Culture as History: The Transformation of American Society in the Twentieth Century* (New York: Pantheon, 1984).

8. See, for example, Hadley Cantril, "The Prediction of Social Events," *Journal of Abnormal and Social Psychology,* 1938, *33*:364–89; Douglas McGregor, "The Major Determinants of the Prediction of Social Events," *Journal of Abnormal and Social Psychology,* 1938, *33*:179–204; Dorwin Cartwright and John R. P. French, Jr., "The Reliability of Life-History Studies," *Character and Personality,* 1939–40, *8*:110–19; Norman A. Polansky, "How Shall a Life-History Be Written?" *Character and Personality,* 1940–1, *9*:188–207; and the note by Alfred L. Baldwin in the *Psychological Bulletin,* 1940, *37*:518–19 on "The Statistical Analysis of the Structure of a Single Personality."

9. Although Murphy was more actively involved in the subfield of social psychology than in personality psychology during this period, he was engaged in writing large portions of what would become his major theoretical statement on personality, *Personality: A Biosocial Approach to Origins and Structure* (New York: Harper, 1947). See Lois Barclay Murphy's discussion of the decade-and-a-half genesis of the text in the chapter "Developing a Theory of Personality," in her *Gardner Murphy: Integrating, Expanding and Humanizing Psychology* (Jefferson, N.C.: McFarland, 1990), pp. 192–215.

10. William James, "Preface," in his *Talks to Teachers on Psychology and to Students on Some of Life's Ideals* (Cambridge, Mass.: Harvard University Press, 1983/1899), p. 4.

11. Horace M. Kallen, "Introduction: The Meaning of William James for 'Us Moderns,'" in William James, *The Philosophy of William James: Selected from His Chief Works* (New York: Modern Library, 1925), p. 53.

12. Lois Barclay Murphy, "Where Are Our Future Abe Lincolns?" in *Children Also Are People,* pamphlet contained in the Lawrence K. Frank Papers, National Library of Medicine, Washington, D.C., p. 36.

13. Gordon W. Allport, "The Psychologist's Frame of Reference," *Psychological Bulletin,* 1940, *37*:1–28, p. 25.

14. John Dewey, "'Consciousness' and Experience," in his *The Influence of Darwin on Philosophy and Other Essays in Contemporary Thought* (New York: Peter Smith, 1951), p. 242. This address was originally given in 1899 before the Philosophic Union of the University of California under the title "Psychology and Philosophic Method."

15. Gordon W. Allport, *The Use of Personal Documents in Psychological Science* (New York: Social Science Research Council, 1942), p. 148.

16. Ibid., p. 190.

17. The evaluation comes from Robert M. Crunden in his *From Self to Society, 1919–1941* (Englewood Cliffs, N.J.: Prentice-Hall, 1972), p. x. In his *Culture as History,* Susman discusses the phrase "the people" in "The People's Fair: Cultural Contradictions of a Consumer Society," p. 212, and the "culture concept" in "Culture of the Thirties," pp. 153–5. On the changing conception of "culture," see also John S. Gilkeson, Jr., "The Domestication of 'Culture' in Interwar America, 1919–1941," in Brown and van Keuren, *The Estate of Social Knowledge.*

18. W. H. Auden, "The Unknown Citizen," *New Yorker,* January 6, 1940, *15*:19. The poem was composed in March of 1939, after Auden had left Great Britain to live in New York City; on the poem's date, see Edward Mendelson, ed., *W. H. Auden: Collected Poems* (London: Faber and Faber, 1976), p. 201; and, on Auden's life, see Humphrey Carpenter, *W. H. Auden: A Biography* (Boston: Houghton Mifflin, 1981).

19. Frank Capra, *The Name above the Title: An Autobiography* (New York: Random House, 1971), p. 186. Capra's "little-guy" heroes were joined on the screen by a grittier breed of protagonists keeping faith with their personal codes of honor, in the tough-guy detectives fashioned by Dashiell Hammett and Raymond Chandler. Perhaps most iconic of all these representations of the individual, however, was Charlie Chaplin's "Little Tramp," especially as realized in *Modern Times.* Lawrence W. Levine has begun an important analysis of the period's films; see his "Hollywood's Washington: Film Images of National Politics during the Great Depression," *Prospects,* 1985, *10*:169–95.

20. James Thurber, *My Life and Hard Times* (New York: Perennial Library, 1933), pp. 9, 2, 12. Thurber explained that humorists are writers who sit "on the edge of the chair of Literature," and whose stock-in-trade was to talk "largely about small matters and smally about great affairs" (pp. 10, 11). Robert Elias, in "James Thurber: The Primitive, the Innocent, and the Individual," *American Scholar,* 1958, *27*:355–63, suggests that "the most interesting aspect of Thurber's artistic development is in terms of his search for a place where the individual can finally reside – or preside" (p. 356).

21. Alan Brinkley, *Voices of Protest: Huey Long, Father Coughlin, and the Great Depression* (New York: Alfred Knopf, 1982), p. xi. Long was a former governor of Louisiana and the state's senator, while Coughlin was a Catholic priest from the Detroit area with a nationwide radio program. During the 1930s, both men became leaders of dissident political movements denouncing the inequalities of wealth and power that modern American capitalism had generated. Coughlin's economic analyses were rife with anti-Semitism.

22. William E. Leuchtenberg, *Franklin D. Roosevelt and the New Deal: 1932–1940* (New York: Harper and Row, 1963), p. 331. For Roosevelt's speech, see his *Papers,* vol. 1, p. 624. During Herbert Hoover's administration, only one man had been needed to handle all the letters the White House received; under Roosevelt it took a staff of fifty to respond to the President's mail; Leuchtenberg, *Franklin D. Roosevelt,* p. 331. For further information, see Russell D. Buhite and David W. Levy, eds., *FDR's Fireside Chats* (Norman: University of Oklahoma Press, 1992).

23. Roosevelt, *Papers,* vol. 5, p. 235; this is from his acceptance of the renomination for the presidency in 1936.

24. The citation is from a 1938 radio address; Roosevelt, *Papers,* vol. 7, p. 586. On Roosevelt and radio, see William Stott, *Documentary Expression and Thirties America* (New York: Oxford University Press, 1973), p. 82. Stott credits Roosevelt's radio success to his personal, friendly, and informal style, an unusual stance at that time for radio addresses by a public figure. Roosevelt, for example, displayed a fondness for ad-libbing, a practice that transgressed one of the medium's cardinal conventions. Stott recounts that, during one broadcast, Roosevelt interrupted his talk to ask for a glass of water, waited for it to be brought, swallowed audibly, and then casually explained that it was quite hot that night in Washington. The gesture drew thousands of sympathetic letters (p. 82).

25. Roland Marchand, *Advertising the American Dream: Making Way for Modernity, 1920–1940* (Berkeley and Los Angeles: University of California Press, 1985), p. 352. These advertising appeals have much in common with how the individual was characterized in popular psychology of the "personality-building, will-power" variety. See, for instance, Richard J. Weiss, *The American Myth of Success: From Horatio Alger to Norman Vincent Peale* (Urbana: University of Illinois Press, 1969); Susman, "'Personality' and the Making of Twentieth Century Culture," in *Culture as History;* and Donald Meyer, *The Positive Thinkers* (Garden City, N.J.: Doubleday, 1965).

26. Marchand, *Advertising,* pp. 353, 357. Allport also noted the deployment of "personalization" in radio broadcasting in Hadley Cantril and Gordon Allport, *The Psychology of Radio* (New York: Harper and Brothers, 1935); see, for example, pp. 72, 96.

27. Marchand, *Advertising,* p. 358.

28. Gordon W. Allport, "The Personalistic Psychology of William Stern," *Character and Personality,* 1936–1937, *5*:231–46, p. 243.

29. Allport, "Frame," pp. 13–14.

30. Gordon W. Allport, "Psychology of Socialism," undated and unpublished typescript (Gordon W. Allport Papers, Harvard University Archives; hereafter Allport Papers). On the distinction between "individuality" and "individualism," see the entry for "Individual," in Raymond Williams, *Keywords: A Vocabulary of Culture and Society* (New York: Oxford University Press, 1976).

31. John Dewey and John L. Childs, "The Underlying Philosophy of Education," in William H. Kilpatrick, ed., *The Educational Frontier* (New York: Century, 1933), pp. 291–2.

32. Dewey, "'Consciousness' and Experience," p. 242.

33. Harold L. Ickes, *The New Democracy* (New York: W. W. Norton, 1934), p. 32.

34. Ibid., p. 32.

35. Emma Goldman, *The Place of the Individual in Society* (Chicago: Free Society Forum, 1940), p. 4. On reading anarchism back into its influential place in American thought, see David DeLeon, *The American as Anarchist: Reflections on Indigenous Radicalism* (Baltimore: Johns Hopkins University Press, 1978); on Goldman, see particularly pp. 94–9. Also pertinent is Deborah J. Coon, "Courtship with Anarchy: The Socio-Political Foundations of William James's Pragmatism," unpublished Ph.D. dissertation, Harvard University, 1988, and "'One Moment in the World's Salvation': Anarchism and the Radicalism of William James," *Journal of American History,* 1996, *83*:70–99.

36. Ralph Barton Perry, *Shall Not Perish from the Earth* (New York: Vanguard Press, 1940), p. 134.

37. John Dewey, *Individualism Old and New* (New York: Minton, Balch, 1930), pp. 18, 81, 83.

38. Lois Barclay Murphy, "Backgrounds of the Exploratory Courses in Psychology," in Lois Barclay Murphy, Eugene Lerner, Jane Judge, and Madeleine Grant, *Psychology for Individual Education* (New York: Columbia University Press, 1942), p. 5.

39. Allport, *Personal Documents,* p. xi.

40. See Jerome S. Bruner and Gordon W. Allport, "Fifty Years of Change in American Psychology," *Psychological Bulletin,* 1940, *37*:757–76, p. 768.

41. Arthur G. Bills, "Changing Views of Psychology as Science," *Psychological Review,* 1938, *45*:377–94, p. 384.

42. The first evaluation is from psychologist Barbara Burks (then at the Carnegie Institute of Washington), in "The College Intelligencer," a publicity circular consisting of testimonial excerpts (Allport Papers). The latter quotation is from a review of Allport's *Personality* by Harold Lasswell in *Ethics,* 1938, *49*:105–7, p. 105.

43. Dwight Chapman, review of *Personality: A Psychological Interpretation* by Gordon W. Allport, in *Sociometry,* 1938, *1*:420–5, p. 425.

44. Allport, *Personal Documents,* p. 54.

45. Ibid., p. 140.

46. Allport, *Personality,* pp. 3–4. Allport noted that "these stages of scientific labor" – abstracting an item of behavior, observing its recurrence in many members of a hypothetical class, using this uniformity as the basis for a generalization or a law, and then conducting repeated tests of empirical verification – "are described repeatedly in treatises on the scientific method" (p. 4). Allport offered the 1925 text, *Essentials of Scientific Method,* by Abraham Wolf, as a typical example.

47. Allport, *Personality,* p. 4 (emphasis added).

48. Ibid., pp. 5, 19.

49. Ibid., p. 5.

50. Ibid., pp. 23, vii.

51. For an interesting contemporary discussion of this point, see Stephen Jay Gould's "The Horn of Triton," in his *Bully for Brontosaurus* (New York: W. W. Norton, 1991).

52. Allport, *Personality,* p. 3.

53. Allport, *Personal Documents,* p. 56.

54. Allport, *Personality,* p. 549.

55. Letter from Gordon Allport to Professor W. E. Leonard dated April 5, 1928 (Allport Papers).

56. Gordon Allport, "The Personalistic Psychology of William Stern," in Gordon W. Allport, *The Person in Psychology* (Boston: Beacon Press, 1968), p. 296. This essay, originally published in 1957, was drawn from Gordon W. Allport, "William Stern: 1871–1938," *American Journal of Psychology,* 1938, *51*:770–4, and "The Personalistic Psychology of William Stern," *Character and Personality,* 1937, *5*:231–46. Allport also wrote of Stern's viewpoints in Allport, *Personality,* particularly pp. 550–8. See also Heinz Werner, "William Stern's Personalistics and Psychology of Personality," *Character and Personality,* 1938–9, *7*:109–25.

57. Relevant work of Stern's would be his three-volume *Person und Sache* (Leipzig: Barth, 1918–24); *Studien zur Personwissenschaft* (Leipzig: Barth, 1930); and *General Psychology from the Personalistic Standpoint* (New York: Macmillan, 1938/1935). For citations to American personalists, see chapter 1. Other aspects of the German philosophical and psychological context in which Stern's work was embedded can be found in Mitchell Ash, *Gestalt Psychology in German Culture, 1890–1967: Holism and the Quest for Objectivity* (Cambridge: Cambridge University Press, 1995); and William Woodward, *From Mechanism to Value: Hermann Lotze in German Medical, Philosophical, and Psychological Thought* (Cambridge: Cambridge University Press, forthcoming).

58. William Stern, "William Stern," in Carl Murchison, ed., *A History of Psychology in Autobiography,* vol. 1 (Worcester, Mass.: Clark University Press, 1930), p. 371.

59. Allport, "William Stern," p. 771.

60. Allport, *Personality,* p. 556.

61. Ibid., pp. 549, 550.

62. William James, "Pragmatism and Humanism," in *Pragmatism,* (Cambridge: Harvard University Press, 1975/1907), p. 126.

63. Ibid., p. 116.

64. William James, *A Pluralistic Universe,* (Cambridge: Harvard University Press, 1977/1909), p. 7.

65. Allport, *Personal Documents,* p. xii.

66. Ibid., p. 79.

67. Ibid., pp. 3–4. Allport named this perspective "dogmatic phenomenology" (p. 4).

68. William James, *Varieties of Religious Experience: A Study in Human Nature,* (Cambridge: Harvard University Press, 1985/1902), p. 492.

69. Allport, *Personal Documents,* p. 5.

70. Ibid., p. 6. The earlier James work that Allport is referencing here is the *Principles of Psychology.*

71. Allport, *Personal Documents,* p. 6.

72. Ibid., pp. 139, 172.

73. Ibid., p. 172.

74. Ibid., p. 141.

75. Ibid., p. 184.

76. Allport, "The Personalistic Psychology of William Stern," p. 244.

77. Allport, *Personal Documents,* p. 143.

78. Ibid., p. 6 (Allport is quoting James).

79. Ibid., p. 6.

80. Ibid., p. 37.

81. Ibid., p. 9.

82. William James, "The Energies of Men," *Philosophical Review,* 1907, *16*:1–20, p. 2; this piece was James's presidential address before the American Philosophical Association. It should be noted that, although not a practicing physician, James did possess medical training. Eugene Taylor makes the interesting argument that James's experimental psychology originated not "from the tradition of Helmholtz, Fechner, and Wundt," but rather from "the French experimental tradition, which blended the clinic with the scientific experiment"; see Eugene Taylor, "New Light on the Origin of William James's Experimental Psychology," in Michael G. Johnson and Tracy B. Henley, eds., *Reflections on The Principles of Psychology: William James after a Century* (Hillsdale, N.J.: Lawrence Erlbaum Associates, 1990), pp. 33, 55–6.

83. See "An Autobiography," in Gordon W. Allport, *The Person in Psychology: Selected Essays* (Boston: Beacon Press, 1968 [originally published 1967]), pp. 381, 378–9, 386. Given Allport's championing of the case study, one might expect him to be sympathetic to psychoanalysis; Allport believed, however, that Freud used case studies merely as typical examples of general phenomena. See Allport, *Personality,* esp. pp. 12–3.

84. For a discussion of the similarities of Murray's work to that of Gardner Murphy and Allport, see Eugene Taylor, "The Case for a Uniquely American Jamesian Tradition in Psychology," in Margaret Donnelly, ed., *Reinterpreting the Legacy of William James* (Washington, D.C.: American Psychological Association, 1992).

85. Josiah Macy, Jr., Foundation, *A Review by the President of Activities for the Six Years Ended December 31, 1936* (Ludwig Kast, president), p. 27 (Kast is quoting here from the Letter of Gift), p. 18. Murphy's first Macy grant was for her research on sympathy, and was awarded to her in 1932 by Kast.

86. On this point, see Ginzburg, "Clues: Roots of an Evidential Paradigm."

87. On the rise of quantification, see, for example, Gail A. Hornstein, "Quantifying Psychological Phenomena: Debates, Dilemmas, and Implications," in Jill G. Morawski, ed., *The Rise of Experimentation in American Psychology* (New Haven, Conn.: Yale University Press, 1988); and Kurt Danziger's chapter, "From Quantification to Methodolatry," in his *Constructing the Subject.* For contemporary data on these trends, see Bruner and Allport, "Fifty Years of Change in American Psychology"; and Allport, "Psychologist's Frame of Reference."

88. Lois Barclay Murphy, *Social Behavior and Child Personality: An Exploratory Study of Some Roots of Sympathy* (New York: Columbia University Press, 1937), p. 187.

89. Ibid., pp. 50, 237.

90. Gordon W. Allport review of Murphy's *Social Behavior* in the *Journal of Abnormal and Social Psychology,* 1938, *33*:538–43, p. 542.

91. Allport, "Psychologist's Frame," pp. 19–20.

92. Murphy, *Social Behavior,* pp. 50, 14.

93. Ibid., pp. 20, 50.

94. Murphy, "Roots," p. 175; Murphy, "Lois Barclay Murphy," p. 98.

95. Murphy, "Roots," pp. 97–8; Murphy, *Gardner Murphy,* p. 168.

96. Murphy, *Gardner Murphy,* pp. 167–71.

97. For a brief overview of the dissemination of Rorschach's methods following his death in 1922, see Morris Krugman, "Out of the Inkwell: The Rorschach Method," *Character and Personality,* 1940–1, *9*:92–110. On the diversity of projective techniques being developed during the 1930s, see Lawrence K. Frank, "Projective Methods for the Study of Personality," *Journal of Psychology,* 1939, *8*:389–413 and Ruth Horowitz and Lois Barclay Murphy, "Projective Methods in the Psychological Study of Children," *Journal of Experimental Education,* 1938–9, *7*:133–40. Frank apparently coined the term "projective methods," while Horowitz and Murphy's article appears to be the first one published in an American journal on the subject.

98. Bruno Klopfer, Morris Krugman, Douglass M. Kelley, Lois Barclay Murphy, and David Shakow, "Shall the Rorschach Be Standardized?" *American Journal of Orthopsychiatry,* 1939, *9*:514–28, p. 519. Each author, as members of a roundtable discussion, has a section.

99. Interview of Lois Barclay Murphy by the author, November 1988.

100. Murphy, "Roots," p. 98.

101. Murphy, *Gardner Murphy,* p. 167; Murphy, "Roots," p. 98. Sarah Lawrence also experimented with using Rorschach tests to assess students' individual characteristics.

102. Klopfer, in Klopfer et al., "Shall the Rorschach," p. 514.

103. Murphy in ibid., pp. 526, 527.

104. William Stern, "Cloud Pictures: A New Method for Testing Imagination," *Character and Personality,* 1937–8, *6*:132–46, p. 133.

105. Ibid., pp. 133–4.

106. Eugene Lerner and Lois Barclay Murphy, "Editorial Foreword," in Eugene Lerner and Lois Barclay Murphy, eds., *Methods for the Study of Young Children, Monographs of the Society for Research in Child Development,* vol. 6, 1941, p. ix.

107. Ibid.

108. Lawrence K. Frank, "Preface," in ibid., pp. vi, v. See also Allport's citing of psychologists' aversion to studying subjective experience as a significant methodological problem in Allport, "Psychologist's Frame of Reference," p. 19.

109. Frank, "Preface," p. vi.

110. Frank, "Projective Methods for the Study of Personality," pp. 403, 391. In regard to the intellectual basis of Frank's assertion about the "idiosyncratic meanings" of stimuli for individuals, Mitchell Ash points out that this claim derives from Kurt Koffka, "Introspection and the Method of Psychology," *British Journal of Psychology,* 1925, *15*:149–61.

111. Lerner and Murphy, *Methods for the Study,* pp. viii, x.

112. Ibid., pp. 8, viii (Frank's article on projective methods is being quoted here).

113. Ibid., p. viii (the Horowitz and Murphy article is being quoted here).

114. Ibid., pp. 11–12. In terms of her own note-taking presence in the room, Murphy felt that "many of the children accepted the idea that the pad and pencil were busy-work and felt actually less under surveillance when they were present than when they were not" (p. 12).

115. Barbara Biber, Lois B. Murphy, Louise P. Woodcock, and Irma S. Black, *Child Life in School: A Study of a Seven-Year-Old Group* (New York: E. P. Dutton, 1942), pp. 16–17. While the observational techniques used at Bank Street were similar to those used at the Sarah Lawrence nursery school, the projective testing procedure used during these Bank Street studies was more formalized, occurring on Friday mornings. The *Child Life* text also includes a chapter by Anna Hartoch and Ernst Schachtel, in which they discuss the methods and principles of the Rorschach technique and how analysts arrive at interpretations of the responses. They also present analyses of the children's specific answers.

116. Ibid., p. 351.

117. Ibid., p. 10.

118. Lerner and Murphy, "Editorial Foreword," p. x (emphasis added). Allport places a central emphasis in *Personality* on the significance of "insight" as a means of self-knowledge; see, for example, pp. 220–5.

119. It should be noted that there was much discussion of the phenomenon of "insight" in Gestalt theory, introduced in English to American psychologists in such works as Kurt Koffka, *The Growth of the Mind: An Introduction to Child-Psychology* (New York: Harcourt, Brace, 1925); and Wolfgang Köhler, *The Mentality of Apes* (New York: Harcourt, Brace, 1925). Although I believe that Allport and the Murphys' resonance with the concept derives from its use in religious contexts, this does not preclude their also being sensitized to it by the discourse of the Gestaltists. This is a question that merits further study.

120. Josiah Royce, *The Sources of Religious Insight* (New York: Charles Scribner's Sons, 1923/1912), pp. 5, 6.

121. Ibid., p. 6.

122. Bills, "Changing Views," pp. 377, 384.

123. Ibid., pp. 386, 383.

124. Frank, "Projective Methods," p. 391.

125. William F. Quillian, "Evolution and Moral Theory in America," in Stow Persons, ed., *Evolutionary Thought in America* (New Haven, Conn.: Yale University Press, 1950), p. 415. For a survey of nineteenth- and twentieth-century theories of emergent evolution, see David Blitz, *Emergent Evolution: Qualitative Novelty and the Levels of Reality* (Dordrecht: Kluwer Academic, 1992). The term was apparently coined by C. Lloyd Morgan; see his *Emergent Evolution* (London: Williams and Norgate, 1923). Emergent evolution theorists by no means agreed on the substance or implications of the doctrine. See, for example, Hans Driesch, "Emergent Evolution," H. Wildon Carr, "Life and Matter," Arthur O. Lovejoy, "The Meanings of 'Emergence' and Its Modes," and W. M. Wheeler, "Emergent Evolution of the Social," all in Edgar S. Brightman, ed., *Proceedings of the Sixth International Congress of Philosophy* (New York: Longmans, Green, 1927); William Morton Wheeler, *Emergent Evolution and the Development of Societies* (New York: W. W. Norton, 1928); and William McDougall, *Modern Materialism and Emergent Evolution* (New York: D. Van Nostrand, 1929). Alfred North Whitehead's widely read *Science and the Modern World* (New York: Macmillan, 1944/1925) displays many points of reference with the principles of emergent evolution and is perhaps the best reference point for this intellectual perspective. For a brief sketch of Whitehead's "organicism," see Donald Worster, *Nature's Economy: The Roots of Ecology* (Garden City, N.Y.: Anchor Books, 1979), especially pp. 317–20.

126. Worster, *Nature's Economy,* p. 322.

127. Sharon E. Kingsland, "Toward a Natural History of the Human Psyche: Charles Manning Child, Charles Judson Herrick, and the Dynamic View of the Individual at the University of Chicago," in Keith R. Benson, Jane Maienschein, and Ronald Rainger, eds., *The Expansion of American Biology* (New Brunswick, N.J.: Rutgers University Press, 1991), p. 217.

128. Ibid., p. 218. As Kingsland notes, the emergent doctrine contained implications for theology and social theory, as in the belief that "the desires and aspirations of humanity are determiners in the operation of the universe on the same footing with physical determiners" (p. 219). For an excellent discussion of emergent evolution in relation to religious thinking, see Stow Persons, "Evolution and Theology in America," in Persons, *Evolutionary Thought in America.* For a contemporary evaluation, see Cornelia Geer Le Boutillier, *Religious Values in the Philosophy of Emergent Evolution* (New York: Columbia University, 1936).

129. Gardner Murphy, *General Psychology* (New York: Harper, 1933), p. 628 (emphasis added).

130. Ibid., p. 639.

131. Gardner Murphy, *A Briefer General Psychology* (New York: Harper, 1935), p. 524.

132. Irwin Edman, "Foreword," in Henri Bergson, *Creative Evolution* (Westport, Conn.: Greenwood Press, 1977/1907), p. x.

133. Letter from Lois Barclay to "Auntie dear," dated December 6, 1920 (Lois Barclay Murphy Papers in the History of Medicine Division, National Library of Medicine, Washington, D.C.).

134. Quoted in Richard J. Bernstein, "Introduction," in James, *A Pluralistic Universe,* p. xxii. This work contains a chapter entitled "Bergson and His Critique of Intellectualism." As Peter Hare explains, James borrowed the term "tychism" from Charles Pierce and understood it to be a doctrine "which represents order as being gradually won and always in the making." See Hare's "Introduction" in James, *Some Problems of Philosophy,* p. xxviii (Hare is quoting James). Also pertinent is James on Harald Höffding; see James's "Preface to Höffding's *Problems of Philosophy,*" in William James, *Essays in Philosophy* (Cambridge, Mass.: Harvard University Press, 1978).

135. Hare, "Introduction," p. xxiii.

136. James, *Some Problems of Philosophy,* p. 29.

137. Bernstein, "Introduction," p. xxv.

138. Gordon W. Allport, "The Productive Paradoxes of William James," in Gordon W. Allport, *The Person in Psychology: Selected Essays* (Boston: Beacon Press, 1968 [the essay was written in 1942 and published the following year]), p. 321.

139. Gordon W. Allport, "Intuition as a Method in Psychology," undated unpublished manuscript, circa 1927 (Allport Papers).

140. H. S. Jennings, "Diverse Doctrines of Evolution, Their Relation to the Practice of Science and of Life," *Science,* 1927, *65*:19–25, p. 19.

141. Quoted on p. 17 of Gordon W. Allport, "The Study of Personality by the Intuitive Method: An Experiment in Teaching from *The Locomotive God,*" *Journal of Abnormal and Social Psychology,* 1929, *24*:14–27.

142. Ibid., pp. 21, 22.

143. Allport, *Personality,* p. 207.

144. Ibid., p. 194.

145. Allport, *Personal Documents,* p. 158.

146. Ibid., p. 159.

147. Ibid. This point about Heine appears to have been taken from an observation made by Social Gospel theologian and personalist philosopher Edgar Brightman, which appeared in his *The Future of Christianity* (New York: Abingdon Press, 1937), pp. 40–2.

148. Gordon W. Allport, "Psychology in the Near Future," undated unpublished manuscript (Allport Papers). The address was delivered in Durham, New Hampshire, to undergraduates; from internal references it appears to be from 1940 or 1941.

149. Allport, *Personal Documents,* pp. 159, 160.

150. Gordon W. Allport, "Critical Report on S.S.R.C. [Social Science Research Council] Predictions of Personal Adjustment," unpublished manuscript dated June 2, 1941 (Allport Papers); Allport, "Psychology in the Near Future."

151. Allport, *Personal Documents,* pp. 160, 188.

152. Lawrence K. Frank, "Structure, Function and Growth," *Philosophy of Science,* 1935, *2*:210–35, pp. 228–9.

153. Ibid., p. 229. See also Lawrence K. Frank, "Causation: An Episode in the History of Thought," *Journal of Philosophy,* 1934, *31*:421–8.

154. Allport, *Personality,* p. 297.

155. Ibid., p. 11.

156. In contemporary philosophy, Nancy Cartwright has explored this aspect of causal explanation, arguing that many valid scientific explanations cannot be covered by laws. Most commonly accepted laws are ceteris paribus laws, holding only in special circumstances, usually ideal conditions. See "The Truth Doesn't Explain Much," *American Philosophical Quarterly,* 1980, *17*:184–9.

157. In *Nature's Capacities and Their Measurement* (Oxford: Oxford University Press, 1989), Nancy Cartwright also uses this analogy for making the point that "laws are a poor stopping-point. It is hard to find them in nature and we are always having to make excuses for them: why they have exceptions – big or little; why they only work for models in the head; why it takes an engineer with a special knowledge of real materials and a not too literal mind to apply physics to reality" (p. 8).

158. Allport, *Personal Documents,* p. 58.

159. Gerd Gigerenzer, Zeno Swijtink, Theodore Porter, Lorraine Daston, John Beatty, and Lorenz Krüger, *The Empire of Chance: How Probability Changed Science and Everyday Life* (Cambridge: Cambridge University Press, 1989), pp. 39, 41–42. Sentiments of this kind by Allport align him with the predominantly German critique of statistical assumptions described by Theodore M. Porter in his book, *The Rise of Statistical Thinking 1820–1900* (Princeton, N.J.: Princeton University Press, 1986). Such critics charged, for example, that "the mean of all the organs and limbs for a particular species of animal would likely fail even to be a viable organism" (p. 172). M. Norton Wise elaborates on this perspective in "How Do Sums Count? On the Cultural Origins of Statistical Causality," in Lorenz Krüger, Lorraine J. Daston, and Michael Heidelberger, eds., *The Probabilistic Revolution: Ideas in History,* vol. 1 (Cambridge, Mass.: MIT Press, 1987). For an insightful discussion of the relation of psychological theory to statistical models, see Gerd Gigerenzer and David J. Murray, *Cognition as Intuitive Statistics* (Hillsdale, N.J.: Lawrence Erlbaum Associates, 1987).

160. Allport, *Personality,* p. 2. Allport also found Kurt Lewin's comments regarding the lawfulness of single cases suggestive; see, for example, p. 363 of *Personality.* On Lewin's attempts to construct ideal-typical situations, see Mitchell G. Ash, "Emigré Psychologists after 1933: The Cultural Coding of Scientific and Professional Practices," in Mitchell G. Ash and Alfons Söllner, eds., *Forced Migration and Scientific Change: Emigré German-Speaking Scientists and Scholars after 1933* (Cambridge: Cambridge University Press, 1996).

161. Allport, *Personality,* p. 21. Evelyn Fox Keller, in *Reflections on Gender and Science* (New Haven, Conn.: Yale University Press, 1985), makes a similar critique, remarking, for example, that "the kinds of order generated or generable by law comprise only a subset of a larger category of observable or apprehensible regularities, rhythms, and patterns" (p. 132). Further discussion can be found on pp. 131–5.

162. Allport, *Personality,* p. 558.

163. Gardner Murphy, *Historical Introduction to Modern Psychology,* rev. ed. (New York: Harcourt, Brace, 1949/1929).

164. Ibid., p. 208.

165. Ibid., p. 209.

Chapter 4

1. The phrase "common public world of nature" is Lawrence K. Frank's. See his article "Projective Methods for the Study of Personality," *Journal of Psychology,* 1939, *8*:389–413, p. 391. For discussions of the "objectivity" concept in historical context in the social sciences, see Robert C. Bannister, *Sociology and Scientism: The American Quest for Objectivity, 1880–1940* (Chapel Hill: University of North Carolina Press, 1987); Peter Novick, *That Noble Dream: The "Objectivity Question" and the American Historical Profession* (Cambridge: Cambridge University Press, 1988); Kurt Danziger, *Constructing the Subject: Historical Origins of Psychological Research* (Cambridge: Cambridge University Press, 1990); Dorothy Ross, *The Origins of American Social Science* (Cambridge: Cambridge University Press, 1991); and Theodore Porter, *Trust in Numbers: The Pursuit of Objectivity in Science and Public Life* (Princeton, N.J.: Princeton University Press, 1995).

2. Gardner Murphy, "The Research Task of Social Psychology," *Journal of Social Psychology,* 1939, *10*:107–20, p. 119.

3. Leon Fink, in his article "'Intellectuals' versus 'Workers': Academic Requirements and the Creation of Labor History," *American Historical Review,* 1991, *96*:395–421, notes that the close association of leading Progressive intellectual figures – and here he mentions such individuals as Jane Addams, Florence Kelley, and John Dewey – with labor circles has gone largely unexplored by historians (p. 419). The scientific activism I will be discussing in this chapter drew sustenance from this same foundational relationship and has similarly been overlooked.

4. Jane Addams, *Democracy and Social Ethics* (New York: Macmillan, 1913/1902), pp. 6–7, 6. Lois Barclay Murphy recalled how, as a young girl in Chicago, "Jane Addams . . . became my idol"; see Lois Barclay Murphy, *Gardner Murphy: Integrating, Expanding and Humanizing Psychology* (Jefferson, N.C.: McFarland, 1990), p. 55.

5. Addams, *Democracy and Social Ethics,* p. 7.

6. Wade Crawford Barclay, "Prayer for Justice to Labor," in Wade Crawford Barclay, ed., *Challenge and Power: Meditations and Prayers in Personal and Social Religion* (New York: Abingdon Press, 1936), p. 52.

7. Harry F. Ward, *The Gospel for a Working World* (New York: Missionary Education Movement of the United States and Canada, 1918), p. 192. See also Harry F. Ward, *Our Economic Morality and the Ethic of Jesus* (New York: Macmillan, 1929).

8. Francis J. McConnell, *The Christian Ideal and Social Control* (Chicago: University of Chicago, 1932), pp. 36–7.

9. Ibid., p. 34.

10. Henry Wallace, "The Social Advantages and Disadvantages of the Engineering-Scientific Approach to Civilization," *Science,* 1934, *79*:1–5, p. 3.

11. Cited in Peter J. Kuznick, *Beyond the Laboratory: Scientists as Political Activists in 1930s America* (Chicago: University of Chicago Press, 1987), p. 40. Kuznick's study is an indispensable starting point in regard to scientific activism. The author was the sociologist Read Bain, writing in the March 1933 issue of *Social Forces.*

12. For an overview of this issue, see George Daniels, "The Pure-Science Ideal and Democratic Culture," *Science,* 1967, *156*:1699–1705. See also Steven Shapin, who, in "Science and the Public," has sketched out the incompatibility that exists between public demands for the accountability of scientists for the knowledge that they produce, and the "autonomy that, scientists said, was the condition for the health of science, its capacity to yield objective knowledge, and, thus, to produce the knowledge upon which technological innovation could be based"; Steven Shapin, "Science and the Public," in R. C. Olby, G. N. Cantor, J. R. R. Christie, and M. J. S. Hodge, eds., *Companion to the History of Modern Science* (London: Routledge, 1990), p. 1004. For a discussion of changing views of science from the point of view of physics, see Daniel J. Kevles, *The Physicists: The History of a Scientific Community in America* (Cambridge, Mass.: Harvard University Press, 1987).

13. Ronald C. Tobey, *The American Ideology of National Science, 1919–1930* (Pittsburgh, Pa.: University of Pittsburgh Press, 1971), p. 229.

14. Wesley Mitchell, "The Public Relations of Science," reprinted in Robert Kargon, ed., *The Maturing of American Science: A Portrait of Science in Public Life Drawn from the Presidential Addresses of the American Association for the Advancement of Science 1920–1970* (Washington, D.C.: American Association for the Advancement of Science, 1974), p. 81 (Mitchell's speech is from 1939). See also chapter 1 in Kuznick, *Beyond the Laboratory.* On p. 64, Kuznick cites a speech given at MIT by President Franklin Roosevelt, in which Roosevelt held science and engineering responsible for the depression's social and economic dislocations; the speech was reprinted in *Science* in 1936, accompanied by a rebuttal from Karl Compton.

15. A task that was taken up by sociologist Robert K. Merton in his "A Note on Science and Democracy," *Journal of Legal and Political Sociology,* 1942, *1*:115–26. A later version of this essay, entitled "Science and Democratic Social Structure," can be found in Robert K. Merton, *Social Theory and Social Structure,* rev. ed. (Glencoe: Free Press, 1968). On Merton, see David A. Hollinger, "The Defense of Democracy and Robert K. Merton's Formulation of the Scientific Ethos," *Knowledge and Society: Studies in the Sociology of Culture Past and Present,* 1983, *4*:1–15 and Everett Mendelsohn, "Robert K. Merton: The Celebration and Defense of Science," *Science in Context,* 1989, *3*:269–89.

16. Mitchell, "Public Relations," p. 89.

17. Robert W. Rydell, "The Fan Dance of Science: American World's Fairs in the Great Depression," *Isis,* 1985, *76*:525–42, p. 542. Other evaluations of this public relations push can be found in Warren Susman, "The People's Fair: Cultural Contradictions of a Consumer Society," in Warren Susman, *Culture as History: The Transformation of American Society in the Twentieth Century* (New York: Pantheon, 1984); and Folke Kihlstedt, "Utopia Realized: The World's Fairs of the 1930s," in Joseph Corn, ed., *Imagining Tomorrow: History, Technology, and the American Future* (Cambridge, Mass.: MIT Press, 1986).

18. Rydell, "Fan Dance," p. 529.

19. Quoted in ibid., p. 531. Rydell relates: "From November 1930 through May 1931, the SAC [Science Advisory Council] produced thirty fifteen-minute nationwide radio broadcasts – required listening in many college science classes – that promoted scientific features at the [Chicago] fair and the gospel of scientific idealism generally" (p. 530).

20. Paul Carter, "Science and the Common Man," *American Scholar,* 1975/76, *45*:778–94, p. 791.

21. On this, see especially Bannister, *Sociology and Scientism;* and Porter, *Trust in Numbers.*

22. Anonymous, "Scientists as Leaders," *New Republic,* September 24, 1930, *64*:140–1, p. 141. The article identifies the man only as "Mr. Ashby," an "English delegate, with Oxford behind him" (p. 140).

23. Ibid., p. 141.

24. Boyd H. Bode, *Progressive Education at the Crossroads* (New York: Newson, 1938), p. 24. Indeed, the 1934–5 course catalog for Sarah Lawrence – where Lois Barclay Murphy was a faculty member – confronted the "ivory tower" question head on. In describing the institution's structure, the catalog advised: "A college situation of this sort, with students of varied background, and faculty who are participating actively in home and civic life, is no ivory tower where a student may retreat from 'real life.' It is rather a sector of real life where a student may secure the most adequate preparation for subsequent participation in and adjustment to a changing world"; quoted in Louis Tomlinson Benezet, *General Education in the Progressive College* (New York: Teachers College, Columbia University, 1943), p. 56.

25. Benjamin Ginzburg, "The Scientist in a Crumbling Civilization," *Social Frontier,* 1934, *1*:13–18. For a taste of the ecumenical use being made of the ivory tower critique, see Eric F. Goldman, "Historians and the Ivory Tower," *Social Frontier,* 1936, *3*:278–80. Adherents of an "objective" approach to history, Goldman remarked, "would never have denied themselves the right to attack prevailing institutions and ideals had they not been convinced that by describing things as they were they were also describing things as they *ought to be.* . . . They were quite ready 'to tell a story and leave philosophy to others,' because they were pleased with the story they had to tell" (p. 279).

26. Ginzburg, "Scientist in a Crumbling Civilization," pp. 13, 14.

27. Ibid., p. 14.

28. Ibid., p. 15.

29. Ibid., p. 18.

30. Murphy, "Research Task," p. 118.

31. Ibid., p. 118.

32. Ibid., pp. 118, 119.

33. Gardner Murphy, "The Growth of Our Social Attitudes," published in Paul Kaufman, ed., *Understanding Ourselves: A Survey of Psychology Today* (Washington, D.C.: Graduate School of the United States Department of Agriculture, 1938). I am working from a typescript of the speech (Gardner Murphy and Lois Barclay Murphy Papers at the Archives of the History of American Psychology, University of Akron [hereafter Murphy Papers]).

34. Ibid.

35. Jerome S. Bruner and Gordon W. Allport, "Fifty Years of Change in American Psychology," *Psychological Bulletin,* 1940, *37*:757–76, p. 774 all. "Fifty Years of Change" presents the results of the statistical survey of journals Allport conducted as the basis for his presidential address, "The Psychologists' Frame of Reference."

36. Bruner and Allport, "Fifty Years of Change," p. 775.

37. Allport, "Frame of Reference," p. 25.

38. Ibid., p. 27.

39. Ibid.

40. Ibid., p. 26.

41. Gordon W. Allport, "Psychology in the Near Future," undated unpublished manuscript (Gordon W. Allport Papers, Harvard University Archives). The address was delivered in Durham, New Hampshire, to undergraduates; from internal references it appears to be from 1940 or 1941.

42. George Hartmann, "Value as the Unifying Concept of the Social Sciences," *Journal of Social Psychology,* 1939, *10*:563–75, p. 566. Hartmann, one of SPSSI's founders and its third president, was a prominent member of the Socialist party. When Hartmann acted on his pacifist beliefs and was fired from Columbia University for his leadership in opposing the United States' entry into World War II, Allport arranged to have him appointed as a visiting lecturer at Harvard and as director of research for the Cambridge-Somerville Youth Study (with which Allport was affiliated); on Hartmann's travails, see Lorenz J. Finison, "The Psychological Insurgency: 1936–1945," *Journal of Social Issues,* 1986, *42*:21–33, p. 25; S. Stansfeld Sargent and Benjamin Harris, "Academic Freedom, Civil Liberties, and SPSSI," *Journal of Social Issues,* 1986, *42*:43–67, pp. 46–7; and Arthur Jenness, "Gordon W. Allport," in David L. Sills, ed., *International Encyclopedia of the Social Sciences* (New York: Free Press, 1979), p. 15.

43. Murphy, "Research Task," p. 120.

44. Gardner Murphy's students covered a number of topics: the relationship between public opinion and the individual; how social norms are established; how socialization during young adulthood influences attitude change. See, for example, Gardner Murphy and Rensis Likert, *Public Opinion and the Individual* (New York: Harper & Row, 1938); Muzafer Sherif, *The Psychology of Social Norms* (New York: Harper and Row, 1936); and Theodore Newcomb, *Personality and Social Change* (New York: Holt, Rinehart and Winston, 1943). A close ally of Murphy's in social psychology at Columbia was Otto Klineberg; Klineberg attacked biological notions of racial inferiority in his *Race Differences* (New York: Harper and Row, 1935), which came out under Murphy's editorship.

45. Murphy and Murphy, *Experimental Social Psychology,* p. 8.

46. Ibid., p. 17.

47. Leonard Doob, review of Murphy et al., *Experimental Social Psychology,* in the *Psychological Bulletin,* 1938, *35*:112–15, p. 112.

48. May, along with Hugh Hartshorne, was author of the famous 1929 study demonstrating the lack of trait-consistency across situations in schoolchildren who were given various opportunities for deceit. In Lois Murphy's biography of Gardner Murphy, she states that although May did initially sign on as coauthor, he later "decided he was too busy to do justice to the task"; see Murphy, *Gardner Murphy,* p. 115. Gardner then asked Lois – who was unexpectedly on hiatus after being fired from Sarah Lawrence as an agitator for the extension of faculty rights – to undertake the review of research on children. Murphy was rehired by the college the following year after the president was fired. On the IHR, see J. G. Morawski, "Organizing Knowledge and Human Behavior at Yale's Institute of Human Relations," *Isis,* 1986, *77*:219–42.

49. Mark A. May, review of Murphy and Murphy, *Experimental Social Psychology,* in the *Psychological Bulletin,* 1933, *30*:776–8, p. 777.

50. Doob, review of Murphy et al., *Experimental Social Psychology,* rev. ed., pp. 113–14, p. 114.

51. Gordon W. Allport, review of Carl Murchison, ed., *Social Psychology,* in *Psychological Bulletin,* 1929, *26*:709–10, p. 710.

52. Addams, *Democracy and Social Ethics,* p. 7.

53. Murphy, *Social Behavior,* p. 3.

54. Ibid., pp. 3, 23.

55. Barbara Biber, Lois B. Murphy, Louise P. Woodcock, and Irma S. Black, *Child Life in School: A Study of a Seven-Year-Old Group* (New York: E. P. Dutton, 1942), p. 13.

56. Ibid., pp. 13, 16.

57. Murphy, *Social Behavior,* pp. 24–5. Murphy also implicated mainstream Protestantism in the stabilization of these social patterns: "Much of the sympathy in the western tradition, emanating from Christianity, has been characterized by a quality of patronizing, a distinct sense of the stronger taking care of the weaker, rather than a group taking care of group members. This pattern of defense of the weak, and the self-consciousness toward them, produced a sledge-hammer reaction against this religion by some who considered it a form of enlightened self-interest on the part of those who wished to maintain their own power by keeping the weak always weak" (ibid., p. 28).

58. Ibid., pp. 26, 29.

59. Biber et al., *Child Life,* p. 14.

60. See, for example, Lois B. Murphy, "Interiorization of Family Experience by Normal Pre-School Children, as Revealed by Certain Projective Methods," *The Psychologists' League Journal,* 1940, *4*:3–4. In this piece Murphy sought to demonstrate how projective techniques such as finger painting and free play with miniature life toys indicated the patterns of experience that had been assimilated by preschool children in relation to their families. It should be noted that the journal in which Murphy published this piece was sponsored by the Psychologists League, an organization that was "to the left, politically, of any other organization of psychologists at the time," including SPSSI; see Lorenz J. Finison, "Unemployment, Politics, and the History of Organized Psychology, II: The Psychologists League, the WPA, and the National Health Program," *American Psychologist,* 1978, *33*:471–7, p. 475.

61. Ruth Horowitz and Lois Barclay Murphy, "Projective Methods in the Psychological Study of Children," *Journal of Experimental Education,* 1938–9, *7*:133–40, p. 136.

62. Ibid., p. 137.

63. Ibid., p. 138.

64. Allport addressed the need for "an efficient Pied Piper" to lead away the "plague of rats" afflicting the profession in "Personality: A Problem For Science or a Problem for Art?" a 1938 essay reprinted in his *Personality and Social Encounter* (Boston: Beacon Press, 1960), p. 4. On the appeal of animal experimentation for his peers, see also Allport's presidential address, "Frame of Reference," and the companion article, "Fifty Years of Research." Reflecting in later years on the place of her research in the 1930s in relation to the work of learning theorists during this period, Lois Murphy remarked: "[Edward] Tolman had more imagination [than Clark Hull] – his idea about the rat pausing to make a choice – well this is all so interesting – if the rat is making a choice and making his own decision about which way he's going to go what about kids?" (interview by the author, November 1988).

65. Murphy and Murphy, "The Influence of Social Situations," p. 1035.

66. Ibid., p. 1036.

67. Ibid., p. 1037.

68. Ibid., pp. 1037–8, 1093.

69. Ibid., p. 1053. Elsewhere Gardner Murphy emphasized the crucial role that these other disciplines could play in contextualizing psychological research questions, arguing that there was "a pressing need for a thorough grounding in the content and the methods of history," for such questions as "the economic evolution of modern Europe must be thoroughly studied, and its implications for the present social psychology of Europe and America grasped, before we can frame significant problems in attitude research or work our way to the formulation of appropriate research techniques"; Murphy, "Research Task," pp. 113–14.

70. Murphy and Murphy, "The Influence of Social Situations," p. 1053.

71. Ibid., p. 1062.

72. Ibid.

73. Ibid., p. 1047.

74. Ibid.

75. Ibid.

76. Murphy and Murphy, *Experimental Social Psychology,* p. 2.

77. Gordon W. Allport, "Report of Conference Called Jan. 18, 1935 (Princeton) by Dr. Lawrence Frank, General Education Board, on Significance of the Concepts of Gestalt Psychology for the Problem of the Development of Personality," typescript notes (Allport Papers). Allport remarked that Frank "despaired of [achieving a] scientific approach, because no single variable can be interpreted in one un[eq]uivocal way," and stated that Murphy [no designation which Murphy] agreed, adding that psychologists "cannot even measure environment, let alone personality, which depends on environment and on other things as well." Allport found these views "pessimistic, simply because they had not distinguished environment from person. One does get dizzy when [the] two are combined" (ibid.)

78. Gordon W. Allport, review of Murphy, *Social Behavior and Child Personality,* in the *Journal of Abnormal and Social Psychology,* 1938, *33*:538–43, pp. 543, 539.

79. Ibid., pp. 538, p. 539. Sociologist Kimball Young, on the other hand, reprimanded Murphy for not being sociological *enough* in a survey review in the *American Journal of Sociology,* 1938–9, *44*:463–9. Young began by allowing that "the laboratory-trained psychologists are gradually learning about the importance of society and culture in human behavior" (p. 463). Among the demurrals issued by Young, however, was one directed at Murphy: "Unfortunately, the author in interpreting her data paid no attention to the theoretical contributions of George H. Mead and John Dewey. For example, much of her material would take on more significance if she had recognized the social nature of the rise of the self" (p. 466).

80. Allport, review of *Social Behavior,* p. 540.

81. Gardner Murphy, "Research Task," p. 112 (Eduard Spranger was a German psychologist of the *Verstehen* school). Murphy was here arguing for psychologists to draw more fully on biology, or, as he put it, "the language of Morgan and Todd, Child, Gasser, and Ranson" (ibid.).

82. Murphy and Murphy, "Influence of Social Situations," p. 1048.

83. Allport, "Frame of Reference," p. 19.

84. Allport, *Personality,* p. viii.

85. Ibid., pp. 37–8. Allport referenced here E. Faris, "The Concept of Social Attitudes," *Journal of Applied Sociology,* 1925, *9*:404–9.

86. Allport, *Personality,* pp. 38–9.

87. Wade Crawford Barclay, "The Nurture of Life," in Wade Crawford Barclay, Arlo A. Brown, Alma S. Sheridan, William J. Thompson, and Harold J. Sheridan, *Life in the Making* (New York: Methodist Book Concern, 1917), p. 11.

88. Gordon W. Allport, "The Spirit of Richard Clarke Cabot," in his *The Person in Psychology: Selected Essays* (Boston: Beacon Press, 1968), p. 374. Allport is quoting from a memorandum accompanying Cabot's will.

89. Murphy, *Social Behavior,* p. 321.

90. Steven Shapin discusses how symbolic representations of the "solitary philosopher" have informed the image of the socially disengaged scientist, in "'The Mind Is Its Own Place': Science and Solitude in Seventeenth-Century England," *Science in Context,* 1990, *4*:191–18.

91. Hartmann, "Value as the Unifying Concept of the Social Sciences," p. 568.

92. Ibid., p. 570.

93. Gordon W. Allport, untitled Chapel Talk from November 15, 1940, using Micah 7:2, 5–9, 19 (Allport Papers).

94. Kuznick, *Beyond the Laboratory,* pp. 177, 181.

95. Lorenz J. Finison, "Psychologists and Spain: A Historical Note," *American Psychologist,* 1977, *32*:1080–4.

96. Lawrence Levine, "Hollywood's Washington: Film Images of National Politics During the Great Depression," *Prospects,* 1985, *10*:169–95, pp. 170, 178, 181. Levine suggests that "this despair can be seen in how often those who champion the cause of the people and the cause of democracy are forced to go beyond the democratic method, are forced to lie to and manipulate the people for their own good" (pp. 181–2).

97. Gordon W. Allport, "What is Fascism?" undated typescript with the notation "1939?" (Allport Papers).

98. Allport, November 15, 1940 Chapel Talk.

99. Ibid.

100. Ibid.

101. Letter to Dr. Krechevsky from Gordon Allport, dated March 6 [1936] (SPSSI Archives, at the Archives of the History of American Psychology, University of Akron [hereafter SPSSI Archives]). Allport had written a number of senators on this score.

102. "Psychology of Socialism," unattributed and undated typescript in Allport's style and with marginal notations in his hand (Allport Papers). From references within the text the piece appears to be from the late 1930s.

103. Gordon W. Allport, "Attitudes," in Carl Murchison, ed., *Handbook of Social Psychology* (Worcester, Mass.: Clark University Press, 1935), p. 806.

104. Ibid., p. 813.

105. Hadley Cantril and Gordon W. Allport, *The Psychology of Radio* (New York: Harper and Brothers, 1935), pp. 9, 8–9, 7–8.

106. Ibid., pp. 8–9.

107. Ibid., pp. 270–1.

108. Ibid., pp. 4, 43, 48.

109. Ibid., p. 48.

110. Murphy, "Research Task," pp. 107, 119 (emphasis added).

111. In "Psychology of Socialism" Allport remarked that "socialism is not a science, but a hope, a longing, a faith, a goal."

112. Murphy, "Research Task," p. 119.

113. Ibid., p. 120.

114. Ibid.

115. Ibid., pp. 119–20.

116. Anonymous, "Society for the Psychological Study of Social Issues Yearbook for 1942: Resistance to Social Change," unpublished and undated typescript (SPSSI Archives).

117. Another factor was perhaps the immense problems that had arisen in publishing what became the 1940 yearbook, *Industrial Conflict: A Psychological Interpretation.* For a discussion of the lengthy, traumatic and divisive attempt to produce that yearbook, see Lorenz J. Finison, "An Aspect of the Early History of the Society for the Psychological Study of Social Issues: Psychologists and Labor," *Journal of the History of the Behavioral Sciences,* 1979, *15*:29–37.

118. Barclay, *Challenge and Power,* p. 11.

119. McConnell, *The Christian Ideal and Social Control,* p. 34.

Chapter 5

1. Rachel Carson, *Under the Sea-Wind: A Naturalist's Picture of Ocean Life* (New York: Simon and Schuster, 1941), pp. xvi–xvii.

2. Henry David Thoreau, *The Writings of Henry David Thoreau: Journal,* vol. 13 (Boston: Houghton Mifflin, 1906), p. 154 (the entry is from February 18, 1860).

3. Lois Barclay Murphy, *Social Behavior and Child Personality: An Exploratory Study of Some Roots of Sympathy* (New York: Columbia University Press, 1937), p. 50; Thoreau, *Writings,* p. 154.

4. Gregg Mitman, *The State of Nature: Ecology, Community, and American Social Thought, 1900–1950* (Chicago: University of Chicago Press, 1992), p. 1. For an overview of recent thinking in the history of American biology, see Ronald Rainger, Keith R. Benson, and Jane Maienschein, eds., *The American Development of Biology* (Philadelphia: University of Pennsylvania Press, 1988); and Keith R. Benson, Jane Maienschein, and Ronald Rainger, eds., *The Expansion of American Biology* (New Brunswick, N.J.: Rutgers University Press, 1991).

5. Sharon E. Kingsland, "Toward a Natural History of the Human Psyche: Charles Manning Child, Charles Judson Herrick, and the Dynamic View of the Individual at the University of Chicago," in Benson et al., *The Expansion of American Biology,* p. 199. Child was one of the biologists that Gardner Murphy urged the members of SPSSI to read; see his "The Research Task of Social Psychology," *Journal of Social Psychology,* 1939, *10*:107–20, p. 112. For an examination of biology pursued from mechanist premises, see Philip J. Pauly's study of Jacques Loeb, *Controlling Life: Jacques Loeb and the Engineering Ideal in Biology* (New York: Oxford University Press, 1987).

6. Evelyn Fox Keller, *A Feeling for the Organism: The Life and Work of Barbara McClintock* (New York: W. H. Freeman, 1983), p. 8. See also Barbara McClintock, *The Discovery and Characterization of Transposable Elements: The Collected Papers of Barbara McClintock* (New York: Garland, 1987).

7. Scott Gilbert, "Cellular Politics: Ernest Everett Just, Richard B. Goldschmidt, and the Attempt to Reconcile Embryology and Genetics," in Rainger et al., *The American Development of Biology,* p. 328.

8. Ernest Everett Just, *The Biology of the Cell Surface* (London: Technical Press, 1939), p. 10. On Just's life and career, see Kenneth R. Manning, *Black Apollo of Science: The Life of Ernest Everett Just* (New York: Oxford University Press, 1983).

9. Stephen Jay Gould, "Mighty Manchester," review of Freeman Dyson's *Infinite in All Directions* in *The New York Review of Books,* October 27, 1988, p. 32.

10. See "So-Called Social Science," in A. L. Kroeber, *The Nature of Culture* (Chicago: University of Chicago Press, 1952), p. 70. The essay is from 1936.

11. Gordon W. Allport, *Personality: A Psychological Interpretation* (New York: Henry Holt, 1937), p. 21.

12. Daniel S. Lehrman, "Behavioral Science, Engineering, and Poetry," in Ethel Tobach, Lester R. Aronson, and Evelyn Shaw, eds., *The Biopsychology of Development* (New York: Academic Press, 1971), pp. 459, 470.

13. Gordon W. Allport, "The Psychologist's Frame of Reference," *Psychological Bulletin,* 1940, *37*:1–28, pp. 18, 19.

14. Murphy, "Research Task," pp. 114, 115. Murphy in fact had a life-long interest in astronomy – his father, Edgar Gardner Murphy, was an amateur astronomer who had written a work of popular astronomy, *A Beginner's Star Book* (New York: G. P. Putnam's Sons, 1912) (Murphy used the pseudonym Kelvin MacKready). Also, during Gardner Murphy's youth, his uncle, William Elkin, was the director of the Yale Observatory. See Lois Barclay Murphy, *Gardner Murphy: Integrating, Expanding and Humanizing Psychology* (Jefferson, N.C.: McFarland, 1990), p. 183.

15. Allport, *Personality,* p. 21.

16. Murphy, *Social Behavior,* pp. 18, 50, 92.

17. William James, "Frederic Myers's Service to Psychology," *Proceedings of the Society for Psychical Research,* 1901, *17*:13–23, p. 14.

18. Ibid., pp. 22, 23.

19. William James, "What Psychical Research Has Accomplished," in *The Will to Believe and Other Essays in Popular Philosophy* (Cambridge, Mass.: Harvard University Press, 1979/1897), p. 239.

20. Ibid., 240.

21. The phrase "discipline emeritus" is from Paul Farber, "The Transformation of Natural History in the Nineteenth Century," *Journal of the History of Biology,* 1982, *15*:145–52, pp. 146–7. On the modernist ends of thinkers such as James, see, especially, James Kloppenberg, *Uncertain Victory: Social Democracy and Progressivism in European and American Thought, 1870–1920* (New York: Oxford University Press, 1986). For further discussions of modernism, see, for example, Malcolm Bradbury and James McFarlane, "The Name and Nature of Modernism," in Malcolm Bradbury and James McFarlane, eds., *Modernism: 1890–1930* (London: Penguin Books, 1976); Stephen Kern, *The Culture of Time and Space, 1880–1930* (Cambridge, Mass.: Harvard University Press, 1983); Raymond Williams, *The Politics of Modernism: Against the New Conformists* (London: Verso, 1989); and Daniel J. Singal, ed., *Modernist Culture in America* (Belmont, Calif.: Wadsworth, 1991).

22. Paul Jerome Croce, "William James's Scientific Education," *History of the Human Sciences,* 1995, *8*:9–27, p. 11; and Eugene Taylor, "Radical Empiricism and the New Science of Consciousness," *History of the Human Sciences,* 1995, *8*:47–60, p. 47. This is a special issue on William James.

23. In 1943 Allport remarked that James's radical empiricism "might have served as the foundations for an American school of phenomenology, but it did not." Gordon Allport, "The Productive Paradoxes of William James," in Gordon Allport, *The Person in Psychology: Selected Essays* (Boston: Beacon Press, 1968), p. 305. For later discussions of this issue, see James M. Edie, *William James and Phenomenology* (Bloomington and Indianapolis: Indiana University Press, 1987); and Max Herzog, "William James and the Development of Phenomenological Psychology in Europe," *History of the Human Sciences,* 1995, *8*:29–46.

24. A suggestive modification of the idea of modernism in relation to the psychology of such figures as John Dewey can be found in David Hollinger, "The Knower and the Artificer, with Postscript 1993," in Dorothy Ross, ed., *Modernist Impulses in the Human Sciences 1870–1930* (Baltimore: Johns Hopkins University Press, 1994). Hollinger's intention in this piece is to point out the enthusiasm of certain intellectuals for organizing "culture around the cognitive capacities of human beings," a perspective that he designates as "cognitive modernism" to distinguish it from a canonical modernism that highlights strategies of artifice (p. 32). Although there is much to recommend Hollinger's formulation, it would obscure those aspects of radical empiricism that I want to emphasize here, particularly the sense of emotionality that informs this stance as a relational form of knowing. The entire volume of which Hollinger's essay is a part is a provocative exploration of the relationship of modernism to the human sciences.

25. Places from which to start investigating romanticism would include Andrew Cunningham and Nicholas Jardine, eds., *Romanticism and the Sciences* (Cambridge: Cambridge University Press, 1990); and Roy Porter and Mikulaš Teich, eds., *Romanticism in National Context* (Cambridge: Cambridge University Press, 1988). See also H. A. M. Snelders, "Romanticism and Naturphilosophie and the Inorganic Natural Sciences 1797–1840: An Introductory Survey," *Studies in Romanticism,* 1970, *9*:193–215; Pierce C. Mullen, "The Romantic as Scientist: Lorenz Oken," *Studies in Romanticism,* 1977, *16*:381–99; Fredrick Amrine, Francis J. Zucker, and Harvey Wheeler, eds., *Goethe and the Sciences: A Reappraisal* (Dordrecht: D. Reidel, 1987); and Dennis L. Sepper, *Goethe Contra Newton: Polemics and the Project for a New Science of Color* (Cambridge: Cambridge University Press, 1988).

26. Lois Barclay Murphy, "Lois Barclay Murphy," in Agnes O'Connell and Nancy

Russo, eds., *Models of Achievement: Reflections of Eminent Women in Psychology* (New York: Columbia University Press, 1983), p. 96.

27. Murphy, *Gardner Murphy,* p. 82.

28. Gardner Murphy, "A Story of Dyadic Thought and Work with Gardner and Lois Murphy as Written by Gardner," unpublished manuscript dated 1968 (Gardner Murphy and Lois Barclay Murphy Papers, Archives of the History of American Psychology, University of Akron [hereafter Murphy Papers]). Lois Murphy's interest in mysticism continued through her years at Vassar and Union Theological Seminary, where she wrote papers on the subject; see Murphy, *Gardner Murphy,* p. 179.

29. Murphy, *Gardner Murphy,* pp. 64, 32.

30. Gardner Murphy, "Gardner Murphy," in E. G. Boring and Gardner Lindzey, eds., *A History of Psychology in Autobiography,* vol. 5 (New York: Appleton-Century-Crofts, 1967), p. 256.

31. Murphy, *Gardner Murphy,* p. 66.

32. Ibid., p. 62.

33. Ibid., p. 35. Lois Barclay Murphy adds that Gardner's "ashes were interred at his request in the King family plot on a high ridge in the Sleepy Hollow Cemetery" (p. 35).

34. Murphy, *Gardner Murphy,* p. 183.

35. Ibid., p. 195.

36. Gordon W. Allport, review of P. M. Symonds, *Diagnosing Personality and Conduct* in the *Journal of Social Psychology,* 1932, *3*:391–7, p. 391.

37. Gordon W. Allport, "An Autobiography," in Allport, *The Person in Psychology,* p. 388.

38. Ibid., pp. 386–7.

39. Ibid., p. 387.

40. William Stern, "William Stern," in Carl Murchison, ed., *A History of Psychology in Autobiography,* vol. 1 (Worcester, Mass.: Clark University Press, 1930), p. 336. For a sense of the specific intellectual and cultural contexts in which Stern participated, see Mitchell Ash, *Gestalt Psychology in German Culture, 1890–1967: Holism and the Quest for Objectivity* (Cambridge: Cambridge University Press, 1995); Anne Harrington, *Reenchanted Science: Holism in German Culture from Wilhelm II to Hitler* (Princeton, N.J.: Princeton University Press, 1996); and Fritz K. Ringer, *The Decline of the German Mandarins: The German Academic Community, 1890–1933* (Cambridge, Mass.: Harvard University Press, 1969).

41. Stern, "William Stern," pp. 336–7.

42. Ibid., p. 345.

43. Gordon W. Allport, "Intuition and the Aesthetic Attitude," unpublished typescript dated December 12, 1927 (Gordon W. Allport Papers, Harvard University [hereafter Allport Papers]); and Gordon W. Allport, "Intuition as a Method in Psychology," unpublished and undated typescript (Allport Papers). The concerns and format of the second paper mark it as a companion to the first; additionally, Allport's comment in the latter work that Eduard Spranger's book *Lebensformen* is "at present being translated" puts it at about 1927 – the English translation, titled *Types of Men,* appeared in 1928. Psychologist William Hunt, who was an undergraduate at Dartmouth at this time, recalled that Allport was just "back from Germany with its stress on the cultural, subjective, and idiomorphic; and the peculiar blending of the intuitive and scientific approaches which was to distinguish his work left a deep impression on me and leavened my somewhat naive behaviorism"; William A. Hunt, "To Live with Judgment," in T. S. Krawiec, ed., *The Psychologists,* vol. 1 (New York: Oxford University Press, 1972), p. 115.

44. Allport, "Intuition as a Method in Psychology."

45. Allport, *Personality,* p. 540. Allport is quoting the psychologist Eduard Spranger.

46. Ibid.

47. Allport, "Intuition and the Aesthetic Attitude."

48. Allport, *Personality,* p. 539. On Dilthey, see H. P. Rickman, *Wilhelm Dilthey: Pioneer of the Human Studies* (Berkeley: University of California Press, 1979); and Rudolf A. Makkreel, *Dilthey: Philosopher of the Human Studies* (Princeton, N.J.: Princeton University Press, 1975).

49. Stern, "William Stern," p. 370. Stern's views on sympathy were also of interest to Lois Barclay Murphy, forming part of the intellectual backdrop to her study of children's sympathy. Murphy cited Stern's work from his 1924 text *The Psychology of Early Childhood,* as well as W. Boeck's 1909 study, *Das Mitleid bei Kindern.*

50. Gordon W. Allport, "The Study of the Undivided Personality," *Journal of Abnormal and Social Psychology,* 1924, *19*:132–41, p. 140. For a sense of the contours of intellectual debate during this period in Germany, see Mitchell G. Ash, "Gestalt Psychology in Weimar Culture," *History of the Human Sciences,* 1991, *4*:395–415.

51. Allport, *Personality,* p. 540.

52. Ibid., pp. 542, 543.

53. Ibid., p. 543.

54. Allport, "Intuition and the Aesthetic Attitude." Mitchell Ash has pointed out to me that what Allport is referencing as being from the *Sturm und Drang* is instead from *Faust.*

55. Allport, *Personality,* p. 543.

56. Thoreau, *Writings,* pp. 154–5.

57. Roy R. Male, "*Sympathy* – a Key Word in American Romanticism," in Kenneth Walter Cameron, ed., *Romanticism and the American Renaissance: Essays on Ethos and Perception in the Age of Emerson, Thoreau, Hawthorne, Melville, Whitman and Poe* (Hartford, Conn.: Transcendental Books, 1977), quoted on p. 21.

58. Ibid.

59. Arthur Schwartz, "The American Romantics: An Analysis," in Cameron, *Romanticism and the American Renaissance,* p. 40.

60. Ibid., p. 39.

61. Walt Whitman, "There Was a Child Went Forth," in *Walt Whitman: Complete Poetry and Collected Prose* (New York: Literary Classics of the United States, Inc., 1982), p. 491. The poem is from the 1891–1892 edition of *Leaves of Grass.* Lois Barclay Murphy ended a recent article with this same segment of Whitman's poem; see Lois Barclay Murphy and Colleen T. Small, "The Baby's World," *Zero to Three: Bulletin of the National Center for Clinical Infant Programs,* 1989, *10*:1–6, p. 6.

62. Male, "*Sympathy* – a Key Word in American Romanticism," quoted on p. 21.

63. Ibid.

64. William James, *A Pluralistic Universe* (Cambridge, Mass.: Harvard University Press, 1977/1909), p. 111 (second emphasis added).

65. Donald Dewsbury, "Comparative Psychology and Ethology: A Reassessment," *American Psychologist,* 1992, *47*:208–15, p. 208.

66. Barbara Laslett, "Unfeeling Knowledge: Emotion and Objectivity in the History of Sociology," *Sociological Forum,* 1990, *5*:413–33, quoted on p. 421. Ogburn's comment is from 1930. For a metatheoretical discussion of this question, see Lorraine Daston, "The Moral Economy of Science," *Osiris,* 1995, *10*:1–24, special issue on "Constructing Knowledge in the History of Science." For a review of the question of emotion as an area of psychological study, see Kenneth Gergen, "Metaphor and Monophony in the 20th-century Psychology of Emotions," *History of the Human Sciences,* 1995, *8*:1–23.

67. Lois Barclay Murphy and Henry Ladd, *Emotional Factors in Learning* (Westport, Conn.: Greenwood Press, 1944), pp. 5–6.

68. Suzanne Clark, *Sentimental Modernism: Women Writers and the Revolution of the Word* (Bloomington and Indianapolis: Indiana University Press, 1991), p. 71. In her text Clark examines the work of Emma Goldman, Edna St. Vincent Millay, Louise Bogan, Kay Boyle, Annie Dillard, and Alice Walker.

69. Allport, "Personalistic Psychology of William Stern," in Allport, *Person in Psychology,* p. 296.

70. Gardner Murphy, "There Is More Beyond," in T. Krawiec, *The Psychologists: Autobiographies of Distinguished Living Psychologists,* vol. 2 (London: Oxford University Press, 1973), p. 323.

71. Clark, *Sentimental Modernism,* pp. 1–3.

72. On this point for literary intellectuals see ibid., p. 4. As Clark cannot resist remarking, "becoming an intellectual in America is sort of like being inducted into the army (or maybe the first grade) and learning not to be a sissy" (ibid., p. 12).

73. The key critical text in regard to such work as Stowe's is Jane Tompkins, *Sensational Designs: The Cultural Work of American Fiction 1790–1860* (New York: Oxford University Press, 1985). Tompkins's work is a revisionist analysis of the views put forth in Ann Douglas, *The Feminization of American Culture* (New York: Alfred A. Knopf, 1977). On the eighteenth-century context of sentimental literature, see, for example, R. F. Brissenden, *Virtue in Distress: Studies in the Novel of Sentiment from Richardson to Sade* (New York: Macmillan, 1974). Brissenden notes the fact that, in Sterne's world, it is "the stoic who is the real enemy in a sentimental journey, not the man who allows himself to feel" (p. 219). See also Michael Bell, *The Sentiment of Reality: Truth of Feeling in the European Novel* (London: George Allen and Unwin, 1983); and Janet Todd, *Sensibility: An Introduction* (London: Methuen, 1986).

74. See, for example, the discussion in Keller, *A Feeling for the Organism,* pp. 197–207.

75. Gardner Murphy and Lois Barclay Murphy, *Experimental Social Psychology* (New York: Harper and Brothers, 1931), pp. 22, 23.

76. Ibid., p. 23.

77. Lehrman, "Behavioral Science, Engineering, and Poetry," p. 462.

78. Ibid., p. 464.

79. Simon Schaffer, "Astronomers Mark Time: Discipline and the Personal Equation," *Science in Context,* 1988, *2*:115–45, pp. 118–19. On the issue of precision, see also M. Norton Wise, ed., *The Values of Precision* (Princeton, N.J.: Princeton University Press, 1995).

80. Eugene Lerner and Lois Barclay Murphy, "Editorial Foreword," in Eugene Lerner and Lois Barclay Murphy, eds., *Methods for the Study of Personality in Young Children,* 1941, *Monographs of the Society for Research in Child Development,* vol. 6, p. x.

81. Lawrence K. Frank, "Preface," in ibid., p. vi. For an extended discussion of this point see Lawrence K. Frank, "Projective Methods for the Study of Personality," *Journal of Psychology,* 1939, *8*:389–413.

82. On this point, see Steven Shapin, "Science and the Public," in R. C. Olby, G. N. Cantor, J. R. R. Christie and M. J. S. Hodge, eds., *Companion to the History of Modern Science* (London: Routledge, 1990); and Schaffer, "Astronomers Mark Time."

83. On the problems inherent in accepting as "self-evident" the status of experimentation, see Steven Shapin and Simon Schaffer, "Understanding Experiment," in their *Leviathan and the Air-Pump: Hobbes, Boyle and the Experimental Life* (Princeton, N.J.: Princeton University Press, 1985), and David Gooding, T. J. Pinch, and Simon Schaffer, eds., *The Uses of Experiment: Studies in the Natural Sciences* (Cambridge: Cambridge University Press, 1989).

84. William Ritter, *The California Woodpecker and I: A Study in Comparative Zoology in Which Are Set Forth Numerous Facts and Reflections by One of Us about Both of Us* (Berkeley: University of California Press, 1938), p. xiii. See also William E. Ritter, "Mechanical Ideas in the Last Hundred Years of Biology," *American Naturalist,* 1938, *72*:315–23.

85. Leila Zenderland, "Education, Evangelism, and the Origins of Clinical Psychology: The Child-Study Legacy," *Journal of the History of the Behavioral Sciences,* 1988, *24*:152–65, p. 157.

86. Barbara Biber, Lois B. Murphy, Louise P. Woodcock, and Irma S. Black, *Child Life in School: A Study of a Seven-Year-Old Group* (New York: E. P. Dutton, 1942), p. 7.

87. Lois Barclay Murphy, "Experiments in Free Play," in Lerner and Murphy, *Methods for the Study of Personality in Young Children,* p. 4. Mitchell Ash remarks that Lewin was critical of the statistical or group data methods favored by American psychologists as early as 1931. The experimental style of Lewin's own work, Ash states, "was characterized from the beginning by attention to the interaction of experimenter and subject as well as the principle that the psychological experiment should attempt to capture the dynamics of real-life situations." Lewin believed that procedures "that retained the essential structural relations of everyday life situations would yield a more believable experimental picture than the manipulation of isolated independent and dependent variables" (Mitchell G. Ash, "Cultural Contexts and Scientific Change in Psychology," *American Psychologist,* 1992, *47*:198–207; the quotations are from p. 203). See also Ash, *Gestalt Psychology in German Culture.* Also pertinent is Lewin's pioneering use of film to try and capture the "life-space;" see Mel van Elteren and Helmut E. Luck, "Kurt Lewin's Films and Their Role in the Development of Field

Theory," in Susan A. Wheelan, Emmy A. Pepitone, and Vicki Abt, eds., *Advances in Field Theory* (Newbury Park: Sage, 1990).

88. Murphy, "Experiments in Free Play," pp. 4–5.

89. Biber et al., *Child Life in School,* pp. 24–5.

90. Ibid., pp. 21, 33, 34.

91. Ibid., p. 46

92. Ibid., p. 37.

93. Ibid., p. 21.

94. Margaret Morse Nice, *Studies in the Life History of the Song Sparrow I: A Population Study of the Song Sparrow, Transactions of the Linnean Society of New York,* 1937, *4*:1–247, p. 1. Nice's research was conducted from 1929 to 1936 and reported in two parts, the second of which is Margaret Morse Nice, *Studies in the Life History of the Song Sparrow II, Transactions of the Linnean Society of New York,* 1943, *6*:1–328.

95. Nice, *Studies in the Life History of the Song Sparrow II,* p. 2.

96. Delacour and Lorenz quoted in Marcia Myers Bonta, "Margaret Morse Nice: Ethologist of the Song Sparrow," in her *Women in the Field: America's Pioneering Women Naturalists* (College Station: Texas A&M University, 1991), pp. 229, 222.

97. Margaret Morse Nice, "Edmund Selous – An Appreciation," *Bird-Banding,* 1935, *6*:90–6, p. 90.

98. On Nice and Thompson, see Margaret W. Rossiter, *Women Scientists in America: Struggles and Strategies to 1940* (Baltimore: Johns Hopkins University Press, 1982), pp. 149, 151. Nice mentions Woolley on p. 80 of her autobiography, *Research is a Passion with Me* (Toronto: Consolidated Amethyst Communications, 1979).

99. In the language studies, Nice used her own daughters as subjects. For examples of Nice's dual research interests, see such works as "A Child Who Would Not Talk," *Pedagogical Seminary and Journal of Genetic Psychology,* 1925, *32*:105–43 and "Study of the Nesting of the Magnolia Warbler," *Wilson Bulletin,* 1926, *38*:185–99. For another instance of the overlap between ornithology and psychology during this period, see Thorleif Schjelderup-Ebbe's contribution, "Social Behavior of Birds," in Carl Murchison, ed., *Handbook of Social Psychology* (Worcester, Mass.: Clark University Press, 1935). Note, for example, Schjelderup-Ebbe's remark that "every bird is a personality" (p. 947). See also Ritter, *The California Woodpecker and I.*

100. But see volume 11 of *Osiris* on "Field Practice" for a general discussion of the field.

101. Morton Deutsch, "Field Theory in Social Psychology," in Gardner Lindzey, ed., *Handbook of Social Psychology* (Cambridge: Addison-Wesley Press, 1954), pp. 181–2. For a similar example that is contemporaneous with the 1930s, see Lawrence K. Frank, "Structure, Function and Growth," *Philosophy of Science,* 1935, *2*:210–35, esp. pp. 221–35. The field metaphor in biology is examined in Donna Jeanne Haraway, *Crystals, Fabrics, and Fields: Metaphors of Organicism in Twentieth-Century Developmental Biology* (New Haven, Conn.: Yale University Press, 1976).

102. Allport, "The Psychologist's Frame of Reference," p. 14.

103. On the fieldworker archetype, see George W. Stocking, Jr., "The Ethnographic Sensibility of the 1920s and the Dualism of the Anthropological Tradition," in George

W. Stocking, Jr., ed., *Romantic Motives: Essays on Anthropological Sensibility* (Madison: University of Wisconsin Press, 1989).

104. Alan Trachtenberg, "Ever – The Human Document," in *America and Lewis Hine: Photographs 1904–1940* (New York: Aperture, 1977), p. 132; Whitman is quoted on p. 132.

105. See Helen Merrell Lynd, *Field Work in College Education* (New York: Columbia University Press, 1945), which is based on Sarah Lawrence's programs. For examples of how fieldwork was used in psychology at Sarah Lawrence, see Lois Barclay Murphy, Eugene Lerner, Jane Judge, and Madeleine Grant, *Psychology for Individual Education* (New York: Columbia University Press, 1942).

106. *The Plow That Broke the Plains* (1936) centered on the Dust Bowl and the exodus of Midwest farm families. For information on Steiner's career see Ralph Steiner, *Ralph Steiner: A Point of View* (Middletown, Conn.: Wesleyan University Press, 1978).

107. Lois Barclay Murphy, Interview with Ms. Ronnie Walker, March 1988 (Sarah Lawrence College Archives).

108. This paper is in the Murphy Papers; the notation indicates that the year is circa 1936–7. Francis G. Peabody, who established the Harvard department of Social Ethics – in which Allport both majored as an undergraduate and taught as an instructor – used a similar logic in arguing for the creation of the Harvard Social Museum, where students of "the sciences which concern themselves with human history and conduct" could adopt the "familiar method of the naturalist," which entailed interpreting nature by seeing, touching, scrutinizing and analyzing. The museum consisted of diagrams, statistical charts, documents, reports, photographs, and exhibits. See Francis G. Peabody, *The Social Museum as an Instrument of University Teaching: A Classified List of the Collections in the Social Museum of Harvard University to February, 1911* (Cambridge, Mass.: Harvard University Press, 1911), p. 1.

109. William Stott, *Documentary Expression and Thirties America* (New York: Oxford University Press, 1973), p. x. See also T. V. Reed's "Unimagined Existence and the Fiction of the Real: Postmodernist Realism in *Let Us Now Praise Famous Men*," *Representations,* 1988, *24*:156–76, in which he points out that the collapse of the nation's economic structure in the 1930s seemed to have "brought down systems of representation with it . . . usher[ing] in the most intense period of documentation, the most exhaustive effort to represent the 'real' in American history" (p. 156). This piece has been reprinted in T. V. Reed, *Fifteen Jugglers, Five Believers: Literary Politics and the Poetics of American Social Movements* (Berkeley: University of California Press, 1992).

110. Warren Susman, "The Culture of the Thirties," in his *Culture as History: The Transformation of American Society in the Twentieth Century* (New York: Pantheon, 1984), p. 159. The essay was originally published in 1983. On the uses that would be made of film as a research tool in the life sciences, see Gregg Mitman, "Cinematic Nature: Hollywood Technology, Popular Culture, and the American Museum of Natural History," *Isis,* 1993, *84*:637–61.

111. Stott, *Documentary Expression and Thirties America,* p. 47.

112. Gordon W. Allport, *The Use of Personal Documents in Psychological Science* (New York: Social Science Research Council, 1942), p. xi.

113. Gordon Allport, "Dewey's Individual and Social Psychology," in his *The Person in Psychology* p. 349. The essay is from 1939.

114. Allport, *Personality,* p. 549.

115. Allport, "Intuition as Method."

116. Gordon W. Allport, "Some Guiding Principles in Understanding Personality," *The Family* (June 1930), p. 127.

117. Allport, "Intuition as Method."

118. Gordon W. Allport and Philip E. Vernon, *Studies in Expressive Movement* (New York: Macmillan, 1933). Allport wrote several major texts with younger colleagues such as Vernon, as well as a number of articles. In this case, as Vernon reports, Allport "usually wrote up the more theoretical and interpretative aspects, myself the more empirical [statistical] side," and I will presume that Allport had sole authorship of the nonstatistical portions of *Studies in Expressive Movement.* See Philip E. Vernon, "The Making of an Applied Psychologist," in T. S. Krawiec, ed., *The Psychologists: Autobiographies of Distinguished Living Psychologists,* vol. 3 (Brandon, Vt.: Clinical Psychology, 1978), p. 307.

119. Gordon W. Allport, "The Standpoint of *Gestalt* Psychology," *Psyche,* 1923–4, *4*:354–61, p. 355. For the question of form as taken up in biology, see Haraway, *Crystals, Fabrics, and Fields,* especially pp. 39–48.

120. Gordon W. Allport review of Symonds, *Diagnosing Personality and Conduct,* p. 392.

121. Allport, "The Study of the Undivided Personality," p. 133.

122. Ibid., pp. 135, 137. The Murphys agreed with Allport on this point, noting that "if we are going to find genuinely *characteristic* aspects of personalities which differ from one person to another and are *relatively invariable* from one situation to another, it will probably have to be the more subtle aspects of *style,* or *manner* in which a given individual expresses his reaction, or the content of his *interests*"; see Lois Barclay Murphy and Gardner Murphy, "The Influence of Social Situations upon the Behavior of Children," in Murchison, *Handbook of Social Psychology,* p. 1048.

123. Goodwin Watson, "Psychology in Germany and Austria," *Psychological Bulletin,* 1934, *10*:755–76, p. 767. An important precedent to this idea of the uniqueness of expressive movement can be found, as Hugh Honour discusses, in the significance that the Romantics had assigned to the artistic "sketch," which was held to be "the least premeditated form of art, in which the painter's or sculptor's feelings might seem to be recorded with spontaneous directness." The sketch was believed to reveal "in the most direct manner possible the individuality of the artist's 'touch,' his or her unique manner of expression"; see Hugh Honour, *Romanticism* (New York: Harper and Row, 1979), p. 17. See also Carlo Ginzburg's article, "Clues: Roots of an Evidential Paradigm," in his *Myths, Emblems, Clues* (London: Hutchinson Radius, 1990).

124. Allport and Vernon, *Studies in Expressive Movement,* pp. v, vii.

125. Ibid., p. viii.

126. Ibid., chap. 4.

127. Ibid., p. 97. *Studies in Expressive Movement* also included an appended chapter by Edwin Powers which discussed matching narrative personality sketches with handwriting analysis (graphology).

128. Ibid., p. ix.

129. Ibid., p. 134.

130. Dorothea Johannsen, review of Allport and Vernon's *Studies in Expressive Movement* in the *Journal of Abnormal and Social Psychology,* 1933–4, *28*:333–4, p. 334.

131. A. A. Roback, review of Allport and Vernon's *Studies in Expressive Movement* in *Character and Personality,* 1932–3, *1*:322–4, p. 323.

132. Gardner Murphy, review of J. L. Moreno, *Who Shall Survive? A New Approach to the Problem of Human Interrelations,* in *Journal of Social Psychology,* 1935, *6*:388–93, p. 391.

133. Gardner Murphy, "The Research Task of Social Psychology," *Journal of Social Psychology,* 1939, *10*:107–20, p. 115.

134. Muzafer Sherif, *The Psychology of Social Norms* (New York: Octagon Books, 1967/1936; originally published by Harper's, under Murphy's editorship), p. 106.

135. Gardner Murphy, "Introduction," in Sherif, *The Psychology of Social Norms,* p. ix.

136. Ibid., p. viii.

137. Ibid., pp. xi–xii.

138. It should be noted that before joining the Columbia department Sherif was a graduate student of Allport's at Harvard.

139. Gardner Murphy, Lois Barclay Murphy, and Theodore Newcomb, *Experimental Social Psychology: An Interpretation of Research upon the Socialization of the Individual,* rev. ed. (New York: Harper and Brothers, 1937), pp. 13, 15.

140. The experiments are described in chapter 4 of René Warcollier, *Experiments in Telepathy* (New York: Harper and Brothers, 1938). Seymour H. Mauskopf and Michael R. McVaugh, in *The Elusive Science: Origins of Experimental Psychical Research* (Baltimore: Johns Hopkins University Press, 1980), also describe these researches on pp. 62–3. Murphy worked with both approaches during the 1930s; see ibid., p. 189. See also his examples in Gardner Murphy, "Things I Can't Explain," *American Magazine* (November 1936), pp. 40–1, 130–2.

141. Mauskopf and McVaugh, *Elusive Science,* pp. 62–3.

142. Gardner Murphy, "Dr. Rhine and the Mind's Eye," *American Scholar,* 1938, *7*:189–200, p. 189.

143. James, "What Psychical Research Has Accomplished," p. 240.

144. Gardner Murphy, "Notes for a Parapsychological Autobiography," *Journal of Parapsychology,* 1957, *21*:165–78, pp. 177–8.

145. Murphy, "Introduction," *Mind to Mind,* p. xii.

146. Farber, "Transformation of Natural History," pp. 146–7.

147. In addition to Kingsland, "Toward a Natural History of the Human Psyche," see also Philip J. Pauly, "The Appearance of Academic Biology in Late Nineteenth-Century America," *Journal of the History of Biology,* 1984, *17*:369–97; and Jan Sapp, "The Struggle for Authority in the Field of Heredity, 1900–1932: New Perspectives on the Rise of Genetics," *Journal of the History of Biology,* 1983, *16*:311–42. For an overview of the reevaluation that is under way among some historians of science see N. Jardine, J. A. Secord, and E. C. Spary, eds., *Cultures of Natural History* (Cambridge, Mass.: Cambridge University Press, 1996), which contains essays examining natural history from the Renaissance to the beginning of the twentieth century; volume 11 of *Osiris* on "Field Practice"; Benson et al., *The Expansion of American Biology;* and Rainger et al., *The American Development of Biology.*

148. See, for example, Olby et al., *Companion to the History of Modern Science,* where the essay on natural history is for the years 1670–1802.

149. Francis B. Sumner, "The Naturalist as a Social Phenomenon," *American Naturalist,* 1940, *74*:398–408, p. 400.

150. Ibid. Sumner wryly notes, however, that "we naturalists may take satisfaction in the knowledge that none of the discoveries which we make in our field lends itself very well to the butchery of our colleagues who speak some other language" (p. 399).

151. Theodosius Dobzhansky, "Are Naturalists Old-Fashioned?" *American Naturalist,* 1966, *100*:541–50, pp. 542, 545. See also Theodosius Dobzhansky, "Taxonomy, Molecular Biology, and the Peck Order," *Evolution,* 1961, *15*:263–4. For a recent discussion of Dobzhansky's work in relation to that of his closest colleagues, see Robert E. Kohler, "Drosophila and Evolutionary Genetics: The Moral Economy of Scientific Practice," *History of Science,* 1991, *29*:335–75; and Robert E. Kohler, *Lords of the Fly: Drosophila Genetics and Experimental Life* (Chicago: University of Chicago Press, 1994). Also of interest is Mark B. Adams, ed., *The Evolution of Theodosius Dobzhansky: Essays on His Life and Thought in Russia and America* (Princeton, N.J.: Princeton University Press, 1991).

152. Perry Miller, *Life of the Mind in America* (New York: Harcourt, Brace and World, 1965); and Perry Miller, *Nature's Nation* (Cambridge, Mass.: Belknap Press, 1967). For a recent example of the interest that cultural historians have taken in the relationships between landscape and the formation of Americans' sense of national identity, see John Sears, *Sacred Places: American Tourist Attractions in the Nineteenth Century* (New York: Oxford University Press, 1989).

153. Barbara Novak, *Nature and Culture: American Landscape Painting 1825–1875* (New York: Oxford University Press, 1980), pp. 16, 17, 4. Note here the hermeneutic approach to nature, an image at odds with the idea that nature's language is "factual" while that of the Bible is "metaphorical." For a more complete analysis on God as Nature in America, see Novak, chapter 1, "The Nationalist Garden and the Holy Book."

154. On the interest in natural history in the nineteenth century, see Nathan Reingold, ed., *Science in Nineteenth Century America: A Documentary History* (New York: Hill and Wang, 1964), p. 162; and Daniel Goldstein, "'Yours For Science': The Smithsonian Institution's Correspondents and the Shape of Scientific Community in Nineteenth-Century America," *Isis,* 1994, *85*:573–99. On Barnum, see Neil Harris, *Humbug: The Art of P. T. Barnum* (Boston: Little, Brown, 1973), p. 3; and John Betts, "Barnum and Natural History," *Journal of the History of Ideas,* 1959, *20*:353–68.

155. Marcel C. LaFollette has done promising spadework on the question of popular images of science; see her survey of popular magazines in *Making Science Our Own: Public Images of Science 1910–1955* (Chicago: University of Chicago Press, 1990).

156. I discuss this point at greater length in a current work-in-progress, "Hearing Natural Knowledge Spoken in the Vernacular: History of Science and the Tale of Luther Burbank."

157. Quoted in Peter Dreyer, *A Gardener Touched with Genius: The Life of Luther Burbank* (Berkeley and Los Angeles: University of California Press, 1985), p. 217. The author is Orland E. White, professor of agricultural biology and director of the

University of Virginia's Blandy Experimental Farm; the passage appears in a letter to W. L. Howard, dated December 21, 1938; see p. 216 and n. 1, p. 278.

158. On this particular political and scientific arena, see, for example, Adrian Desmond, "Artisan Resistance and Evolution in Britain, 1819–1848," *Osiris,* 1987, *3*:77–110.

159. For an analysis of the ways in which boundaries are constructed between scientists and nonscientists see Thomas Gieryn, "Boundary-Work and the Demarcation of Science from Non-Science: Strains and Interests in Professional Ideologies of Scientists," *American Sociological Review,* 1983, *48*:781–95. For a brief discussion of the issues at stake in setting off science from the public see Shapin, "Science and the Public."

Chapter 6

1. Ruth Benedict, *Patterns of Culture* (Boston: Houghton Mifflin, 1934), p. 278.

2. Eugene W. Lyman, "Religious Education for a New Democracy," *Journal of Religion,* 1923, *3*:449–57, p. 449.

3. On this point, see Morton and Lucia White, *The Intellectual versus the City: From Thomas Jefferson to Frank Lloyd Wright* (Cambridge, Mass.: Harvard University and MIT Press, 1962); Thomas Bender, *Community and Social Change in America* (Baltimore: Johns Hopkins University Press, 1978); and Lawrence Levine, *Highbrow/Lowbrow: The Emergence of Cultural Hierarchy in America* (Cambridge, Mass.: Harvard University Press, 1988).

4. George Lipsitz, *The Sidewalks of St. Louis: Place, People, and Politics in an American City* (Columbia: University of Missouri Press, 1991), p. 2.

5. William James, *Some Problems of Philosophy* (Cambridge, Mass.: Harvard University Press, 1979/1911), p. 48.

6. Quoted in Ann Douglas, *Terrible Honesty: Mongrel Manhattan in the 1920s* (New York: Farrar, Straus, and Giroux, 1995), p. 118.

7. Ibid., pp. 116, 118.

8. Gordon Allport, "The Productive Paradoxes of William James," in Gordon W. Allport, *The Person in Psychology: Selected Essays* (Boston: Beacon Press, 1968 [the essay is from 1943]), p. 320.

9. William James, "Are We Automata?" in William James, *Essays in Psychology* (Cambridge, Mass.: Harvard University Press, 1983 [the essay is from 1879]), p. 51. This passage also appears in James's *Principles of Psychology,* in the "Stream of Thought" chapter. See William James, *Principles of Psychology,* vol. 1 (Cambridge, Mass.: Harvard University Press, 1981/1890), p. 277. On the dynamic and perspectival nature of James's thinking, see also David E. Leary, "William James and the Art of Human Understanding," *American Psychologist,* 1992, *47*:152–60; and Richard J. Bernstein, "Introduction," in William James, *A Pluralistic Universe* (Cambridge, Mass.: Harvard University Press, 1977/1909).

10. William Graebner, *The Engineering of Consent: Democracy and Authority in Twentieth-Century America* (Madison: University of Wisconsin Press, 1987), pp. 3, ix.

11. Quoted in David Bakan, "Behaviorism and American Urbanization," *Journal of the History of the Behavioral Sciences,* 1966, *2*:5–28, p. 10 all (Bakan is quoting from Watson's 1928 text, *The Ways of Behaviorism*).

12. Ibid., pp. 10, 11 (emphasis added [Bakan is quoting from the 1924 edition of Watson's *Psychology from the Standpoint of a Behaviorist*]).

13. P. W. Bridgman, *The Intelligent Individual and Society* (New York: Macmillan, 1938), p. 1 both. On Bridgman's life and work, see Maila L. Walter, *Science and Cultural Crisis: An Intellectual Biography of Percy Williams Bridgman, 1882–1961* (Stanford, Calif.: Stanford University Press, 1990).

14. Bridgman, *Intelligent Individual,* pp. 1–2.

15. Crane Brinton, review of Robert Lynd's *Knowledge for What?* in *Saturday Review,* May 6, 1939, pp. 3–4, 4. For discussions of where Lynd fell politically, see Henry Etzkowitz, "The Americanization of Marx: *Middletown* and *Middletown in Transition,*" *Journal of the History of Sociology,* 1979–80, *2*:41–54; and S. M. Miller, "Struggle for Relevance: The Lynd Legacy," *Journal of the History of Sociology,* 1979–80, *2*:58–62.

16. Robert Lynd, *Knowledge for What?* (Princeton, N.J.: Princeton University Press, 1939), pp. 105, 49, 242, 110, 230–1.

17. Ibid., pp. 4, 111, 243, 244.

18. William H. Kilpatrick, "The New Adult Education," in William H. Kilpatrick, ed., *The Educational Frontier* (New York: Century, 1933), pp. 151, 152.

19. George Coe, "The Public Mind," in Harris Franklin Rall, ed., *Religion and Public Affairs: In Honor of Bishop Francis John McConnell* (New York: Macmillan, 1937), pp. 193–4.

20. Harris Franklin Rall, "Social Change," in Rall, *Religion and Public Affairs,* p. 220.

21. Walter Rauschenbusch, *Christianity and the Social Crisis* (New York: Macmillan, 1907), p. 421.

22. Francis J. McConnell, *The Christian Ideal and Social Control* (Chicago: University of Chicago, 1932), pp. 131–2, 140.

23. Ibid., pp. 14, 15.

24. Boyd H. Bode, "The Confusion in Present-Day Education," in Kilpatrick, *The Educational Frontier,* p. 29.

25. Ibid., pp. 29–30.

26. Ibid., pp. 30–1.

27. See Cornelia V. Christenson, *Kinsey: A Biography* (Bloomington: Indiana University Press, 1971).

28. Stephen Jay Gould, "Of Wasps and WASPS," in his *The Flamingo's Smile: Reflections in Natural History* (New York: W. W. Norton, 1985), pp. 159–60.

29. Regina Markell Morantz, "The Scientist as Sex Crusader: Alfred C. Kinsey and American Culture," *American Quarterly,* 1977, *29*:563–89, p. 569.

30. Alfred Kinsey, "Individuals," reprinted in Christenson, *Kinsey,* p. 7 all (the speech was originally delivered on June 5, 1939). For another example of this biological interest in diversity and democracy, see John Beatty, "Dobzhansky and the Biology of Democracy: The Moral and Political Significance of Genetic Variation," in Mark B. Adams, ed., *The Evolution of Theodosius Dobzhansky: Essays on His Life and Thought in Russia and America* (Princeton, N.J.: Princeton University Press, 1991).

31. Kinsey, "Individuals," p. 9.

32. Guenter H. Lenz, "'Ethnographies': American Culture Studies and Postmodern Anthropology," *Prospects,* 1991, *16*:1–40, p. 1.

33. For an overview of the issues, see Edward A. Purcell, Jr., *The Crisis of Democratic Theory: Scientific Naturalism and the Problem of Value* (Lexington: University Press of Kentucky, 1973); Peter Novick, *That Noble Dream: The "Objectivity Question" and the American Historical Profession* (Cambridge: Cambridge University Press, 1988); and James Kloppenberg, "Objectivity and Historicism: A Century of American Historical Writing," *American Historical Review,* 1989, *94*:1011–30.

34. Richard Handler, "Boasian Anthropology and the Critique of American Culture," *American Quarterly,* 1990, *42*:252–73, p. 253. See also John S. Gilkeson, Jr., "The Domestication of 'Culture' in Interwar America, 1919–1941," in JoAnne Brown and David K. van Keuren, eds., *The Estate of Social Knowledge* (Baltimore: Johns Hopkins University Press, 1991).

35. For a discussion of this point, see Clifford Geertz, "Us/Not-Us: Benedict's Travels," in his *Works and Lives: The Anthropologist as Author* (Stanford, Calif.: Stanford University Press, 1988). Of equal importance, see Barbara A. Babcock, "'Not in the Absolute Singular': Re-Reading Ruth Benedict," *Frontiers,* 1992, *12*:39–77.

36. Ruth Benedict, review of Edward Westermarck's *Ethical Relativity* in the *New York Herald Tribune,* June 26, 1932, p. 3.

37. F. H. Matthews, "The Revolt Against Americanism: Cultural Pluralism and Cultural Relativism as an Ideology of Liberation," *Canadian Review of American Studies,* 1970, *1*:4–31, p. 22.

38. Clifford Geertz, "Anti Anti-Relativism," *American Anthropologist,* 1984, *86*:263–78. A number of philosophers and sociologists during the last decades have been exercised over the "threat" that they believe relativist perspectives present for such concepts as rationality and realism. For a sample of some aspects of this debate see Martin Hollis and Steven Lukes, eds., *Rationality and Relativism* (Oxford: Basil Blackwell, 1982); and Robert Nola, ed., *Relativism and Realism in Science* (Dordrecht: Kluwer, 1988). Other relevant collections are Ernan McMullin, ed., *Construction and Constraint: The Shaping of Scientific Rationality* (Notre Dame: University of Notre Dame Press, 1988); and Diederick Raven, Lieteke van Vucht Tijssen, and Jan de Wolf, eds., *Cognitive Relativism and Social Science* (New Brunswick, N.J.: Transaction Publishers, 1992).

39. Geertz, "Benedict's Travels," pp. 105, 106, 107.

40. Ibid., p. 106.

41. See, for example, Ruth Benedict, *Race: Science and Politics* (New York: Modern Age, 1940), and *Race and Cultural Relations: America's Answer to the Myth of a Master Race* (Washington, D.C.: National Education Association, 1942).

42. Benedict, *Patterns,* pp. 23, 24.

43. Ibid., pp. 36, 232.

44. Ibid., p. 249. Benedict provided a judgmental example of her own, when she deplored the existence within American culture patterns of those family men, jurists, and business executives who, as "arrogant and unbridled egoists" pursued "courses of action [that] are often more asocial than those of the inmates of penitentiaries. In terms of the suffering and frustration that they spread about them there is probably no comparison" (p. 277).

45. The allusion to Carl Becker here is deliberate. See Carl Becker, "Everyman His Own Historian," *American Historical Review,* 1932, *37*:221–36.

46. See, for example, Robert Faris, *Chicago Sociology, 1920–1932* (Chicago: University of Chicago Press, 1967); Edward Shils, "Tradition, Ecology, and Institution in the History of Sociology," *Daedalus,* 1970, *99*:760–825; Fred Matthews, *Quest for an American Sociology: Robert Park and the Chicago School* (Montreal: McGill-Queen's Press, 1977); and David J. Lewis and Richard L. Smith, *American Sociology and Pragmatism: Mead, Chicago Sociology, and Symbolic Interaction* (Chicago: University of Chicago Press, 1980).

47. Lucy Salmon has an entry in Edward T. James, Janet Wilson James, and Paul S. Boyer, eds., *Notable American Women, 1607–1950* (Cambridge, Mass.: Belknap Press, 1971) by Violet Barbour; for Salmon's life, see also Louise Fargo Brown, *Apostle of Democracy: The Life of Lucy Maynard Salmon* (New York: Harper and Brothers, 1943). Some of Salmon's social views can be found in, for example, Lucy Maynard Salmon, *Progress in the Household* (Boston: Houghton, Mifflin, 1906). Information on Wylie and Mills can be found, respectively, in Elisabeth Woodbridge Morris, ed., *Miss Wylie of Vassar* (New Haven, Conn.: Yale University Press, 1934); and Herbert Elmer Mills, *College Women and the Social Sciences* (Freeport, N.Y.: Books for Libraries, 1971/1934). Wylie's interest in what she termed the "social affiliations" of literature can be seen in her *Social Studies in English Literature* (Boston: Houghton Mifflin, 1916); the phrase is from p. i. Examples of relativistic stances can be found in such work as Gertrude Buck, *The Social Criticism of Literature* (New Haven, Conn.: Yale University Press, 1916).

48. For Rourke, see Joan H. Rubin, *Constance Rourke and American Culture* (Chapel Hill: University of North Carolina Press, 1980); a key work of Rourke's is *American Humor: A Study of the National Character* (New York: Harcourt, Brace, 1931). See also her works on popular culture such as *Davy Crockett* (New York: Harcourt, Brace, 1934); and *Audubon* (New York: Harcourt, Brace, 1936). For Ware, see Ellen Fitzpatrick, "Caroline F. Ware and the Cultural Approach to History," *American Quarterly,* 1991, *43*:173–98; a key work of Ware's is *Greenwich Village 1920–1930: A Comment on American Civilization in the Post-War Years* (New York: Harper and Row, 1965/1935); see also Caroline Ware, ed., *The Cultural Approach to History* (New York: Columbia University Press, 1940). There are two major biographies of Ruth Benedict: Judith Schachter Modell, *Ruth Benedict: Patterns of a Life* (Philadelphia: University of Pennsylvania Press, 1983); and Margaret M. Caffrey, *Ruth Benedict: Stranger in This Land* (Austin: University of Texas Press, 1989).

49. The Vassar College records for Lois Barclay indicate that she took English courses for each of her four years, among them Laura Wylie's class in nineteenth-century British poets. Wylie as an intellectual influence is mentioned in Gardner Murphy, "A Story of Dyadic Thought and Work with Gardner and Lois Murphy as Written by Gardner," unpublished manuscript dated 1968 (Gardner Murphy and Lois Barclay Murphy Papers, Archives of the History of American Psychology, University of Akron). In a letter to the author dated June 22, 1992, Murphy states, "Tho' I didn't take a course with [Lucy Salmon], I *was* very much influenced (maybe a little awed, too!").

50. Laura Wylie, "What Can Be Done about It?" reprinted in Morris, *Miss Wylie of Vassar,* p. 136.

51. Lynn D. Gordon, *Gender and Higher Education in the Progressive Era* (New Haven, Conn.: Yale University Press, 1990), p. 132 all. Gordon subtitles her chapter on Vassar, "Women with Missions." See also Salmon's parables on how her privileged students should begin erasing class differences in her 1906 work, *Progress in the Household.*

52. Constance Warren, *A New Design for Women's Education* (New York: Frederick A. Stokes, 1940), p. 252. See also Constance Warren, "Self Education: An Experiment at Sarah Lawrence College," *Progressive Education,* 1934, *11*:267–70 and Louis Tomlinson Benezet, *General Education in the Progressive College* (New York: Teachers College, Columbia University, 1943). Five volumes explicating the teaching philosophy at Sarah Lawrence were published during this period: Ruth L. Munroe, *Teaching the Individual* (New York: Columbia University Press, 1942); Lois Barclay Murphy, Eugene Lerner, Jane Judge, and Madeleine Grant, *Psychology for Individual Education* (New York: Columbia University Press, 1942); Esther Raushenbush, *Literature for Individual Education* (New York: Columbia University Press, 1942); Lois Barclay Murphy and Henry Ladd, *Emotional Factors in Learning* (New York: Columbia University Press, 1944); and Helen Merrell Lynd, *Field Work in College Education* (New York: Columbia University Press, 1945). This type of egalitarianism was also incorporated into the structure of the Bank Street Schools, with which Murphy was associated as well; on this aspect of Bank Street, see Joyce Antler, *Lucy Sprague Mitchell: The Making of a Modern Woman* (New Haven, Conn.: Yale University Press, 1987), p. 328.

53. Lucy M. Salmon, *What is Modern History?* (Poughkeepsie, N.Y.: Vassar College, 1917), p. 28.

54. Buck, *The Social Criticism of Literature,* p. 31.

55. Ibid., p. 48. For a brief discussion of Rourke as a student of Buck and Wylie's, see p. 11 of Rubin, *Constance Rourke.*

56. Constance Rourke, "Vassar Classrooms: 'English J' and 'Romanticism,'" in Morris, *Miss Wylie of Vassar,* p. 72. Rubin calls attention to this point in *Constance Rourke,* p. 11

57. Ruth Benedict, "Mary Wollstonecraft," in Margaret Mead, ed., *An Anthropologist at Work: Writings of Ruth Benedict* (Boston: Houghton Mifflin, 1959), p. 491 (Benedict is quoting Wollstonecraft). The piece is an unpublished manuscript circa 1914–15. Benedict, *Patterns,* p. 2.

58. Richard Weiss, "Ethnicity and Reform: Minorities and the Ambience of the Depression Years," *Journal of American History,* 1979, *66*:566–85, p. 567.

59. See, for example, Harvard Sitkoff, *A New Deal for Blacks: The Emergence of Civil Rights as a National Issue* (New York: Oxford University Press, 1978); John B. Kirby, *Afro-Americans in the Roosevelt Era: Liberalism and Race* (Knoxville: University of Tennessee Press, 1980); and Nancy J. Weiss, *Farewell to the Party of Lincoln: Black Politics in the Age of FDR* (Princeton, N.J.: Princeton University Press, 1983).

60. Weiss, "Ethnicity," p. 566, quoted on p. 568. A favorite anecdote about Franklin Roosevelt, apparently apocryphal, has him opening a meeting of the Daughters of the American Revolution with the words: "Greetings, fellow immigrants!" On the burgeoning scientific debate about race during this period, see Hamilton Cravens, *The Triumph of Evolution: The Heredity-Environment Controversy, 1900–1941* (Baltimore: Johns Hopkins University Press, 1988).

61. Carey McWilliams, *Louis Adamic and Shadow-America* (Los Angeles: Arthur Whipple, 1935), p. 77.

62. Weiss, "Ethnicity," p. 569.

63. McWilliams, *Louis Adamic,* p. 76. Caroline Ware, "Cultural Groups in the United States," in Caroline Ware, ed., *The Cultural Approach to History* (Port Washington, N.Y.: Kennikat Press, 1940), p. 73.

64. Weiss, "Ethnicity," p. 578. See also David Hollinger's essays, "Ethnic Diversity, Cosmopolitanism, and the Emergence of the American Liberal Intelligentsia," and "Democracy and the Melting Pot Reconsidered," in his *In the American Province: Studies in the History and Historiography of Ideas* (Baltimore: Johns Hopkins University Press, 1985).

65. William Stott, *Documentary Expression and Thirties America* (New York: Oxford University Press, 1973), quoted on p. 110.

66. Ibid., pp. 110–11.

67. Jerry Mangione details the WPA writers' project in *The Dream and the Deal: The Federal Writers' Project, 1935–1943* (Boston: Little, Brown, 1972). For reference to *The Negro of Virginia,* see pp. 258–61; for *Lay My Burden Down,* p. 263; and for the "Living-Lore Unit," pp. 271–2. For an overview of the arts programs, see Jane De Hart Mathews, "Arts and the People: The New Deal Quest for a Cultural Democracy," *Journal of American History,* 1975, *62*:316–39.

68. Matthews, "The Revolt Against Americanism," pp. 5, 8, 7.

69. Interview with the author, November 1988.

70. Lois Barclay Murphy, "Emotional Development and Guidance in Nursery School and Home," *Childhood Education,* 1936, *12*:306–11, p. 307.

71. Ibid.

72. Ibid., p. 306.

73. Lois Barclay Murphy, "The Nursery School Contributes to Emotional Development," *Childhood Education,* 1940, *16*:404–7, pp. 404, 405.

74. Ibid., p. 405.

75. Ralph Steiner, *Ralph Steiner: A Point of View* (Middletown, Conn.: Wesleyan University Press, 1978), p. 9.

76. Ibid.

77. Ronald K. Goodenow, "The Progressive Educator, Race and Ethnicity in the Depression Years: An Overview," *History of Education Quarterly,* 1975, *15*:365–94, p. 368.

78. Lois Barclay Murphy, *Social Behavior and Child Personality: An Exploratory Study of Some Roots of Sympathy* (New York: Columbia University Press, 1937), p. 50.

79. Ibid., p. 49.

80. Barbara Biber, Lois B. Murphy, Louise P. Woodcock, and Irma S. Black, *Child Life in School: A Study of a Seven-Year-Old Group* (New York: E. P. Dutton, 1942), p. 7.

81. Ibid.

82. Gardner Murphy and Lois Barclay Murphy, "The Influence of Social Situations upon the Behavior of Children," in C. Murchison, ed., *A Handbook of Social Psychology* (Worcester, Mass.: Clark University Press, 1935), p. 1092.

83. Gordon W. Allport, "The Psychologist's Frame of Reference," *Psychological Bulletin,* 1940, *37*:1–28, p. 18.

84. Allport, "Frame," p. 24. Allport also identified these frames as being dependent on the "historical positions" of individuals; see Gordon W. Allport, "The Functional Autonomy of Motives," *American Journal of Psychology,* 1937, *50*:141–56, p. 142.

85. Gordon W. Allport, *The Use of Personal Documents in Psychological Science* (New York: Social Science Research Council, 1942), pp. 168, 167. "Functional autonomy" refers to one of Allport's own theoretical constructions and is discussed below.

86. Gordon W. Allport and H. S. Odbert, *Trait-Names: A Psycho-lexical Study, Psychological Monographs,* 1936, *47*:1–171. Odbert was a graduate student of Allport's.

87. Ibid., p. 2.

88. Ibid., p. 3.

89. Ibid., p. 14.

90. Ibid., p. 15.

91. Gordon W. Allport, *Personality: A Psychological Interpretation* (New York: Henry Holt, 1937), p. 192.

92. Allport and Odbert, *Trait-Names,* p. 10.

93. Allport, *Personality,* p. 321.

94. Ibid., p. 194.

95. Ibid., p. 204.

96. J. P. Guilford, review of Allport and Odbert, *Trait-Names: A Psycho-lexical Study* in the *American Journal of Psychology,* 1936, *48*:673–80, pp. 676, 674. See also Ross Stagner's reviews in *Journal of Applied Psychology,* 1936, *20*:522–3, and *Character and Personality,* 1936–7, *5*:255–7. Both men presented claims on behalf of factor theory as a way out of Allport's dilemma and also suggested that the common traits that they believed were in fact shared by individuals would be found to be grounded in physiology or genetics.

97. J. P. Guilford, review of Allport, *Personality,* in *Journal of Abnormal and Social Psychology,* 1938, *33*:414–20, pp. 416, 419.

98. Ibid., p. 416.

99. Ibid., p. 419.

100. Gardner Murphy, *An Historical Introduction to Modern Psychology* (London: Kegan Paul, 1929), p. 414.

101. Gardner Murphy and Lois Barclay Murphy, *Experimental Social Psychology* (New York: Harper and Brothers, 1931), p. 446 (emphasis added).

102. Gardner Murphy, "Editorial Foreword," *Sociometry,* 1937, *1*:5–7, p. 6.

103. Gardner Murphy, "The Research Task of Social Psychology," *Journal of Social Psychology,* 1939, *10*:105–54, p. 113.

104. Ibid.

105. Murphy, "Editorial Foreword," pp. 5–6.

106. Gardner Murphy, Lois Barclay Murphy and Theodore Newcomb, *Experimental Social Psychology,* rev. ed. (New York: Harper and Brothers, 1937), p. 761.

107. Ibid., pp. 761, 762.

108. Gardner Murphy, "Personality and Social Adjustments," *Social Forces,* 1937, *15*:472–6, p. 472.

109. Ibid., pp. 473–4.

110. Allport, *Personality,* p. 193.

111. Ibid., pp. 193–4.

112. Ibid., p. 194.

113. Murphy et al., *Experimental Social Psychology,* p. 17. See also such remarks by Biber et al., in *Child Life in School,* as the statement that "it is indeed futile to hunt for universal human trends" (p. 7). The authors maintained that "there can be no universal description of stages of development. They must be regarded as functions of a cultural complex" (p. 8).

114. Leonard Doob, review of Murphy et al., *Experimental Social Psychology,* in *Psychological Bulletin,* 1938, *35*:112–15, p. 115.

115. Jane Addams, *Democracy and Social Ethics* (New York: Macmillan, 1913/1902), pp. 176, 177.

116. Rall, "Social Change," p. 217.

117. Letter from Gordon W. Allport to Mr. [Louis] Adamic dated January 16, 1942 (Gordon W. Allport Papers, Harvard University Archives).

Conclusions

1. William James, "The One and the Many," in *Pragmatism* (Cambridge, Mass.: Harvard University Press, 1975/1907), p. 71.

2. Franklin D. Roosevelt, *The Public Papers and Addresses of Franklin D. Roosevelt,* vol. 1 (New York: Random House, 1938), p. 659; Franklin D. Roosevelt, *The Public Papers and Addresses of Franklin D. Roosevelt,* vol. 2 (New York: Random House, 1938), p. 12; Roosevelt, *Public Papers,* vol. 1, p. 646. The first citation is from Roosevelt's acceptance of the Democratic nomination in 1932; the second is from his first inaugural in 1933; the third is from a 1932 campaign address.

3. For a recent analysis of the public prominence achieved by the psychological profession in post–World War II America, see Ellen Herman, *The Romance of American Psychology: Political Culture in the Age of Experts* (Berkeley: University of California Press, 1995). Herman emphasizes the imperative among psychological experts to press for "a larger jurisdiction" for the field of psychology, and the enthusiastic response of policy makers and the public to this initiative.

4. Gardner Murphy, ed., *Human Nature and Enduring Peace* (Boston: Houghton Mifflin, 1945); and Gardner Murphy, *In the Minds of Men: The Study of Human Behavior and Social Tensions in India* (New York: Basic Books, 1953).

5. Otto Klineberg, "Otto Klineberg," in Gardner Lindzey, ed., *A History of Psychology in Autobiography* (Englewood Cliffs, N.J.: Prentice-Hall, 1974), pp. 174–5. Klineberg also served as Director of Applied Social Sciences at UNESCO from 1953 to 1955 (p. 176).

6. Cf. *American Psychologist,* 1967, *22*:496, Invited Distinguished Address by Martin Luther King, Jr., "The Role of the Behavioral Scientist in the Civil Rights Movement." For a detailed discussion of the Clarks, see Ben Keppel, *The Work of Democracy: Ralph Bunche, Kenneth B. Clark, Lorraine Hansberry and the Cultural Politics of*

Race (Cambridge, Mass.: Harvard University Press, 1995); see also Herman, *Romance of American Psychology.*

7. Kenneth B. Clark, "Introduction," in Gordon Allport, *The Nature of Prejudice,* 25th Anniversary Edition (Reading, Mass: Addison-Wesley, 1979/1954), p. ix.

8. Robert Coles, *Children of Crisis,* 5 vols. (Boston: Little, Brown, 1967–80); Roger Barker and Herbert Wright, *One Boy's Day: A Specimen Record of Behavior* (New York: Harper's, 1951 [the Harper's series was under Gardner Murphy's editorship]); *Midwest and Its Children: The Psychological Ecology of an American Town* (Evanston, Ill.: Row and Peterson, 1954); and Roger Barker, *The Stream of Behavior: Explorations of Its Structure and Content* (New York: Appleton-Century-Crofts, 1963).

9. On humanistic psychology see Roy José DeCarvalho, *The Founders of Humanistic Psychology* (New York: Praeger, 1991); and Herman, *The Romance of American Psychology.* On the rise of clinical psychology, see also Herman, *Romance of American Psychology.* For an introduction to cognitive psychology, see Howard Gardner, *The Mind's New Science: A History of the Cognitive Revolution* (New York: Basic Books, 1985).

10. Jerome Bruner, *In Search of Mind: Essays in Autobiography* (New York: Harper and Row, 1983), p. 30 (Bruner also mentions Kurt Koffka's *Principles of Gestalt Psychology* as being influential in this regard), pp. 36, 76.

11. Gordon W. Allport, "The Study of Personality by the Intuitive Method: An Experiment in Teaching from *The Locomotive God,*" *Journal of Abnormal and Social Psychology,* 1929, *24*:14–27, p. 20.

12. Philip Kitcher and Wesley C. Salmon indicate that, with respect to scientific explanation, the hegemony of logical empiricism in the 1950s and 1960s is aptly symbolized by Carl Hempel's *Aspects of Scientific Explanation and Other Essays in the Philosophy of Science;* see their "Preface" in Philip Kitcher and Wesley C. Salmon, eds., *Scientific Explanation: Minnesota Studies in the Philosophy of Science,* vol. 13 (Minneapolis: University of Minnesota Press, 1989), p. xiii. Likewise, Laurence Smith argues that the transplanted logical positivism of the Vienna Circle and the Berlin Society came to be widely regarded in the United States as "*the* philosophy of science, rather than as one approach among others;" cf. his *Behaviorism and Logical Positivism: A Reassessment of the Alliance* (Stanford, Calif.: Stanford University Press, 1986), p. 14.

13. Peter Novick, *That Noble Dream: The "Objectivity Question" and the American Historical Profession* (Cambridge: Cambridge University Press, 1988), p. 546.

14. Thomas S. Kuhn, *The Structure of Scientific Revolutions,* 2d ed. (Chicago: University of Chicago Press, 1970/1962), p. viii. Kuhn's analysis, of course, launched a torrent of soul-searching by numerous social scientists, all looking for the holy grail of "paradigmhood," which they believed would remedy their "immature" status. For discussions of applications of Kuhn's framework to particular social sciences, see, for example, Douglas Lee Eckberg and Lester Hill, Jr., "The Paradigm Concept and Sociology: A Critical Review," in Gary Gutting, ed., *Paradigms and Revolutions: Appraisals and Applications of Thomas Kuhn's Philosophy of Science* (Notre Dame: University of Notre Dame Press, 1980); and R. I. Watson, "Psychology: A Prescriptive Science," *American Psychologist,* 1967, *22*:435–43.

15. For a thoughtful rebuttal to this point of view, see Mitchell G. Ash, "Historicizing Mind Science: Discourse, Practice, Subjectivity," *Science in Context,* 1992, *5*:193–207. Sandra Harding, in *Whose Science? Whose Knowledge? Thinking from Women's Lives*

(Ithaca, N.Y.: Cornell University Press, 1991), makes the provocative contention that, "on scientific grounds, as well as for moral and political reasons, those social sciences that are most deeply critical and most comprehensively context-seeking can provide the best models for all scientific inquiry, including physics" (p. 98); she also touches on this issue in *The Science Question in Feminism* (Ithaca, N.Y.: Cornell University Press, 1986).

16. Reconsideration of the relations between the social and the natural sciences has received recent historical attention. See I. Bernard Cohen, ed., *The Natural Sciences and the Social Sciences: Some Critical and Historical Perspectives* (Dordrecht: Kluwer, 1994); and I. Bernard Cohen, *Interactions: Some Connections between the Natural Sciences and the Social Sciences* (Cambridge, Mass.: MIT Press, 1994).

17. It is interesting to note recent discussions of such issues as complexity, open-endedness, and disorder. Physicist Murray Gell-Mann has given the concept of "complexity" a high profile in such ventures as his general-interest text, *The Quark and the Jaguar: Adventures in the Simple and the Complex* (New York: W. H. Freeman, 1994). See also physicist Freeman Dyson's remarks regarding the distinction between scientists who are "unifiers" and those who are "diversifiers"; Freeman J. Dyson, "Manchester and Athens," in his *Infinite in All Directions* (New York: Harper and Row, 1988), p. 36. Philosopher of science John Dupré has also produced a provocative analysis of related concerns in *The Disorder of Things: Metaphysical Foundations of the Disunity of Science* (Cambridge: Harvard University Press, 1993). And a number of scholars from within science studies have begun considering these themes; see the recent volume of essays edited by Peter Galison and David J. Stump, *The Disunity of Science: Boundaries, Contexts, and Power* (Stanford, Calif.: Stanford University Press, 1996).

18. Clifford Geertz, "Blurred Genres: The Refiguration of Social Thought," in his *Local Knowledge: Further Essays in Interpretive Anthropology* (New York: Basic Books, 1983), p. 19.

19. Ibid., p. 20.

20. William James, "Is Life Worth Living?" in *The Will to Believe and Other Essays in Popular Philosophy* (Cambridge, Mass.: Harvard University Press, 1979/1897), p. 49.

21. Ibid., p. 50.

Index